Springer Theses

Recognizing Outstanding Ph.D. Research

Aims and Scope

The series "Springer Theses" brings together a selection of the very best Ph.D. theses from around the world and across the physical sciences. Nominated and endorsed by two recognized specialists, each published volume has been selected for its scientific excellence and the high impact of its contents for the pertinent field of research. For greater accessibility to non-specialists, the published versions include an extended introduction, as well as a foreword by the student's supervisor explaining the special relevance of the work for the field. As a whole, the series will provide a valuable resource both for newcomers to the research fields described, and for other scientists seeking detailed background information on special questions. Finally, it provides an accredited documentation of the valuable contributions made by today's younger generation of scientists.

Theses may be nominated for publication in this series by heads of department at internationally leading universities or institutes and should fulfill all of the following criteria

- They must be written in good English.
- The topic should fall within the confines of Chemistry, Physics, Earth Sciences, Engineering and related interdisciplinary fields such as Materials, Nanoscience, Chemical Engineering, Complex Systems and Biophysics.
- The work reported in the thesis must represent a significant scientific advance.
- If the thesis includes previously published material, permission to reproduce this must be gained from the respective copyright holder (a maximum 30% of the thesis should be a verbatim reproduction from the author's previous publications).
- They must have been examined and passed during the 12 months prior to nomination.
- Each thesis should include a foreword by the supervisor outlining the significance of its content.
- The theses should have a clearly defined structure including an introduction accessible to new PhD students and scientists not expert in the relevant field.
- Indexed by zbMATH.

More information about this series at http://www.springer.com/series/8790

Hsiang-Hsi Kung

Collective Excitations in the Antisymmetric Channel of Raman Spectroscopy

Doctoral Thesis accepted by Rutgers University, USA

Springer

Hsiang-Hsi Kung
Quantum Matter Institute
University of British Columbia
Vancouver
BC, Canada

ISSN 2190-5053 ISSN 2190-5061 (electronic)
Springer Theses
ISBN 978-3-030-89331-6 ISBN 978-3-030-89332-3 (eBook)
https://doi.org/10.1007/978-3-030-89332-3

This Springer imprint is published by the registered company Springer Nature Switzerland AG
The registered company address is: Gewerbestrasse 11, 6330 Cham, Switzerland

Supervisor's Foreword

In his Ph.D. work, Dr. Hsiang-Hsi (Sean) Kung made a number of breakthrough contributions to condensed matter physics, including the first observation of broken reflection symmetry in the hidden order phase of the enigmatic heavy fermion material URu_2Si_2, and the discoveries of a novel collective mode – the chiral spin wave – and a novel composite particle – the chiral exciton – in a three-dimensional topological insulator Bi_2Se_3, which adds to the potential of topological insulators as a platform for photonics, optoelectronics, and spintronics devices. Each of the above listed discoveries merits an award on its own; to have three of them within one Ph.D. thesis is an unprecedented fit.

Besides the Springer Thesis Award, his dissertation work has received various recognition from prestigious institutions, including the Northeastern Association of Graduate Schools' Outstanding Doctoral Dissertation Award, and the American Physical Society's Richard L. Greene Dissertation Award in Experimental Condensed Matter or Materials Physics. These achievements highlight the importance of this work and its impact on the broad condensed matter physics community.

Distinguished Professor of Physics
Rutgers University, Piscataway, NJ, USA

Girsh Blumberg

Preface

Raman spectroscopy is an inelastic light scattering technique, where a monochromatic light source, typically a laser, is used to exchange energy with the system and thus perturbs the material under study into an excited state. In this process, the energy quanta that the electromagnetic field lose to or gain from the system correspond to specific energy levels that are characteristic to the material, such as lattice vibration modes, aka phonons. That is, each peak in the energy spectrum of the scattered light corresponds to a specific collective excitation of normal modes in the system. Therefore, we gain insights into the material properties by studying the energy spectrum of the scattered light, which is called a Raman spectrum to honor the main discoverer of this effect, Sir C.V. Raman.

One of the most powerful advantages of Raman scattering is that the energy spectra are independent of the incident light energy and only reflect the spectra of the system's excited states. Therefore, experimentalists can pick the excitation energy that maximizes the Raman signal of interest. This effect is also known as resonant Raman scattering, which is done by exciting electrons into an unoccupied band and thus creating an electron–hole pair in the system. There is finite probability that the electron–hole pair decay into an excited state instead of the ground state by emitting a slightly less energetic photon. Another unique property of Raman scattering is its symmetry sensitivity. By using polarized light, one can selectively probe part of the Raman tensor, which in turn reflects collective excitations with a specific symmetry.

My thesis research focuses on using low temperature polarization resolved Raman spectroscopy to identify optically excited collective modes in strongly correlated electron systems and three-dimensional (3D) topological insulators. In particular, we are interested in collective modes present in the pseudovector symmetry channel, such as the A_{2g} representation of D_{4h} group. Although generally forbidden for phononic excitations, it was noted by early workers of the field that contribution from antisymmetric tensors ($A_{2(g)}$ irreducible representation) may arise under special circumstances. Historically, the excitations in the antisymmetric channel have been referred to as the "chiral excitations" due to their intimate relationship with the spin chiral operator. For example, antisymmetric scattering was predicted for Dirac spin-liquid state in the spin-1/2 kagome system but yet unobserved to date.

Here in this monograph, we discovered two examples of electronic excitations arising from the antisymmetric "chiral" channel. More precisely, we observe sharp $A_{2(g)}$ collective modes in two entirely different nonmagnetic systems: heavy fermion metal URu_2Si_2 and 3D topological insulator Bi_2Se_3. Observation of an antisymmetric collective mode in nonmagnetic systems is highly unusual and rare, and the mechanism is different for different cases. This monograph is dedicated to explore the properties, mechanisms, and physical implications from these two new systems. Due to the recently rising interests on the chiralities in condensed matter physics, we hope that this dissertation will intrigue new insights into the possibility of probing chiral excitations with optical methods.

Vancouver, BC, Canada Hsiang-Hsi Kung

Acknowledgment

This work is dedicated to
the great minds and tireless workers
who strove to understand
the mysterious case of the "hidden order".

I am in debt to many people throughout the course of this monograph, and it is impossible to make a comprehensive list to give proper credits to them all. But, I would like to start by thanking my good friend and thesis advisor, Professor Girsh Blumberg, who patiently and tirelessly guided me in the past five years. It is not uncommon that Ph.D. advisors provide good professional guidance to graduate students. But, Girsh has uplifted the standards for Ph.D. advisory to another level. Besides the useful discussions and insightful comments he offered during this monograph, which would have been impossible to complete without many of his ideas, I am most grateful for his teachings on research attitude and career building. He has always demonstrated high degree of self-discipline and self-motivation toward research—I hope I have learned some of that from him. He showed me that if we push ourselves, we can often achieve more than we originally thought—but, if and only if we start pushing. Because in academia as well as in life, nothing will ever happen unless we try.

I am very grateful to have worked with many talented people in Rutgers, who taught me a lot in many aspects: I would like to thank Professor Kristjan Haule, for patiently teaching me the basics of heavy fermions and strongly correlated materials, helping me improve the writings in the manuscript, and aiding me in many ways in the past 5 years; Professor Piers Coleman and Professor Premi Chandra, for the insightful discussions we had about URu_2Si_2 and also for their friendly support throughout my Ph.D. studies; I am also very thankful to the members of the "Blumberg Lab": Dr. Wei-Lu Zhang, for selflessly sharing her laughter and time when I needed them; Shangfei Wu and Mai Ye, for the lengthy but helpful debates about Raman scattering; Alex Lee and Viktor Krapivin, for their company and help during experiments; and Brian Dennis and Dr. Verner Thorsmølle, for teaching me how to properly run and log experiments. I am also very thankful to the good friendships I built up with several Rutgers graduate students and postdocs, in particular Alejandro "Ale" Gomez and Aniket "Po'boy" Patra, for listening to my whining and teaching me to laugh out the stress in life with good friends.

I am fortunate to have collaborated with many great people around the world: I am very thankful to Professor Mydosh from the Leiden University, for tirelessly helping me improve the manuscript and carefully explaining to me the Kondo physics in URu_2Si_2; Professor Bauer at the LANL and Professor Baumbach at the FSU Tallahassee, for providing me with the best crystals of URu_2Si_2; Professor Maple, Dr. Ran, and Dr. Kanchanavatee from the UCSD, for very generously providing me large pieces of $URu_{2-x}Fe_xSi_2$ single crystals and being very forgiving for my slow progress; Professor Cheong (Rutgers) and Professor Wang, for generously sharing with me a very rare batch of ultraclean Bi_2Se_3 single crystal; Professor Oh (Rutgers) and his group members, Matthew, Nikesh, Maryam, and Jisoo, for helping me better understand the basics of 3D topological insulators through many discussions; Professor Smirnov and his student Zhengguang Lu from the FSU Tallahassee, for tirelessly helping my experiments in the MagLab; Professor Merlin and his student Ibrahim Boulares from the University of Michigan, and Professor Kemper from the NC State University who very patiently taught me many things about the surface phonon modes and surface resonance in Bi_2Se_3; and Professor Maslov and Dr. Maiti from the University of Florida who very patiently helped me understand the physics of chiral spin modes in Bi_2Se_3.

Last but not the least, my families who endured my capricious temper during the past 5 years, and probably way before that. My special love to Lingwei, for her selfless sacrifices and for displaying great patience with me.

Regretfully, I have to cut off the 10-page-long list I had in mind at here and express my final gratitude toward the financial supports: The thesis work presented here was done under the support of the NSF grants DMR-1104884 and DMR-1709161 and the US DOE—Basic Energy Sciences grant DE-SC0005463. The financial support of the predoctoral University and Louis Bevier Graduate Fellowship is also acknowledged.

Hsiang-Hsi "Sean" Kung

Parts of This Thesis Have Been Published in the Following Journal Articles

1. H.-H. Kung, R.E. Baumbach, E.D. Bauer, V.K. Thorsmølle, W.-L. Zhang, K. Haule, J.A. Mydosh, G. Blumberg, Chirality density wave of the 'hidden order' phase in URu_2Si_2. Science **347**, 6228 (2015). https://doi.org/10.1126/science.1259729
2. H.-H. Kung, S. Ran, N. Kanchanavatee, V. Krapivin, A. Lee, J.A Mydosh, K. Haule, M.B. Maple, G. Blumberg, Analogy between the 'Hidden Order' and the orbital antiferromagnetism in URu_2Si_2. Phys. Rev. Lett. **117**, 227601 (2016). https://doi.org/10.1103/PhysRevLett.117.227601
3. H.-H. Kung, S. Maiti, X. Wang, S.-W. Cheong, D.L. Maslov and G. Blumberg, Chiral spin mode on the surface of a topological insulator. Phys. Rev. Lett. **119**, 136802 (2017). https://doi.org/10.1103/PhysRevLett.119.136802
4. H.-H. Kung, M. Salehi, X. Wang, I. Boulares, A.F. Kemper, N. Koirala, M. Brahlek, P. Lostak, C. Uher, R. Merlin, S.-W. Cheong, S. Oh, G. Blumberg, Surface vibrational modes of the topological insulator Bi_2Se_3 observed by Raman spectroscopy. Phys. Rev. B **95**, 245406 (2017). https://doi.org/10.1103/PhysRevB.95.245406
5. H.-H. Kung, A.P. Goyal, D.L. Maslov, X. Wang, A. Lee, A.F. Kemper, S.-W. Cheong, G. Blumberg, Observation of chiral surface excitons in a topological insulator Bi_2Se_3. Proc. Natl. Acad. Sci. **116**, 4006 (2019). https://doi.org/10.1073/pnas.1813514116

Contents

List of Abbreviations

AFM	Antiferromagnetism
ARPES	Angle resolved photoemission spectroscopy
BCT	Body centered tetragonal
BS	Bismuth selenide (Bi_2Se_3)
BZ	Brillouin zone
CDW	Charge density wave
CP	Circularly polarized
DFT	Density functional theory
DMFT	Dynamical mean field theory
DOS	Density of state
FM	Ferromagnetism
FWHM	Full width at half maximum
FS	Fermi surface
HO	Hidden order
IR	Infrared
IRR	Irreducible representation
KK	Kramers–Kronig
PL	Photoluminescence
QEP	Quasi-elastic peak
RRR	Residual resistivity ratio
SC	Superconductivity
SDW	Spin density wave
SOC	Spin–orbit coupling
ST	Simple tetragonal
STM	Scanning tunneling microscopy
TI	Topological insulator

TMD Transition metal dichalcogenide
TSS Topological surface state
TR Time resolved
TRIM Time reversal invariant momentum
TRS Time reversal symmetry
URS Uranium ruthenium silicide (URu_2Si_2)
UV Ultraviolet

Chapter 1
Introduction

Abstract Polarization resolved Raman spectroscopy is a powerful tool as it simultaneously offers exceptional symmetry and energetic resolution (~0.1 meV), making it particularly suitable for studying low energy collective modes in systems with nontrivial electron correlation. In the condensed matter physics, Raman spectroscopy has been crucial in understanding superconductivity by studying symmetry of the Cooper pair breaking peak, electron–phonon interaction in semiconductors by studying the plasmon–polaritons, and charge density waves by studying the amplitude modes.

In this chapter, I will first introduce the generic theory of polarization resolved Raman spectroscopy in solids. Focus will be given to electronic Raman scattering, where light scatters with the electronic quantum mechanical states. Minimal group theory will be used to justify the symmetry analysis of polarization resolved Raman scattering data. Then, I will discuss how Raman scattering can be used to probe the "chiral excitations," i.e., collective excitations out of a chiral ground state, by measuring inelastic light scattering in the pseudovector-like antisymmetry channel.

1.1 The Scope of This Thesis

In this thesis, we will focus on using low temperature Raman spectroscopy to identify optically excited collective modes in strongly correlated electron systems and three-dimensional (3D) topological insulators. In particular, we are interested in collective modes presenting in the pseudovector-like symmetry channel, e.g., the A_{2g} representation of D_{4h} group, which is antisymmetric with respect to in-plane reflections. Previously, such excitations are primarily seen in magnetic materials where the time reversal symmetry is broken [1]. Here, we present two examples where A_{2g} collective modes arise from time reversal symmetry preserved ground states.

H.-H. Kung, *Collective Excitations in the Antisymmetric Channel of Raman Spectroscopy*, Springer Theses, https://doi.org/10.1007/978-3-030-89332-3_1

In the first example, we study a heavy fermion metal $URu_{2-x}Fe_xSi_2$, which holds long standing interest in the strongly correlated electron community due to an emergent long range order it exhibits at low temperature and $x < 0.1$, known as the "hidden order" (HO). By studying the temperature dependent Raman response function in all allowed symmetry channels for various iron concentrations x, we found evidences of broken symmetries and a possible order parameter that explains the low temperature phase diagram. In the low Fe concentration HO phase, we observed a sharp in gap mode with A_{2g} symmetry and small leakage into the A_{1g} symmetry. Our results show strong indication of the local reflection symmetries broken at the uranium site in the HO phase [2]. As we increase x, the energy of an A_{2g} collective mode decreases to almost zero at the phase boundary and recovers in a well-known antiferromagnetic (AFM) phase. This is a direct evidence that both HO and AFM phases are related to an A_{2g} type order parameter, which arranges the uranium-$5f$ orbitals into orders breaking either local chirality or time reversal symmetry [3].

In the second example, we study a 3D topological insulator Bi_2Se_3, which has a rhombohedral lattice with the D_{3d} point group symmetry. At the crystal surface characterized by the C_{6v} point group, strong spin orbit coupling and time reversal symmetry conspire to form topologically protect Dirac states with chiral spin texture. While the existence and spin texture of the surface states are mostly understood for Bi_2Se_3 through photoemission and scanning tunneling spectroscopies, very few examples of collective excitations from the Dirac surface states have so far been identified, and therefore interactions between the Dirac particles are yet unclear. In our study, we tune the excitation energy into resonance between two surface bands to enhance the signal contributed by the Dirac electrons. As photon energy tuned to 1.8 eV, we observed a sharp collective mode with energy of about 150 meV appearing below the edge of a gapped continuum, in the pseudovector-like A_2 symmetry of the C_{6v} group. By comparing the data with calculations, we identify this peak as the transverse collective spin mode of surface Dirac fermions: a collective spin–flip excitation from the lower to upper Dirac cone [4]. The chiral properties of the surface states could be further explored by measuring the photoluminescence (PL) spectra. With the excitation photon energy tuned to around 2.7 eV, we observe the appearance of a PL peak centered at 2.26 eV, where the intensity is almost perfectly circularly polarized. By comparing the excitation profile with band structure calculations, we identify this PL peak as 2D excitonic bound states composed of surface massless electrons and surface massive holes, both subject to strong spin–orbit coupling which locks their spins and momenta into chiral textures. Due to the chiral structure, the surface excitons emit circularly polarized PL that preserves the angular momenta of the incident light [5]. These are new and direct optical measurements of the dynamical response from the 2D Dirac fermions in topological materials. Both the spin mode and the chiral exciton survive even at room temperature with decreased intensity, which is likely due to the more available decay channels through interacting with surface phonons [6]. The robustness of the chiral collective modes suggests potential applications in room temperature magnonics and optoelectronic devices.

1.1.1 Brief History

We will start with a brief detour on the history of discovery of inelastic light scattering. Newtonian optics (geometrical optics) states that light always travels in a straight line in empty space and can only be perceived by eyes when directly pointed at from the source. However, interaction with medium can deviate light from its optical path, which is often referred to as diffusion of light. We know by now that the diffusion of light is nothing more than the electromagnetic wave interacting with microparticles in the medium, which behave as tiny oscillating dipoles under the influence of the external field. Thus according to Maxwell's theory of electromagnetism, the scattered light can be deflected into any direction away from the optical path. Depending on whether or not the scattered light maintains the same wavelength as the incident light λ_I, the scattering process can be further separated into elastic and inelastic scatterings.

The elastic scattering of light from dilute particles where λ_I is much smaller than average particle separation has been known to scientists for a long time. However, systematic study was not done until the late nineteenth century by Tyndall. Depending on the nature of the scatterers, the elastic scattering from dilute medium can be further separated into two main categories: when the scatterers have diameter comparable or larger than the light wavelength, they are referred to as Tyndall effect, such as in suspending colloidal liquid; otherwise, they are generally referred to as Rayleigh effect, such as scattering from air molecules. Notice that all of the above treatments on light scattering assume each microparticle as harmonic oscillator, with the dipole moment of each particle, oscillates at the same frequency of the incident light, $\omega_I = 2\pi c/\lambda_I$. Thus the particles always radiate with the same wavelength as the incident light.

The scattering of light by dense medium, such as fluids, requires a different explanation. Since the particle separation is small compared to λ_I there exists coherence between light scattered by different particles, which can no longer be assumed as independent dipolar emitters in Rayleigh's treatments. A straightforward summation over N identical particles leads to destructive interference in every direction except along the forward light path [7], which is obviously inconsistent with our everyday experience of light penetrating water. For example, one can clearly see sunlight beams underwater in a swimming pool, where the light propagation is orthogonal to the observer's viewing direction. This cannot be attributed simply to impurities or dust in the solution, since the elastic line is always present even at large angles to the optical path, and in the purest solutions [8]. It was later pointed out by Landau and Placzek in 1934 that light scattering from local pressure (density) and entropy (temperature) fluctuations will avoid total cancellation of intensity in destructive interference and leads to nonzero intensity for the elastic scattering component away from the optical path [7, 9]. Note that this unshifted component is usually referred to as the "Rayleigh line" in literatures, but its origin is actually quite different from the Rayleigh effect that is causing the blue sky. More detailed discussion can be found in chapter 8 of Ref. [7].

Progress in understanding the inelastic scattering spectra is first made by Brillouin in 1922, who predicted that light interaction with propagating pressure (density) fluctuation in liquid or solids due to sound waves can lead to modification of scattered light wavelength [10]. Using the newly accepted quantum theory of lattice vibrations by Debye, i.e., phonons [11], Brillouin predicted two inelastic peaks displaced symmetrically on both sides of the elastic Rayleigh line, with energy separations related to the sound speed of the medium. The experimental observation was first reported for scattering of light by various pure and transparent liquids in 1930 [8], and the effect was since dubbed Brillouin scattering.

Shortly after Brillouin's thesis, Smekal predicted in 1923 that scattering of light by a two-level system will lead to oscillatory change of the medium's polarizability that causes inelastic sidebands in the spectra of scattered light [12]. The frequency shift of the sidebands from the excitation light frequency will be exactly the "quanta" of the two-level energy difference divided by Plank's constant. A few years later, Kramers and Heisenberg developed a quantum mechanical version of Smekal's theory using the "correspondence principle," describing inelastic light scattering of atoms [13]. The complete quantum mechanical treatment of light will be given by Paul Dirac two years later [14, 15], and thus the Raman scattering from a simple system is predicted and well understood.

Around the same time of Smekal's prediction, Raman and coworkers in Calcutta, India were studying the light transmission through water, because they do not believe that the "blueness" of the sea is due to reflection of the sky [16]. In the process of their study, they discovered sharp inelastic lines in the scattering spectra of several transparent organic liquids and distilled water. Raman spent the next few years studying this phenomenon and realized that not only the radiation is highly polarized, but also the energy shifts of some new lines are the same as the known absorption lines in infrared spectra. In 1928, Raman finally made the announcement that this is the discovery of a new type of "secondary radiation," where "*the incident quantum of radiation is partially absorbed by the molecule and that the unabsorbed part is scattered*" [17, 18]. Raman's discovery was repeated in quartz and several other materials later that year by R.J. Wood with an improved setup [19].

Raman was soon awarded Nobel prize in 1930 due to the discovery of the new type of light scattering now named after him. However, little is known that around the same time in Moscow, Landsberg and Mandelstam also reported similar effect in quartz [20], and it was correctly interpreted as the inelastic light scattering predicted by Smekal in 1923. Detailed historical account of the discovery in Moscow is given in Ref. [21].

The greatest triumph and significance for the discovery of Raman effect were perhaps due to its successful applications in studying of very low energy excitations in materials. These "ultra red" excitations would otherwise be very difficult to investigate experimentally by absorption spectroscopy due to its long wavelength. Raman effect translates the wavelength of these excitations into the visible range, or any other wavelength convenient to the material and experimental setup, by choosing an appropriate incident light energy. Also, in molecules or crystals with inversion symmetry, some of the vibration normal modes are forbidden by symmetry in infrared absorption, which could be observed in Raman scattering.

Another interesting implication of Raman scattering is in the development of quantum theory. By 1913, the great success of early quantum mechanics in explaining blackbody radiation, heat capacity, and the emission/absorption spectra of simple atoms has left very little doubt on the validity of quantizing energy levels in matters. However, this only deals with the emission and absorption of light by materials, and nothing to do with the nature of the light itself. Therefore, the quantization of the electromagnetic field proposed by Einstein in 1905 remains a debated issue for many years [22]. In 1916, Millikan finally, rather unwillingly, admitted that his experimental data of photoelectric effect is fully consistent with Einstein's theory [22], which would imply that the energy of electromagnetic wave is quantized with a quanta proportional to its frequency, i.e., $\hbar\omega$. In 1923, the observation of wavelength change in X-ray scattering by electrons in vacuum (Compton effect) finally convinced most people the quantum description of electromagnetic waves [23]. However, both experiments were done with high energy photon (extreme UV to hard X-ray), leaving the generality of the "particle description" of light at question. The Raman effect on the other hand is an extremely general and direct demonstration to the quantum nature of electromagnetic waves. As Robert W. Wood aptly commented on the Raman effect in 1928, "it appears to me that this very beautiful discovery, which resulted from Raman's long and patient study of phenomena of light scattering, is one of the most convincing proofs of the quantum theory of light which we have at the present time" [19].

1.2 Inelastic Light Scattering

We discussed in the previous section that Raman scattering is a type of "secondary radiation" [17], i.e., the emission of electromagnetic waves from matters exposed to an external field—the *primary radiation*—which could be either a beam of particles or electromagnetic waves. If the primary radiation is photon, then there are generally speaking two types of photonic secondary radiation that are observed in solids, i.e., the luminescence and light scattering. While most of the photons are scattered elastically in the light–matter interaction, there is a small fraction of scattered photon having a different energy than the absorbed. The photon is said to scatter *inelastically* with the system, i.e., either loses or gains energy through creating or absorbing a *quanta* of collective excitation, respectively. This inelastic process is coined the *Raman effect*, where the energy difference is known as the *Raman shift*. Traditionally, the Raman shift is often given in the unit of wavenumber (cm^{-1}) or frequency (THz), where 1 THz=33.356 cm^{-1}. The wavenumber is a convenient energy scale, comparable to the thermal fluctuations ($1\,k_B = 0.695\,\text{cm}^{-1}$/K) and the Zeeman energy ($1\,\mu_B$=0.4668 cm^{-1}/T), but almost an order of magnitude smaller than the most commonly used milli-electron volt (1 meV=8.065 cm^{-1}).

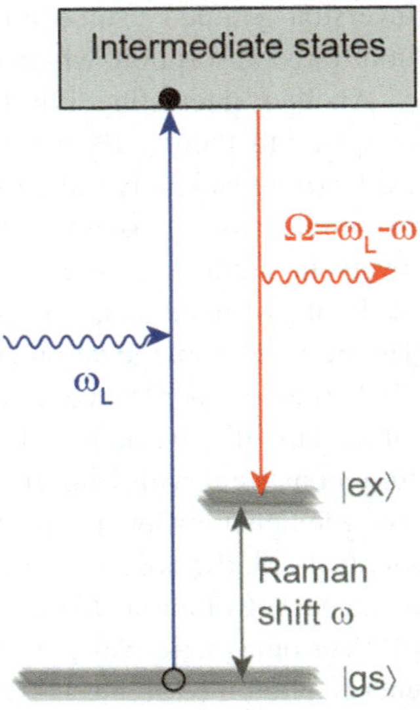

Fig. 1.1 Illustration for the Stokes process, where the incident laser energy is ω_L, and the excited state energy is ω

In order to further distinguish inelastic light scattering from the luminescence, we have to look into the processes of the secondary radiation. In the following, we introduce the 3-step model of secondary radiation typically used to explain the Raman effect [24], also illustrated in Fig. 1.1. First, the incident (primary) photon is absorbed by the material to promote an interband transition into an "intermediate state," which specifies an electron–hole (e–h) pair with the electron in a state of the conduction band and the hole in a state of the valence band. Note that the perturbation theory in quantum mechanics does not require the incident or scattered photon energy to match any of the intermediate state energies in order to acquire nonzero transition probability. If the photon energies are far away from any of the intermediate states, then the transition is said to be *virtual*. This "nonresonant" case is the simplest form of Raman scattering generally discussed in most textbooks and literature [7, 25]. Immediately after the optical absorption and before the band thermalization is fully established, i.e., redistributing electron populations throughout k-space to reach the Fermi–Dirac distribution with a new effective electronic temperature, the e–h pair exchanges energy with the system through inelastic processes such as creating/absorbing phonons, magnons, or other lower energy e–h pairs. Lastly, the e–h pair recombines to emit a "scattered" photon with different energy. Note that these 3 steps can occur in any time orderings.

The differentiation between Raman scattering and luminescence happens in the second step. Generally speaking, if the phase relation of the emitted photons to the absorbed ones is not simply determined by the emission/absorption of collective modes, i.e., the phonons, magnons, plasmons, etc., but containing other unspecified inelastic processes, then we can separate the secondary radiation into two parts. The essentially instantaneous coherent scattering of the e–h pair with elementary excitations is referred to as *Raman scattering*, and the incoherent processes are known as the *hot luminescence* [26]. These two processes are in principle inseparable if the photon energy matches the intermediate states, i.e., the resonant condition. In some cases, due to the different relaxation processes involved, hot luminescence can still be distinguished from resonant Raman scattering by time resolved experiments [26]. Such differentiation is not present for nonresonant case, where no luminescence is expected as all processes occur virtually and no phase decoherence happens. In reality for most common materials, due to the complexity and broadening of the electronic bands, the visible light excitation almost always falls in the *near resonance* regime, where both nonresonant and resonant Raman scattering, as well as photoluminescence, contribute to the spectra. Therefore, one has to examine the excitation profile and polarization dependence to distinguish these different contributions and to gain understanding of the intermediate states involved in the secondary radiation [27, 28], as we will discuss in more detail in Chap. 4.

It is worth noting that the electron and hole will eventually thermalize independently with the system to regain the Fermi–Dirac distribution, and if momentum conservation allows, the electron and hole may recombine by emitting a less energetic photon, i.e. (cold) photoluminescence. Depending on the band structure, the photoluminescence can sometimes have much longer wavelength than the incident photon. In most metals, the thermalization process is fast ($\ll$ 1 ps), and the cold luminescence is generally very broad and featureless in energy. In the typical interesting energy range of Raman scattering, i.e., about 100 meV, cold photoluminescence is a weak almost constant and unpolarized background that can be subtracted from the measured intensity. Due to unscreened Coulomb interaction in insulators and semiconductors, the electron-hole pair often thermalizes to form an exciton with well-defined energy levels [29], which decay back to the ground state by radiating photoluminescence sharply center around the dipole allowed transitions. In this case, the decay time is dominated by the exciton lifetime, which can be as long as nanoseconds and results in very sharp features in the luminescent spectra.

For historical reasons, inelastic light scattering spectra obtained using a Fabry–Pérot interferometer are generally referred to as Brillouin scattering, whereas spectra recorded with grating spectrometers are called Raman scattering. In the rest of this chapter, we will focus on the basic theory of Raman scattering. We will assume that the incident and scattered photon energies are much larger than the Raman shift, but small enough such that the photon momentum is negligible compared to the electronic Brillouin zone.

1.2.1 Theory of Raman Scattering

Depending on the emitted photon energy in an inelastic scattering process, the final state of the system could attain either higher or lower energy than the initial state, known as the Stokes and anti-Stokes processes, respectively. That is, the scattered photon frequency in a Stokes process is $\omega_s = \omega_I - \omega$, where ω_I is the frequency of the incident photon, and ω is the Raman shift [Fig. 1.1]. In the anti-Stokes process, the scattered photon frequency is then $\omega_{as} = \omega_I + \omega$. In principle, the quasi momenta should also be conserved during scattering, i.e., $\mathbf{k_s} = \mathbf{k_I} - \mathbf{q}$ for the Stokes process and $\mathbf{k_{as}} = \mathbf{k_I} + \mathbf{q}$ for the anti-Stokes process, where $\mathbf{q}$ is the momentum transfer from the photon. However, the photon typically employed in Raman spectroscopy has wavelength of about 0.4–1 μm, which is about 1000 times larger than the lattice parameters of most crystals. Therefore, the photon momentum is about 1000 times smaller than the reciprocal lattice vectors, and thus the momentum transfer between incident and scattered photon, $\mathbf{q}$, can be taken as zero for all practical considerations. In other words, Raman scattering only probes collective modes near the Brillouin zone center.

The experimental observable in scattering experiments is called "differential scattering cross section" (or spectral differential scattering cross section):

$$\frac{d^2\sigma}{d\Omega d\omega_s} = \frac{scattered\ particle\ flux}{incident\ particle\ flux}; \tag{1.1}$$

note that Ω is the solid angle of the collection cone *inside* the sample. Physically, the differential scattering cross section measures the rate energy transfer from the incident beam into a scattered beam within energy range of $d\omega_s$ and detected within the solid angle $d\Omega$, through (Stokes) scattering with a medium. Therefore, we need to do a conversion from the calculated *energy flux* per unit area, $\frac{d^2 I_s}{d\Omega d\omega_s}\frac{1}{I_I}$, into the *particle flux* that we measure [7].

$$\frac{d^2\sigma}{d\Omega d\omega_s} = \frac{\omega_I A}{\omega_s}\frac{d^2 I_s}{d\Omega d\omega_s}\frac{1}{I_I}, \tag{1.2}$$

where A is the effective scattering area. For light scattering experiments, the photon energy flux is given by the magnitude of Poynting vector [7].

$$\frac{d^2 I_s}{d\Omega d\omega_s}\frac{1}{I_I} = \frac{\omega_s^4 V L n_s}{8\pi^2\epsilon_0 c^3}|\epsilon_s \cdot P_s|^2 \frac{1}{2\epsilon_0 c n_I |E_I|^2}, \tag{1.3}$$

where V is the sample volume and E_I is the incident field amplitude, both will eventually cancel with the "power spectrum" $|\epsilon_s \cdot P_s|^2$. P_s is the light induced polarization in the sample, $P_s = \epsilon_0 \chi \cdot \epsilon_I E_I$, where χ is the dynamical charge susceptibility tensor, and ϵ_s and ϵ_I are directional vector for scattered and incident light, respectively. n_I and n_s are the refractive index of the sample at the wavelength of incident and scattered light, respectively. c is speed of light, and L is effective penetration depth of light. Therefore, the scattering cross section can be simplified as

$$\frac{d^2\sigma}{d\Omega d\omega_s} = \frac{\omega_I \omega_s^3 v n_s}{(4\pi\epsilon_0)^2 c^4} V |\epsilon_s \cdot \chi \cdot \epsilon_I|^2, \tag{1.4}$$

where $v = AL$ is the effective scattering volume. In reality, the evaluation of susceptibility should be taken as thermal dynamic average of all finite energy states,

$$|\epsilon_s \cdot \chi \cdot \epsilon_I|^2 \rightarrow \langle |\epsilon_s \cdot \chi \cdot \epsilon_I|^2 \rangle \equiv \frac{1}{V} \sum_f | \langle f | \epsilon_s \cdot \chi \cdot \epsilon_I | i \rangle |^2, \tag{1.5}$$

with $|i\rangle$ and $|f\rangle$ the initial and final states, respectively.

In principle, all of the physics in the scattering event are encrypted in the last term of Eq. (1.4). Namely, the inelastic light scattering probes the dynamical fluctuations of the charge susceptibility due to the perturbation of the incident photon field. The fluctuation could be due to its coupling to lattice vibrations (phonons) [30] or to magnetic excitations [31]. Equation (1.4) is often referred to as the *macroscopic* theory of light scattering. In general, this is not very useful for *ab initio* calculation of the Raman scattering cross section in crystals, which requires a *microscopic* theory.

Microscopic calculation of the scattering cross section is based on time dependent perturbation theory (the "Fermi's Golden Rules") [14] and the "minimal coupling" (excluding electron spin) Hamiltonian of electromagnetic (EM) field to the electrons,

$$H = H_0 + H_{int} = \frac{(\mathbf{p} + e\mathbf{A})^2}{2m} = \frac{\mathbf{p}^2}{2m} + \frac{e^2}{2m}\mathbf{A} \cdot \mathbf{A} + \frac{e}{m}\mathbf{p} \cdot \mathbf{A}, \tag{1.6}$$

where the first term H_0 describes the kinetic motion of the electron in the medium, and the second and third terms are the interaction Hamiltonian H_{int} that governs the coupling between the electron and EM field. Here we have used the Coulomb gauge, $\nabla \cdot \mathbf{A} = 0$, to simplify the equation. Proper microscopic treatment requires rewriting H_{int} with the vector potential in the quantum operator form [7, 30]. The annihilation and creation operators for a photon with energy ω_k, wave vector $\mathbf{k}$, and polarization ϵ_k are

$$\mathbf{A}(\mathbf{r}) = \sum_k \sqrt{\frac{\hbar}{2\epsilon_0 n^2 V \omega_k}} [\mathbf{a}_k e^{i\mathbf{k}\cdot\mathbf{r}} + \mathbf{a}_k^\dagger e^{-i\mathbf{k}\cdot\mathbf{r}}] \epsilon_k. \tag{1.7}$$

Plugging into Eq. (1.6) and omitting nonlinear terms, the second term becomes [32]

$$H_{AA} = \frac{e^2}{2m} \sum_{k,k'} \left(\frac{\hbar}{2\epsilon_0 n_k n_{k'} V \sqrt{\omega_k \omega_{k'}}} \right) \left[\mathbf{a}_k \mathbf{a}_{k'}^\dagger e^{i(\mathbf{k}-\mathbf{k}')\cdot\mathbf{r}} + \mathbf{a}_k^\dagger \mathbf{a}_{k'} e^{-i(\mathbf{k}-\mathbf{k}')\cdot\mathbf{r}} \right] (\epsilon_k \cdot \epsilon_{k'}). \tag{1.8}$$

To calculate the contributions of Eq. (1.6) in the inelastic light scattering, we need to use the time dependent perturbation theory [14], or more precisely, Fermi's Golden Rule for transition rate between initial state $|i\rangle$ and final state $|f\rangle$,

$$\Gamma_{if} = 2\pi \sum_{i,f} p_i |T|^2 \delta(\omega_I - (\omega_s + \omega_{if})), \tag{1.9}$$

where $\omega_{if} \equiv (E_f - E_i)/\hbar$ is the excitation frequency, commonly called *Raman shift*, and we will simply denote it as ω in the following. $p_i = \exp(-\beta E_i)$ is the probability of being in the initial state, and T is the matrix element that encodes all of the information about interactions between the scattering medium and light. The spectral cross section is proportional to the transition rate by scattered photon density of states per unit solid angle, $\frac{V}{(2\pi)^3} \int d^3\mathbf{k}_s = \frac{V}{8\pi^3}\frac{n_s^3\omega_s^2}{c^3}$, and then normalize to the incident photon flux. We will drop the prefactors in the following since they are not relevant to this experimental thesis study, and rigorous derivations exist in many books and reviews [7, 27, 30].

The scattering matrix element T can be evaluated perturbatively assuming the interaction Hamiltonian H_{int} is weak compared to H_0,

$$T = \langle f|H_{int}|i\rangle + \sum_m \frac{\langle f|H_{int}|m\rangle \langle m|H_{int}|i\rangle}{\hbar(\omega_i - \omega_m)}$$

$$+ \sum_{m,m'} \frac{\langle f|H_{int}|m\rangle \langle m|H_{int}|m'\rangle \langle m'|H_{int}|i\rangle}{\hbar^2(\omega_i - \omega_m)(\omega_i - \omega_{m'})} + \mathcal{O}(H_{int}^4). \tag{1.10}$$

Inserting H_{AA} term of Eq. (1.6) into Eq. (1.10) and keeping only the leading order, the matrix element for scattering with a collection of particles is

$$T \sim \langle f|\sum_j e^{i\mathbf{q}\cdot\mathbf{r}_j}|i\rangle\, (\boldsymbol{\epsilon}_I \cdot \boldsymbol{\epsilon}_s), \tag{1.11}$$

where $\mathbf{q} = \mathbf{k}_I - \mathbf{k}_s$ is the momentum transfer. In typical Raman scattering experiments, the used wavelength of light is much larger than the lattice parameters, i.e., $\mathbf{q} \approx 0$ compared to the Brillouin zone size, and therefore the exponential term is the order of unity (dipole approximation). Thus, in scattering by electronic transitions, the matrix element is only appreciable with the same initial and final states while conserving the momentum, i.e., the *intraband* transitions. In fact, it can be shown that the cross section of this term is equivalent to the single band electron density–density correlation function close to the Brillouin zone center (at $\mathbf{q} \approx 0$), measuring excitations of conduction electrons and plasmons in crystals [7, 32]. This is why Raman scattering from electrons is often referred to as probing the charge density fluctuation. Following the same argument, Raman response vanishes at zero energy, because there is no charge "fluctuation" in the static limit. Note that H_{AA} does not contribute to the "crossed polarization" scatterings due to the term $(\boldsymbol{\epsilon}_I \cdot \boldsymbol{\epsilon}_s)$ and therefore will never contribute to the antisymmetric symmetry channels, e.g., A_{2g} in D_{4h} group.

The third term in Eq. (1.6) can be written as

$$H_{PA} = \frac{e}{m}\sum_{k}\sqrt{\frac{\hbar}{2\epsilon_0 n_k^2 V\omega_k}}[\mathbf{a}_k e^{i\mathbf{k}\cdot\mathbf{r}} + \mathbf{a}_k^{\dagger} e^{-i\mathbf{k}\cdot\mathbf{r}}](\mathbf{p}\cdot\epsilon_k). \tag{1.12}$$

This term obviously does not conserve photon number and thus requires second order perturbation in Eq. (1.10) as an elastic light scattering process. The matrix element then reads

$$\begin{aligned} T \sim \sum_m \Bigg[&\frac{\langle f|\sum_j(\mathbf{p}_j\cdot\epsilon_s)e^{-i\mathbf{k}_s\cdot\mathbf{r}_j}|m\rangle\,\langle m|\sum_j(\mathbf{p}_j\cdot\epsilon_I)e^{i\mathbf{k}_I\cdot\mathbf{r}_j}|i\rangle}{\omega_i-\omega_m+\omega_I} \\ &+\frac{\langle f|\sum_j(\mathbf{p}_j\cdot\epsilon_I)e^{i\mathbf{k}_I\cdot\mathbf{r}_j}|m\rangle\,\langle m|\sum_j(\mathbf{p}_j\cdot\epsilon_s)e^{-i\mathbf{k}_s\cdot\mathbf{r}_j}|i\rangle}{\omega_i-\omega_m-\omega_s}\Bigg]. \end{aligned} \tag{1.13}$$

Note that the matrix element now contains electronic momenta and thus allows scattering between different electronic states, i.e., *interband* transitions [Fig. 1.2b]. Moreover, the incident and scattered photon polarizations need not be parallel as in the H_{AA} term. Therefore, the H_{PA} term is more significant for almost all collective excitations in crystals [7].

We can further draw relation between the experiment (Raman cross section) and the theory (matrix element) by rewriting the T-matrix into correlation functions. Define the single particle states, $|\alpha\rangle$, $|\alpha'\rangle$, $|\beta\rangle$, and $|\beta'\rangle$, and the manybody creation operator $c_\alpha^{\dagger}$. We can combine and rewrite the T-matrices from H_{AA} and H_{PA} [33],

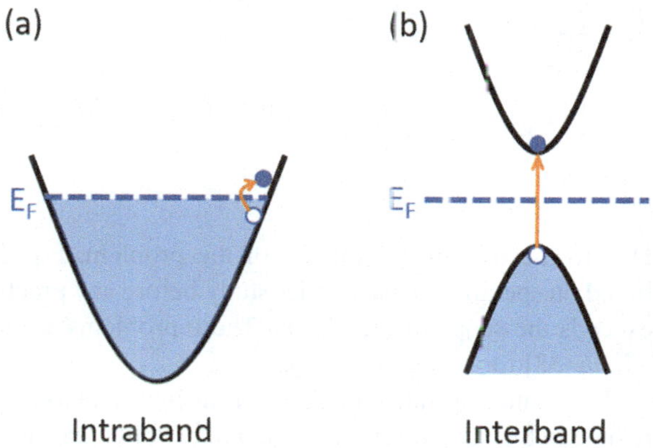

Fig. 1.2 Illustrations of (**a**) intraband and (**b**) interband transitions

$$T \sim (\epsilon_I \cdot \epsilon_s) \sum_{\alpha,\beta} \rho_{\alpha,\beta}(\mathbf{q}) \langle f | c^\dagger_\alpha c_\beta | i \rangle$$
$$+ \frac{1}{m_e} \sum_m \sum_{\alpha,\alpha',\beta,\beta'} p_{\alpha,\alpha'}(\mathbf{q}_s) p_{\beta,\beta'}(\mathbf{q}_I) \tag{1.14}$$
$$\times \left[\frac{\langle f | c^\dagger_\alpha c_{\alpha'} | m \rangle \langle m | c^\dagger_\beta c_{\beta'} | i \rangle}{\omega_i - \omega_m + \omega_I} + \frac{\langle f | c^\dagger_\beta c_{\beta'} | m \rangle \langle m | c^\dagger_\alpha c_{\alpha'} | i \rangle}{\omega_i - \omega_m - \omega_s} \right],$$

where $\rho_{\alpha,\beta}(\mathbf{q}) = \langle \alpha | e^{i\mathbf{q}\cdot\mathbf{r}} | \beta \rangle$ is the electron density matrix element, and $p_{\alpha,\alpha'}(\mathbf{q}_{\mathbf{I},\mathbf{s}}) = \langle \alpha | \mathbf{p} \cdot \epsilon_{I,s} e^{\pm i \mathbf{q}_{I,s}\cdot\mathbf{r}} | \alpha' \rangle$ is the momentum matrix element.

Shastry and Shraiman further argued that if we only consider correlation effects for the electronic bands with energy comparable to the photon energies, ω_I and ω_s, which are typically around 2 eV, then the interaction Hamiltonian can be rewritten as

$$H_{int} = \frac{e^2}{2\hbar^2 c^2} \sum_{\alpha\beta} A_\alpha \gamma_{\alpha\beta} A_\beta + \frac{e}{\hbar c} \mathbf{j} \cdot \mathbf{A}, \tag{1.15}$$

where $\gamma_{\alpha\beta}(\mathbf{q}) = \sum_{\mathbf{k}} \frac{\partial E(\mathbf{k})}{\partial k_\alpha \partial k_\beta} c^\dagger(\mathbf{k} + \mathbf{q}/2) c(\mathbf{k} - \mathbf{q}/2)$ is usually called Maxwell's stress tensor, relating the energy to vector potential, and $j_\alpha(\mathbf{q}) = \sum_{\mathbf{k}} \frac{\partial E(\mathbf{k})}{\partial k_\alpha} c^\dagger(\mathbf{k} + \mathbf{q}/2) c(\mathbf{k} - \mathbf{q}/2)$ is the current operator. Then, the T-matrix can be further generalized to strongly correlated materials [33–35],

$$T \sim \sum_{\alpha\beta} \epsilon^\alpha_I \epsilon^\beta_s$$
$$\times \left[\langle f | \gamma_{\alpha,\beta}(\mathbf{q}) | i \rangle + \sum_m \left(\frac{\langle f | j_\beta(\mathbf{q}_s) | m \rangle \langle m | j_\alpha(\mathbf{q}_I) | i \rangle}{\omega_i - \omega_m - \omega_I} + \frac{\langle f | j_\alpha(\mathbf{q}_I) | m \rangle \langle m | j_\beta(\mathbf{q}_s) | i \rangle}{\omega_i - \omega_m + \omega_s} \right) \right]. \tag{1.16}$$

Due to the overall complexity of the problem, Eq. (1.16) needs to be simplified based on specific systems under study before any practical calculation and therefore exceeds the scope of this thesis. These problems are nicely discussed in Refs. [33, 34, 36–38] for several systems.

The scattering rate calculated from Fermi's Golden Rule in Eq. (1.9) is further related to the autocorrelation function of "effective charge density" operator, $\rho(x, t)$

$$S(\omega, q) = \hbar r_0^2 \frac{\omega_s}{\omega_I} \Gamma_{if} \sim \sum_{\mathbf{k},\mathbf{k}'} \int_{-\infty}^{\infty} \langle c^\dagger_{\mathbf{k}}(t) c_{\mathbf{k}+\mathbf{q}}(t) c^\dagger_{\mathbf{k}'+\mathbf{q}}(0) c_{\mathbf{k}'}(0) \rangle \, e^{i\omega t} dt, \tag{1.17}$$

which is often referred to as the dynamical electron "effective density–density" correlation function [33]. The fluctuation–dissipation theorem further relates the measured fluctuation (correlation function) to the dissipation of the system (linear response function). The cross section of the Stokes process at zero temperature is simply proportional to the imaginary part of the linear response function (dissipation), $\chi''(\omega, q)$, where ω is the photon energy transfer and q is the momentum transfer. That is, the Stokes intensity will be entirely determined by the *spontaneous scattering* due to quantum fluctuations. Notice that this term is purely quantum mechanical and is irrelevant to the details of the response function under consideration [39].

At finite temperature, the thermal fluctuation will result in an additional term in the correlation function, leading to the *stimulated scattering*. This term is related to the nature of the perturbation, and for Raman scattering, it is determined by the Bose–Einstein factor, $n(\omega, T) = 1/(\exp[\frac{\hbar\omega}{k_B T}] - 1)$. Thus the cross section for the Stokes process in Raman scattering is related to the Raman response function $\chi''(\omega, q)$ by

$$\frac{d^2\sigma}{d\Omega d\omega} \sim S(\omega, q) = [1 + n(\omega, T)]\chi''(\omega, q). \tag{1.18}$$

The details of optical corrections for converting this cross section into the scattering intensity measured by the spectrometer *outside* the sample will be given in the next chapter, Sect. 2.2.3.

For materials without *long range ordered* magnetism, the scattering cross section for the Stokes and anti-Stokes processes is related by the principle of detailed balance [7]:

$$\frac{(d^2\sigma/d\Omega d\omega)_{AS}}{(d^2\sigma/d\Omega d\omega)_S} \sim \frac{n(\omega, T)}{1 + n(\omega, T)} = \exp\left[\frac{-\hbar\omega}{k_B T}\right], \tag{1.19}$$

assuming that the energy shift is much less than the incident laser frequency, $\omega \ll \omega_L$. Therefore, by measuring the ratio of scattering cross section for the Stokes and anti-Stokes processes, we can in principle determine the sample temperature and thus estimate the effect of laser heating.

1.2.2 Nonresonant Raman Scattering

As we discussed earlier in the beginning of this section, if both the incident and scattered photon energies are much smaller than the intermediate state energy, $\omega_{I,s} \ll |\omega_i - \omega_m|$, then the scattering is called "non-resonant," and the intermediate states are called "*virtual* states." If we further limit to *intraband* transitions in weakly interacting systems [Fig. 1.2a], then the Raman cross section can be simplified to the following form [33]:

$$\frac{d^2\sigma}{d\omega d\Omega} \sim |\epsilon_I \cdot \mathbf{M}^{-1} \cdot \epsilon_s|^2 \sum_{\mathbf{k},\mathbf{k}'} \int_{-\infty}^{\infty} \langle c_{\mathbf{k}}^{\dagger}(t) c_{\mathbf{k}+\mathbf{q}}(t) c_{\mathbf{k}'+\mathbf{q}}^{\dagger}(0) c_{\mathbf{k}'}(0) \rangle \, e^{i\omega t} dt, \quad (1.20)$$

where $\mathbf{M}^{-1}$ is called the inverse "effective mass tensor" [40],

$$\mathbf{M}^{-1} \equiv \frac{\partial^2 E(\mathbf{k})}{\partial k_\alpha \partial k_\beta} = \frac{1}{m_e}\delta_{\alpha\beta} + \frac{1}{m_e^2}\sum_m \left(\frac{\langle i|p_\beta|m\rangle \langle m|p_\alpha|i\rangle + \langle i|p_\alpha|m\rangle \langle m|p_\beta|i\rangle}{\omega_i - \omega_m} \right). \quad (1.21)$$

This is known as the "effective mass" approximation. This is very useful in metals or large band gap insulators, where the conduction band width or band gap is commonly over 3 eV, such that no *real* interband transitions need to be considered when visible light is used. Within the approximation, by utilizing the symmetry properties of the Raman cross section that we will discuss later, one can even relate symmetry resolved Raman response function to different parts of the Fermi surface [33]. In most metallic materials, unless there exists a singularity in the joint density of states for vertical transitions of visible wavelength, the effective mass approximation is always a good starting point for understanding the Raman scattering spectra.

Up to this point, the Raman scattering cross section has been discussed quite generally without restricted to any specific systems. In the following, we will briefly discuss the most common situations for Raman scattering in solids, i.e., light scattering with lattice vibrations.

1.2.3 Raman Scattering of Phonon

The theory of inelastic light scattering by molecule or lattice vibrations is first developed by Placzek [30], who realized that the vibration of atoms around its equilibrium position creates a time dependent electric polarization, like an oscillating dipole, which as we known leads to emission of electromagnetic waves. In solids, the dominant Raman scattering process is always the one in which the photons interact indirectly with the phonons via the electrons [25]. The photon field interacts with the "deformation potential" induced by the lattice vibrations and thus the energy shift by the phonon energies.

Microscopically, the lattice vibrations modify the dynamical electric susceptibility of the crystal, $\chi(u, t)$ [41]. Note that this is usually called polarizability, $\alpha(u, t)$, for molecule vibrations. Expanding the susceptibility around the equilibrium position of nuclei in the normal coordinates,

$$\chi(u, t) = \chi(0) + (\mathbf{u}(t) \cdot \nabla)\chi(u, t)|_{u=u(0)} + \mathcal{O}(\mathbf{u}^2), \quad (1.22)$$

where $\mathbf{u}(t) = \mathbf{u}_0 e^{-i\omega t} + \mathbf{u}_0^* e^{i\omega t}$ is the lattice displacement due to a phonon mode with frequency ω. The first term in $\chi(u, t)$ does not modify the frequency of perturbation (elastic scattering), and thus we only keep the second term linear in $\mathbf{u}$. Then the time dependent susceptibility is of the form $\chi(t) = \chi_0 e^{-i\omega t} + \chi_0^* e^{i\omega t}$. Assume that the incoming photon has electric field $\mathbf{E} = \mathbf{E}_0 e^{-i\omega_I t} + \mathbf{E}_0^* e^{i\omega_I t}$, where ω_I is the photon frequency. Then, the induced dipole moment is

$$\mathbf{P}(t) = [\chi_0 \cdot \mathbf{E}_0 e^{-i(\omega_I+\omega)t} + \chi_0^* \cdot \mathbf{E}_0^* e^{i(\omega_I+\omega)t}] + [\chi_0^* \cdot \mathbf{E}_0 e^{-i(\omega_I-\omega)t} + \chi_0 \cdot \mathbf{E}_0^* e^{i(\omega_I-\omega)t}]. \tag{1.23}$$

The first term is oscillating at a slightly lower frequency and thus corresponds to the Stokes process, and the second term is responsible for the anti-Stokes process. The quantum mechanical derivation of phonon scattering turns out to be much more complicated, involving third order perturbation that includes the electron–phonon coupling and neglects the H_{AA} term [7, 25, 42, 43]. However, the conclusion remains the same with simple semi-classical arguments [43, 44]. Therefore we will not reproduce them in this thesis.

The Raman active phonon typically has a Lorentzian lineshape [45]. However, when a sharp phonon mode with energy $\hbar\omega_0$ interacts with a broad electronic continuum, the interference between the transition amplitudes for electronic continuum T_e and the phonon mode T_p results in a "Fano lineshape" [44, 46],

$$I(\omega) = \frac{\pi\rho(\omega)T_e^2(\omega_0 - \omega - VT_p/T_e)^2}{(\omega_0 - \omega + V^2R(\omega))^2 + (\pi V^2\rho(\omega))^2}. \tag{1.24}$$

The electron density of states, $\rho(\omega) = \mathrm{Im}[G(\omega)]$, is usually assumed to be some broad function or even a constant, where $G(\omega) = \sum_e 1/(\omega_\epsilon - \omega - i\Gamma)$ is the unperturbed Green's function for the electronic continuum. The electron–phonon interaction is V, and $R(\omega) = \mathrm{Re}[G(\omega)]$. Therefore, the Fano effect is a phonon mode selective probe for studying the electron–phonon coupling in metallic systems, which is especially useful when studying the mechanism of superconductivity in some compounds [47, 48].

1.3 Symmetry Selection Rules in Raman Scattering

The symmetry properties of the scattering cross section, i.e., measured polarization dependences, are determined by the symmetry properties of the Raman response function, $\chi''(\omega)$. Quite generally, the Hamiltonian of a system commutes with the symmetry operators of the crystal's factor group (space group without translational symmetry operators). In other words, the Hamiltonian should transform as the full symmetric irreducible representation of the crystal's factor group. This is called Neumann's Principle [49], first stated by Franz E. Neumann and then later popularized by his student Woldermar Voigt. It was later generalized by Pierre Curie [50], stating that a physical phenomenon may exist and only exist in a medium that possesses the same symmetry characteristic of the phenomenon or that of a subgroup of that symmetry.

The eigenfunctions of a Hamiltonian can be projected to form basis functions for the irreducible representations of the group. For a certain point group under considerations, let the initial state $|i\rangle$ in a scattering process have symmetry Γ_i and the final state $|f\rangle$ have symmetry Γ_f. The symmetry of the excitation, $|f\rangle\langle i|$, is then equal to $\Gamma_f \otimes \Gamma_i^*$. Let the electric field vector for the incident and scattered light transform as Γ_{ϵ_I} and Γ_{ϵ_s}, respectively. Then, a very general group theoretical statement is that only excitations that satisfy the following relation are allowed [7, 51]:

$$\Gamma_f \otimes \Gamma_i^* \subset \Gamma_{\epsilon_I}^* \otimes \Gamma_{\epsilon_s}. \tag{1.25}$$

Since the algebra and symmetry properties of all irreducible representations for all 32 point groups in 3D are documented [52, 53], we can easily find out the allowed excitation symmetries in a crystal.

Consider now the point group D_{4h}, suitable for the lattice symmetry of URu_2Si_2, which we will discuss in Chap. 3. The electric field vector transforms as Γ_5^- (E_u) along a-axis and Γ_2^- (A_{2u}) along c-axis. Then, $\Gamma_5^- \otimes \Gamma_5^- = \Gamma_1^+ \oplus \Gamma_2^+ \oplus \Gamma_3^+ \oplus \Gamma_3^+$ (in the Bethe notation), or equivalently in the Mulliken notation, $A_{1g} \oplus A_{2g} \oplus B_{1g} \oplus B_{2g}$. This suggests that only excitations with the symmetry of these four irreducible representations are allowed for inelastic light scattering from the ab-surface. In order to access the Γ_5^+ (E_g) excitations, we need to have $\Gamma_5^- \otimes \Gamma_2^- = \Gamma_5^+$, that is, scattering from the ac-surface with incident and scattered light polarization orthogonal to each other (crossed polarization).

The superscript plus sign in the Bethe notation or the subscript g in the Mulliken notation denotes parity even (stands for *gerade*, meaning *even* in German). Similarly, the superscript minus sign and the subscript u mean parity odd (*ungerade*). This distinguish is of course dropped for noncentrosymmetric crystals or 2D wallpaper groups, as we will see in the following for Bi_2Se_3. One immediately notices that, for crystals with inversion symmetry, the multiplication of any two *ungerade* representation will always result in gerade representations. Therefore, only at most half of the irreducible representations are allowed in Raman scattering. This is indeed the case, and these excitations are called Raman active modes. The excitations with *ungerade* symmetries are accessible by absorption or emission spectroscopies, such as infrared spectroscopies.

In Bi_2Se_3, an interesting scenario happens where both excitations from the bulk and from the surface need to be considered [Chap. 4]. The bulk lattice has symmetry of the D_{3d} group, which follows from a similar exercise as above, that only excitations with the symmetries $\Gamma_3^- \otimes \Gamma_3^- = \Gamma_1^+ \oplus \Gamma_2^+ \oplus \Gamma_3^+$ are allowed from ab-plane scattering. Note here that scattering from the ac-plane does not yield any extra symmetry channels. The surface states in Bi_2Se_3 belong to a different symmetry group, C_{6v} [54], which obviously does not have inversion symmetry. The electric field vector in ab-plane transforms as Γ_5 (E_1) and as Γ_1 (A_1) along the c-axis. Thus, the accessible symmetry channels from the ab-plane are $\Gamma_5 \otimes \Gamma_5 = \Gamma_1 \oplus \Gamma_2 \oplus \Gamma_6$

$(A_1 \oplus A_2 \oplus E_2)$, and the Γ_5 (E_1) symmetry channel is only accessible from the *ac*-surface. Note that in this group, Γ_3 and Γ_4 symmetries will remain silent for two photon scattering process and will only be accessible by nonlinear optics.

The above relations are usually referred to as the Raman selection rules, which determine the allowed excitations symmetries. However, the symmetry properties of each irreducible representation impose much stronger rules on the Raman susceptibility tensors [51], which relates the cross sections for each symmetry channel to the polarizations of light. The bilinear dependence of Raman susceptibility on electric field polarization, $\epsilon_s \cdot \chi \cdot \epsilon_I$, implies the susceptibility tensor to be rank-2. A very general and pedagogical recipe for obtaining the susceptibility tensors for each symmetry channel of the 32 point groups is given by Callen in Ref. [55], and the results for Raman scattering are tabulated first by L.N. Ovander [56] and then cited in several monographs as the "Raman tensors" [7, 27].

We will not repeat the tables here but merely use group D_{4h} as an example for calculating the polarization dependence of cross sections using the Raman tensors.

$$A_{1g} : \begin{bmatrix} a & & \\ & a & \\ & & b \end{bmatrix} ; A_{2g} : \begin{bmatrix} & c & \\ -c & & \\ & & \end{bmatrix} ; B_{1g} : \begin{bmatrix} d & & \\ & -d & \\ & & \end{bmatrix} ; B_{2g} : \begin{bmatrix} & e & \\ e & & \\ & & \end{bmatrix} , \tag{1.26}$$

where the empty spaces in the 3 by 3 matrices are zeros, and the columns and rows are x, y, and z Cartesian coordinates in the *laboratory frame*. The cross section of, for example, the A_{2g} symmetry is then proportional to $|\epsilon_s^x \cdot \mathbf{R} \cdot \epsilon_I^y|^2 \sim c^2$, where $\mathbf{R}$ is the Raman tensor, $\epsilon^x = [1, 0, 0]$, and $\epsilon^y = [0, 1, 0]$. Thus the excitations with A_{2g} symmetry will be contained in the XY scattering geometry. One would need to do this repeatedly for all symmetry channels and for all scattering geometries, to find out the expected polarization dependence of the excitations. Note that due to time reversal symmetry, the circularly polarized light changes chirality on back reflection. Thus, $\epsilon_I^R = [1, -i, 0]/\sqrt{2}$ and $\epsilon_s^R = [1, i, 0]/\sqrt{2}$. As we will see in Chaps. 3 and 4, the circularly polarized light is important to separate A_{2g} symmetry from the other excitations [33, 57].

1.4 Chiral Excitations in Solids

As we introduced above, the phononic Raman cross section is identified as proportional to the electric susceptibility tensor, which is restricted to real and symmetric form [30]. Thus, time reversal symmetry forbids light scattering through the fully antisymmetric Raman tensors, such as collective modes with A_2 or A_{2g} symmetry [58]. Historically, the excitations in the antisymmetric channel are referred to as the "chiral excitations" due to their intimate relationship with the spin chiral operator $\mathbf{s_1} \cdot (\mathbf{s_2} \times \mathbf{s_3})$ [34, 36, 59]. For example, antisymmetric scattering was predicted for Dirac spin-liquid state in the spin-1/2 kagome system [36], but yet unobserved to date.

It was noted by Placzek at the very end of his book that contribution from antisymmetric tensors ($A_{2(g)}$ irreducible representation) may arise due to breakdown of the polarizability description near resonance conditions [30]. Indeed, this was observed in resonant Raman scattering of heme proteins, such as ferrocytochrome-c [60–62]. Besides resonance scattering, antisymmetric nonresonant vibrational Raman scattering is allowed if one goes beyond the Born–Oppenheimer approximations, such as in systems with exceptionally strong electron–phonon coupling or in the proximity of a phonon mediated phase transition [63]. Generally speaking, magnetic Raman scattering could also couple to the antisymmetric tensor [7]. This is because both the magnetic moments and antisymmetric $A_{2(g)}$ representations transform as a pseudovector (in the lowest order). Many examples exist where sharp magnon or two-magnon peaks were observed in the antisymmetric channel [1, 7, 31]. In magnetic materials, phonon scattering symmetry selection rules are partly relaxed due to Faraday rotation, which also lead to nonzero antisymmetric Raman matrix elements [64, 65].

For nonmagnetic and non-resonance Raman scattering from crystalline samples, much less examples of nonzero signal with A_{2g} symmetry are known. To the best of our knowledge, all of the examples come from electronic Raman scatterings. In general, antisymmetric excitations are allowed for electronic *interband* transitions [56]. For example, transitions between crystal field split states can contribute to the antisymmetric channel. Indeed, the very first observation of Raman scattering in the antisymmetric channel was reported in 1968 from crystal field split states of embedded magnetic ions (Eu^{3+}:YAG) [66] and latter found in several rare earth compounds [67, 68]. However, this is not a common observation, other than very localized $4f$ electrons. Because most systems have full symmetric ground states, and thus an $A_{2(g)}$ interband transition would require a fully antisymmetric $A_{2(g)}$ excited state, which is typically not possible with few exemptions.

The most famous example is perhaps in the insulating cuprates, where an $A_{2(g)}$ symmetry peak slightly below the charge transfer gap edge was observed [32, 69, 70]. This is later understood as excitonic interband transition between B_{1g} and B_{2g} symmetrized copper 3d orbital states [32]. Such transition is in some sense also a crystal field excitation, but the band width is much larger than in $4f$ systems, and therefore the peak is also 2 orders of magnitude broader. An alternative explanation to the peak was given by Khveshchenko and Wiegmann as dynamic chiral spin fluctuations [59], which is a collective zero mode being pulled below the charge transfer gap edge by the exchange interaction J. In this view, the excited state could be thought of as a bound state between the double occupant and the hole sites, where the $A_{2(g)}$ symmetry originates from the local "circular current" induced by the charge transfer.

More recently, antisymmetric continuum was reported from the inter Dirac cones excitations in graphene [71, 72]. Note that the electronic Raman cross section of intraband transitions in metals is usually related to the inverse effective mass tensor, which is always symmetric. That is, Raman scattering cross section of metals typically does not contain any signal in the fully antisymmetric $A_{2(g)}$ channel. However, the excitations in semi-metallic graphene provided an exception due to its peculiar helical band structure, i.e., the electron momentum locked to pseudo-spin. Similar examples of spin-momentum lock-in can be found in many topological materials [Chap. 4] and therefore provide interesting platforms to look for antisymmetric Raman excitations in future works.

Here in this monograph, we add two more examples of electronic excitations from antisymmetric channel to the above list. More precisely, we observe sharp $A_{2(g)}$ collective modes in two entirely different nonmagnetic systems, heavy fermion metal URu_2Si_2 and 3D topological insulator Bi_2Se_3. Observation of an antisymmetric collective mode in nonmagnetic systems is highly unusual and rare, and the mechanism is different for each case. This monograph is dedicated to explore the properties, mechanism, and physical implications from these two new systems.

References

1. M.G. Cottam, D.J. Lockwood, *Light Scattering in Magnetic Solids* (Wiley, New York, 1986)
2. Kung et al., Science **347**, 6228 (2015). https://doi.org/10.1126/science.1259729
3. Kung et al., Phys. Rev. Lett. **117**, 227601 (2016). https://doi.org/10.1103/PhysRevLett.117.227601
4. Kung et al., Phys. Rev. Lett. **119**, 136802 (2017). https://doi.org/10.1103/PhysRevLett.119.136802
5. Kung et al., Proc. Natl. Acad. Sci. U.S.A. **116**, 4006 (2019). https://doi.org/10.1073/pnas.1813514116
6. Kung et al., Phys. Rev. B **95**, 245406 (2017). https://doi.org/10.1103/PhysRevB.95.245406
7. W. Hayes, R. Loudon, *Scattering of Light by Crystals* (Courier Corporation, Chelmsford, 2012)
8. E. Gross, Change of wave-length of light due to elastic heat waves at scattering in liquids. Nature **126**, 201 (1930)
9. L. Landau, G. Placzek, The structure of undisplaced scattered lines. Physik Z. Sowjetunion **5**, 172 (1934)
10. L. Brillouin, Diffusion of light and X-rays by a transparent homogeneous body. Ann. Phys.(Paris) **17**, 88 (1922)
11. P. Debye, Zur Theorie der spezifischen Wäarmen. Annalen der Physik **344**(14), 789–839 (1912). https://doi.org/10.1002/andp.19123441404
12. A. Smekal, Zur Quantentheorie der Dispersion. Naturwissenschaften **11**(43), 873–875 (1923). https://doi.org/10.1007/BF01576902
13. H.A. Kramers, W. Heisenberg, Über die Streuung von Strahlung durch Atome. Zeitschrift für Physik **31**(1), 681–708 (1925). https://doi.org/10.1007/BF02980624
14. P.A.M. Dirac, The quantum theory of the emission and absorption of radiation. Proc. R. Soc. Lond. A Math. Phys. Eng. Sci. **114**(767), 243–265 (1927). https://doi.org/10.1098/rspa.1927.0039
15. P.A.M. Dirac, The quantum theory of dispersion. Proc. R. Soc. Lond. A: Math. Phys. Eng. Sci. **114**(769), 710–728 (1927). https://doi.org/10.1098/rspa.1927.0071
16. C.V. Raman, On the molecular scattering of light in water and the colour of the sea. Proc. R. Soc. Lond. A **101**(708), 64–80 (1922). https://doi.org/10.1098/rspa.1922.0025
17. C.V. Raman, K.S. Krishnan, A new type of secondary radiation. Nature **121**, 501–502 (1928). https://doi.org/10.1038/121501c0

18. C.V. Raman, A new radiation. Indian J. Phys. **2**, 387–398 (1928)
19. R.W. Wood, Wave-length shifts in scattered light. Nature **122**, 349 (1928). https://doi.org/10.1038/122349d0
20. G. Landsberg, L. Mandelstam, Eine neue Erscheinung bei der Lichtzerstreuung in Krystallen. Naturwissenschaften **16**(28), 557–558 (1928). https://doi.org/10.1007/BF01506807
21. I.L. Fabelinskiĭ, Seventy years of Raman scattering. Physics-Uspekhi **41**(12), 1229 (1998)
22. R.A. Millikan, A direct photoelectric determination of Planck's "h". Phys. Rev. **7**, 355–388 (1916). https://doi.org/10.1103/PhysRev.7.355
23. A.H. Compton, A quantum theory of the scattering of X-rays by light elements. Phys. Rev. **21**, 483–502 (1923). https://doi.org/10.1103/PhysRev.21.483
24. A. Pinczuk, E. Burstein, Fundamentals of inelastic light scattering in semiconductors and insulators, in *Light Scattering in Solids I* (Springer, New York, 1983), pp. 23–78
25. R. Loudon, Theory of the first-order Raman effect in crystals. Proc. R. Soc. Lond. A Math. Phys. Eng. Sci. **275**(1361), 218–232 (1963). https://doi.org/10.1098/rspa.1963.0166
26. Y.R. Shen, Distinction between resonance Raman scattering and hot luminescence. Phys. Rev. B **9**, 622–626 (1974). https://doi.org/10.1103/PhysRevB.9.622
27. M. Cardona, Resonance phenomena, in *Light Scattering in Solids II*, ed. by M. Cardona, G. Güntherodt (Springer, Berlin, 1982), pp. 45–49
28. R. Planel, C.B. Guillaume, Optical orientation of excitons, in *Optical Orientation*, Chap. 8, ed. by F. Meier, B.P. Zakharchenya (Elsevier Science Ltd, Amsterdam, 1984), p. 353
29. R.S. Knox, *Theory of Excitons*, vol. 5 (Academic, New York, 1963)
30. G. Placzek, The Rayleigh and Raman scattering, in *Berkeley, Calif: Lawrence Radiation Laboratory* (1959). Translated from the Handbuch der Radiologie, ed. by E. Marx, Leipzig, Akademische Verlagsgesellschaft VI, 2, 209–374 (1934)
31. P.A. Fleury, R. Loudon, Scattering of light by one- and two-magnon excitations. Phys. Rev. **166**, 514–530 (1968). https://doi.org/10.1103/PhysRev.166.514
32. D.V. Salamon, Large-shift Raman scattering in insulating cuprate materials. PhD thesis. University of Illinois at Urbana-Champaign, 1994
33. T.P. Devereaux, R. Hackl, Inelastic light scattering from correlated electrons. Rev. Mod. Phys. **79**, 175–233 (2007). https://doi.org/10.1103/RevModPhys.79.175
34. B.S. Shastry, B.I. Shraiman, Theory of Raman scattering in Mott-Hubbard systems. Phys. Rev. Lett. **65**, 1068–1071 (1990). https://doi.org/10.1103/PhysRevLett.65.1068
35. B.S. Shastry, B.I. Shraiman, Raman scattering in Mott-Hubbard systems. Int. J. Mod. Phys. B **5**, 365–388 (1991). https://doi.org/10.1142/S0217979291000237
36. W.-H. Ko et al., Raman signature of the U(1) Dirac spin-liquid state in the spin-$\frac{1}{2}$ Kagome system. Phys. Rev. B **81**, 024414 (2010). https://doi.org/10.1103/PhysRevB.81.024414
37. A.M. Shvaika et al., Resonant enhancement of inelastic light scattering in strongly correlated materials. Phys. Rev. Lett. **93**, 137402 (2004). https://doi.org/10.1103/PhysRevLett.93.137402
38. A.M. Shvaika et al., Electronic Raman scattering in correlated materials: a treatment of nonresonant, mixed, and resonant scattering using dynamical mean-field theory. Phys. Rev. B **71**, 045120 (2005). https://doi.org/10.1103/PhysRevB.71.045120
39. P. Coleman, *Introduction to Many-Body Physics* (Cambridge University Press, Cambridge, 2015). https://doi.org/10.1017/CBO9781139020916
40. N.W. Ashcroft, N.D. Mermin, *Solid State Physics*. HRW International editions. Holt, Rinehart and Winston, 1976
41. M. Born, K. Huang, *Dynamical Theory of Crystal Lattices*. Oxford Classic Texts in the Physical Sciences (Clarendon Press, Oxford, 1954)
42. R.F. Wallis, M. Balkanski, *Many-body Aspects of Solid State Spectroscopy* (North-Holland Physics, New York, 1986)
43. M.V. Klein, Vibrational Raman scattering from crystals, in *Dynamical Properties of Solids VI: The Modern Physics of Phonons*, ed. by G.K. Horton, A.A. Maradudin (Elsevier, New York, 1990), pp. 65–128

44. M.V. Klein, Electronic Raman scattering, in *Light Scattering in Solids I*, ed. by M. Cardona, G. Güntherodt (Springer, Berlin, 1983), pp. 169–172
45. R Loudon, Theory of the resonance Raman effect in crystals. J. Phys. France **26**(11), 677–683 (1965). https://doi.org/10.1051/jphys:019650026011067700
46. U. Fano, Effects of configuration interaction on intensities and phase shifts. Phys. Rev. **124**, 1866–1878 (1961). https://doi.org/10.1103/PhysRev.124.1866
47. I. Ohana, M.S. Dresselhaus, S. Tanuma, Resonant Raman effect and Fano distortion in the stage-2 graphite donor intercalation compound C/Rb. Phys. Rev. B **43**, 1773–1776 (1991). https://doi.org/10.1103/PhysRevB.43.1773
48. P. Zhou et al., Raman-scattering study of the electron-phonon interaction in M_3C_{60} (M=K,Rb). Phys. Rev. B **48**, 8412–8417 (1993). https://doi.org/10.1103/PhysRevB.48.8412
49. F.E. Neumann, O.E. Meyer, *Vorlesungen über die Theorie der Elasticität der festen Körper und des Lichtäthers: gehalten an der Universität Königsberg*, vol. 4 (BG Teubner, Leipzig, 1885)
50. P. Curie, Sur la symétrie dans les phénomènes physiques, symétrie d'un champ électrique et d'un champ magnétique. Journal de Physique Théorique et Appliquée **3**(1), 393–415 (1894)
51. J.F. Nye, *Physical Properties of Crystals: Their Representation by Tensors and Matrices* (Oxford science publications, Clarendon Press, New York, 1985)
52. G.F. Koster. *Properties of the Thirty-Two Point Groups*. Massachusetts Institute of Technology Press Research Monograph (M.I.T. Press, Cambridge, 1963)
53. E.B. Wilson, J.C. Decius, P.C. Cross, *Molecular Vibrations: The Theory of Infrared and Raman Vibrational Spectra*. Dover Books on Chemistry (Dover Publications, New York, 2012)
54. J. Li, J.J. Tu, J.L. Birman, Symmetry predicted transitions in 3D topological insulators. Solid State Commun. **163**, 11–14 (2013). http://dx.doi.org/10.1016/j.ssc.2013.03.010
55. H. Callen, Crystal symmetry and macroscopic laws. Am. J. Phys. **36**(8), 735–748 (1968). https://doi.org/10.1119/1.1975102
56. L.N. Ovander, The form of the Raman tensor. Opt. Spectrosc **9**, 302 (1960)
57. J. Nestro, T.G. Spiro, Circularly polarized Raman spectroscopy: Direct determination of antisymmetric scattering in the resonance Raman spectrum of ferrocytochrome c. J. Raman Spectrosc. **1**(6), 539–550 (1973). https://doi.org/10.1002/jrs.1250010604
58. F.-C. Liu, A.D. Buckingham, Antisymmetric resonant vibrational Raman scattering. Chem. Phys. Lett. **207**(4), 325–331 (1993). http://dx.doi.org/10.1016/0009-2614(93)89008-6
59. D.V. Khveshchenko, P.B. Wiegmann. Raman scattering and anomalous current algebra in Mott insulators. Phys. Rev. Lett. **73**, 500–503 (1994). https://doi.org/10.1103/PhysRevLett.73.500
60. T.G. Spiro, T.C. Strekas, Resonance Raman spectra of hemoglobin and cytochrome c: inverse polarization and vibronic scattering. Proc. Natl. Acad. Sci. **69**(9), 2622–2626 (1972)
61. L.D. Barron, *Molecular Light Scattering and Optical Activity*, 2nd edn. (Cambridge University Press, Cambridge, 2004)
62. H. Hamaguchi, The resonance effect and depolarization in vibrational Raman scattering, in *Advances in Infrared and Raman Spectroscopy*, ed. by R.J.H. Clark, R.E. Hester (Wiley-Heyden, London, 1985), pp. 273–310
63. F. Liu, Antisymmetric nonresonant vibrational Raman scattering. J. Phys. Chem. **95**(19), 7180–7185 (1991). https://doi.org/10.1021/j100172a017
64. Z. Peng-Xiang, T. An, Influence of Faraday rotation on the selection rule of phonon Raman scattering in magnetic materials. J. Raman Spectrosc. **14**(5), 326–329 (1983). https://doi.org/10.1002/jrs.1250140506
65. R.A. Yadav, Raman scattering in magnetic materials. J. Chem. Sci. **102**(5), 629–634 (1990). https://doi.org/10.1007/BF03040790
66. J.A. Koningstein, O.S. Mortensen, Experimental Observation of an antisymmetric Raman scattering tensor. Nature **217**(5127), 445–446 (1968). https://doi.org/10 1038/217445a0
67. H. Rho, M.V. Klein, P.C. Canfield, Polarized Raman scattering studies of crystal-field excitations in ErNi2B2C. Phys. Rev. B **69**, 144420 (2004). https://doi.org/10.1103/PhysRevB.69.144420

68. H. Rho, M.V. Klein, P.C. Canfield, Crystal-field excitations and spin-phonon interactions in $DyNi_2B_2C$: Raman scattering study. Phys. Rev. B **82**, 064423 (2010). https://doi.org/10.1103/PhysRevB.82.064423
69. R. Liu et al., Novel Raman-active electronic excitations near the charge-transfer gap in insulating cuprates. Phys. Rev. Lett. **71**, 3709–3712 (1993). https://doi.org/10.1103/PhysRevLett.71.3709
70. P.E. Sulewski et al., Observation of chiral spin fluctuations in insulating planar cuprates. Phys. Rev. Lett. **67**, 3864–3867 (1991). https://doi.org/10.1103/PhysRevLett.67.3864
71. E. Riccardi et al., Gate-dependent electronic Raman scattering in graphene. Phys. Rev. Lett. **116**, 066805 (2016). https://doi.org/10.1103/PhysRevLett.116.066805
72. O. Kashuba, V.I. Fal'ko, Signature of electronic excitations in the Raman spectrum of graphene. Phys. Rev. B **80**, 241404 (2009). https://doi.org/10.1103/PhysRevB.80.241404

Chapter 2
Experimental Setup

Abstract Since the days when C.V. Raman and coworkers discovered inelastic light scattering using Calcutta sunlight as the source in the 1920s (Raman, Indian J Phys 2:387–398, 1928), many improvements have been done on Raman spectroscopy apparatus in the past century. Raman scattering setup has evolved into very diverse forms, from portable or even handheld Raman spectrometers widely used by forensic scientists and minefield workers to gigantic state-of-the-art triple-stage grating spectrometers used in research labs. The data collection efficiency has also improved many orders of magnitude, alongside with better spectral resolution and polarization optics. This is due to improvements on almost every single element: laser light source, polarization optics, aberration corrected lens, holographic blazed gratings, off-axis parabolic mirrors, high precision slits, back illuminated CCD detectors, and many other factors (Hayes and Loudon, Scattering of light by crystals. Courier Corporation, Chelmsford, 2012; Palmer, Loewen, Diffraction grating handbook. Newport Corporation, New York, 2005). Given the diversity and breadth of this topic, it is impossible to have a complete account of all involved instrumentations. In this chapter, I will introduce the basic elements common to most low temperature polarization resolved Raman spectroscopic research labs, but focus will be given to the setups we employ here at Rutgers. Then I will discuss the data acquisition, error sources, and data analysis techniques we used in this monograph.

2.1 Raman Scattering Setup

2.1.1 Laser

Although C.V. Raman originally used sun light as excitation source for his pioneering experiments of inelastic light scattering [1], it is more desirable to use monochromatic light source, where the energy is as sharply peaked as possible (monochromaticity). Before the advent of both ruby laser and He–Ne laser in 1960, gas discharge lamps were the only choice of light source for Raman scattering. However, since the cross section of Raman scattering is usually very small,

H.-H. Kung, *Collective Excitations in the Antisymmetric Channel of Raman Spectroscopy*, Springer Theses, https://doi.org/10.1007/978-3-030-89332-3_2

Table 2.1 The list of laser wavelength and typical output power in the Kr^+ laser used in our lab (Coherent Innova 302C), under "single line" configuration, operating at TEM_{00} mode (except for the 752 nm line) under the listed current. Notice that this is the "optimized operating current" that outputs maximum power and was measured by Coherent technician immediately after the replacement of a new plasma tube. The typical operating current we use is around 32–35 A, and the output power is less than listed below

Wavelength (nm)	Energy (eV)	Power (mW)	Operating current (A)
799.3	1.55	103	36
752.5	1.65	200	36
676.4	1.83	190	41
647.1	1.92	760	41
568.2	2.18	340	41
530.9	2.34	440	45
520.8	2.38	272	45
482.5	2.57	155	45
476.2	2.60	195	45
413.1	3.00	650	45
406.7	3.05	415	45
350.7	3.54	500	45

extremely long exposure time is needed to observe the Raman lines, limiting its applicability. In the accounts of Landsberg, Raman scattering in quartz requires more than 15 h exposure time and sometimes more than 100 h [2]. Therefore, it is desirable to have a much more intense and monochromatic light source. The appearance of laser nicely solves both problems and has become an essential tool for most of the Raman scattering experiments nowadays.

For most of the experiments in this thesis, we use a Kr^+ ion laser (Coherent Innova 302C) that allows several lasing lines from UV to NIR range, where the most prominent laser lines are listed in Table 2.1. Notice that the "operating current" tabulated in this table is the "optimized value" that outputs maximum power and was measured by coherent technician immediately after the replacement of a new plasma tube. The typical operating current we use is around 32 A, and the output power is a little less than listed here.

Ion laser is one of the most popular choices for Raman scattering due to its stability, large output power, and monochromaticity. Among them, Ar+ and Kr+ ion lasers are the most popular choices, because both of them have multiple lasing lines in the visible spectrum, convenient for conventional Raman spectroscopy.

In an ion laser, the laser light is produced in the "plasma tube" [Fig. 2.1], which is essentially a sealed cylinder with low pressure Kr gas. During operation, an electric field is supplied to maintain the Kr gas ionized and in a highly excited state, creating a "population inversion" for some atomic states, where the number of particles in a higher energy state exceeds the number of particles in a lower energy state. The energy stored within the population inverted state is eventually released through cascade of "stimulated photon emission," where all ions coherently radiate photons

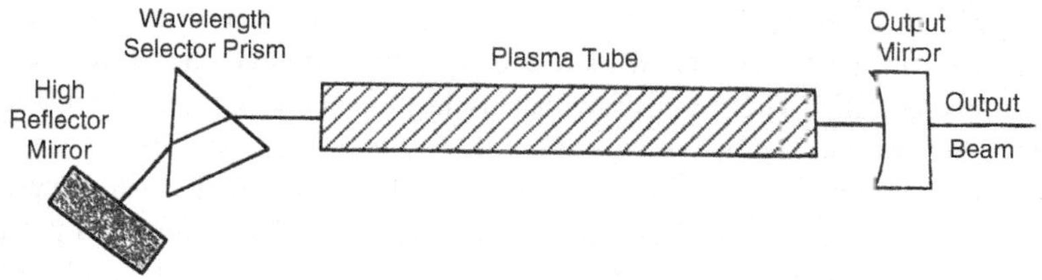

Fig. 2.1 Illustration of the resonance cavity in typical gas ion lasers. The ends of the plasma tube are sealed with "Brewster windows," not shown in the drawing

with the same energy and same phase and thus producing very intense, collimated, and coherent radiation.

The ends of the plasma tube are sealed with "Brewster windows." These are quartz crystals with surface normal oriented at Brewster's angle relative to the light propagation direction. With this geometry, only light with p-polarization can pass through the window and trigger stimulated emission. Thus the output laser light is polarized in the direction that experiences zero loss through the Brewster windows.

Finally, two mirrors are placed outside the plasma tube and on the optical axis, such as drawn in Fig. 2.1. These two mirrors form an optical cavity that stores the electromagnetic energy escaped from the plasma tube. Both the "high reflector" and "output mirror" are quartz coated with dielectric materials to achieve its desired reflectivity within a certain frequency range. While the high reflector has practically 100% reflectivity, the output mirror is designed to have reflectivity slightly less than perfect to allow part of the light intensity to escape the cavity and form the output beam.

Due to the very large number of states in the plasma tube, many laser lines and even much more plasma lines are simultaneously emitting. This is however inconvenient for Raman scattering experiments, where we would like to use monochromatic excitation. This is achieved through placing a "selector prism" in the optical path, known as the "single line" configuration, shown in Fig. 2.1. Light through the prism is dispersed such that only photon wavelength within a small window may reach the high reflector flat mirror and be send back into the optical cavity for further amplification. Thus, by adjusting the prism orientation, we can easily switch between several laser lines.

Apart from the large size, expensive maintenance, requirements for large electric power and water cooling, the main drawback of ion gas lasers is the nuisance of "plasma lines." These are the incoherent, non-lasing, emission lines of the ion plasma. The intensity of the plasma lines are typically comparable to, or even stronger than, many Raman spectroscopic features we are interested in. Furthermore, as an ion laser ages, the once optimized tube condition will unavoidably change slightly, and therefore some weaker laser lines will stop emission after a few years.

Another popular genre is the diode lasers (semiconductor lasers), which due to its high efficiency can be made very compact with appreciable laser output power without water cooling. Diode lasers with any wavelength across the entire visible spectrum can be found with relatively low price and therefore becoming increasingly popular. The main drawback is its poor collimation, which requires additional lenses outside the laser cavity.

Another common laser source is the solid state laser, which uses solid crystals as lasing medium instead of the plasma tube in ion lasers. The lasing medium needs to be optically pumped by another source, usually a diode laser. One of the most popular solid state lasers is based on neodymium-doped yttrium aluminum garnet (Nd:YAG) crystal, which has a 532 nm line often used for "laser pointers." The first ever built optical laser, the "Ruby laser," is also a type of solid state laser. In our lab, we have an air-cooled 532 nm solid state laser with about 50 mW output power from JDS Uniphase Corporation, a 445 nm laser diode with about 5 mW output power from Lilly Electronics, and a 462 nm laser diode with about 500 mW output power from Thorlabs.

Ti:Sapphire and dye lasers are the most common types of coherent light source with tunable wavelength we have in our lab, however both requiring pumping from another laser source. The author does not use either of the tunable lasers in this monograph.

2.1.2 Optical Setup

Figure 2.2 is an illustration of the basic optical setup for Raman spectroscopy in the "pseudo backscattering" at Rutgers. Notice that the dimensions are not all drawn in scale. All elements are fixed on a large optical table, but adjustments can be easily made depending on the needs of the particular experiment. Several reflective high

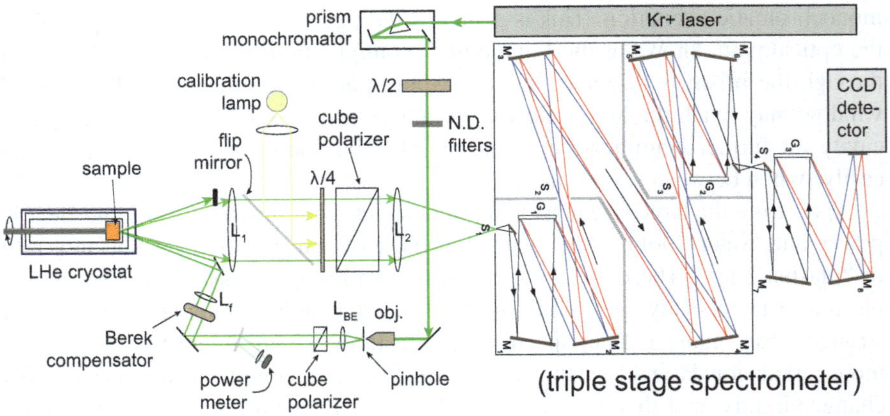

Fig. 2.2 Illustration of the optical setup for Raman spectroscopy at Rutgers. Note that the dimensions are not drawn in scale

power neutral density filters are placed in the optical path to reduce laser power if needed. A complimentary "micro-Raman" setup sharing the optical table, laser source, and spectrometer is not related to this monograph and thus is not shown here.

As we explained in the previous section, the ion lasers typically contain a large amount of plasma lines in the visible range and need to be filtered out as early as possible. In our lab, we place a tunable prism type monochromator immediately behind the laser and then pass the dispersed light through a 5 μm diameter pinhole about 1.5 m behind the monochromator. An objective (obj. in Fig. 2.2) and a thin lens (L_{BE}) are placed across the pinhole as a beam expander, which magnifies the 1.5 mm diameter beam output from the laser to about 10 mm, such that we can conveniently focus into a smaller spot (about 50 μm) on the sample. The pinhole size is chosen to match the diffraction limit of the objective, $D = \frac{4\lambda f}{\pi d}$, where λ is the laser wavelength, $f = 11$ mm is the objective focal length, and $d \approx 1.5$ mm is the laser beam diameter. This combined setup also serves as a "spatial filter" that significantly suppresses the unwanted plasma lines in the laser and also produces a better Gaussian beam profile that propagates along the optical path.

A focusing lens, L_f, is used to focus the beam onto sample. Depending on different operations, this can be an aspheric lens that results in a spherical spot of about 50 μm spot diameter on the sample, or a cylindrical lens that results in a significantly elongated spot, or even a Mitutoyo long working distance objective that produces smaller spots. For most practical usage, it is sufficient to tilt an aspheric lens slightly (about 20°) to produce an elongated elliptical spot with larger area, thus reducing the power density (laser heating) on the sample.

To avoid the elastic reflection from being collected, we employ the "quasi-backscattering" geometry as shown in Fig. 2.2, where the reflected beam is outside the collection cone and blocked. The incident light is at a small angle (about 17°) with respect to the sample surface normal. In all materials studied in this work, the refractive indices are high enough such that the light propagation vector inside the crystals is always less than 10°, which is negligible for all practical concerns.

The secondary emission from the sample is collected by an achromatic lens with $f/2$. After passing the polarization optics ("analyzers"), the light is focused onto the spectrometer's entrance slit by another achromatic lens. A large 50 mm diameter flip mirror is placed in front of the cube polarizer to allow fast switching to calibration lamps as source. The details of the polarization optics, spectrometers, and calibration lamps will be discussed in the following sections.

Polarization Optics

The polarization of laser output is determined by the Brewster windows of the plasma tube, which in our setup is always vertically polarized (out-of-plane direction in Fig. 2.2). Thus a broadband $\lambda/2$ waveplate is placed behind the monochromator to rotate the laser polarization into the direction suitable for the experiment. By design, $\lambda/2$ waveplate has the property of rotating the linear polarization of light by 2θ relative to the input polarization, in the plane perpendicular to the optical path, where θ is the angle between input polarization and the waveplate's "slow axis." A broadband Brewster type polarization cube is placed behind L_{BE} to ensure that the degree of linear polarization on the sample is as close to 100% as possible.

Notice that a Berek compensator is basically an adjustable retardation waveplate, and therefore it can also operate in the $\lambda/2$ mode. In this case, the $\lambda/2$ waveplate is unnecessary and can be replaced by the Berek compensator drawn in Fig. 2.2. However, the degree of polarization in a Berek compensator critically depends on its proper alignment and therefore is quite time consuming and troublesome. In most cases, we only use the Berek compensator to produce circularly polarized light. When put into $\lambda/4$ mode with $\theta = 45°$, the Berek compensator transforms linearly polarized light into circular polarization. The advantage of using a Berek compensator over broadband $\lambda/4$ waveplate is its flexibility in wavelength (200–1600 nm) and polarization of incident light. A Berek compensator can be adjusted to produce almost perfect circularly polarized light for a wide range of wavelength, whereas any broadband waveplate suffers finite deviation from $\lambda/4$ retardation, except within some small wavelength window. But because of this flexibility, one has to be careful when aligning the Berek compensator to avoid introducing ellipticity in the output light. This is done by adjusting 4 degrees of freedom (assuming the light beam passes through the center of Berek compensator): 1 rotating axis around optical path ("orientation"), the 2 tilting axis orthogonal to the optical axis, and the retardation of the compensator.

The typical procedures for Berek alignment are the following: First, one has to properly align the compensator to be normal to the optical path by adjusting the 2 tilting degrees of freedom. This is done by looking at the back reflection from the compensator surface when retardation is set to 0. When properly aligned, the reflection should coincide with input beam. One can try to rotate the compensator ("orientation"), and the back reflection spot should not move with changing orientation. Next, one needs to align the slow axis of the Berek compensator with respect to the linear polarization of the incident light, which is determined by the cube polarizer behind L_{BE}. Luckily, the slow axis is usually marked on the compensator mount when bought. As long as the orientation of the cube polarizer is known, it is trivial to orient Berek's slow axis at 45° to the incident light. Usually, one can buy a factory pre-aligned polarizer with markings on the mount and then use it as a reference to align all other linear polarizers in the lab. In typical Raman spectroscopy labs, *all* linear polarizers' orientation is aligned with respect to the gratings' ruling direction, which is typically normal to the plane of incidence, thus normal to the optics table. After the Berek orientation is set to 45°, one rotates the retardation to $\lambda/4$, which is documented for commonly used wavelengths in a pre-calculated table (provided by the company from which the compensator is bought from).

Due to many unavoidable errors in all the previous operations, it is important to check and fine-tune the retardation at this point. This is done by putting a reflective mirror, e.g., Ag coated, behind the Berek compensator and looking at the back reflection on a "half screen." This screen should be placed between LBE and the cube polarizer and should be positioned such that it blocks exactly half of the incident beam. Needless to say, the reflective mirror has to be tilted in a way such that the back reflected beam has a tiny overlap on the "half screen." If the output beam after Berek is perfect circularly polarized, the mirror reflected beam through the Berek will be a linearly polarized light with polarization orthogonal to the cube polarizer (combined action of reversed helicity by mirror reflection, and the opposite retardation by passing through Berek in the opposite direction) and thus should produce zero intensity on the "half screen."

In the light collection path (lighter green in Fig. 2.2), a broadband ("zero order") $\lambda/4$ waveplate, marked by $\lambda/4$ in Fig. 2.2, is used instead of a compensator because the Raman signal often spans a wide wavelength range. This waveplate must have clear aperture comparable to the collection lens, L_1, which is 38 mm in our lab. Therefore, we use a 50 mm clear aperture mica-type $\lambda/4$ waveplate. The slightly larger diameter ensures that all scattered lights collected by L_1 pass through the waveplate, so that the intensity can be consistently compared between experiments. Sheet mica has the spectacular property of relatively flat retardation profile within the visible spectrum with high transmission and can be made into large diameter. Another popular choice is polymer waveplate, which has lower transmission but can be made into very large size with much lower prices.

The cube polarizer behind the $\lambda/4$ waveplate is often referred to as the "analyzer," with the clear aperture of 50 mm. The cube polarization is oriented in the direction that optimizes the grating throughput in the spectral regime of interest.

Spectrometer

In our lab, we use a custom designed Czerny–Turner triple stage spectrometer by Acton Research Corporation, to collect and disperse the secondary emission from the sample. All of the gratings and slit positions in the spectrometer are motorized and computer controlled. The basic structure and optical paths inside the spectrometer are illustrated in Fig. 2.2 [3], where the black lines denote the optical path of undispersed "white light," the red and blue lines denote the paths of light with longer and shorter wavelength, respectively.

In this design, the dispersion of incident light is entirely achieved by reflective planar gratings, which is basically a more advanced version of double slit interference (in the first order): $\sin\alpha + \sin\beta = \lambda/d$, with α and β being the incident and diffracted light angles (from grating surface normal, thus having opposite signs) for a particular light wavelength λ, and d is the averaged distance between grating grooves [Fig. 2.3]. With elementary trigonometry, we see that the above "grating equation" (diffraction equation for first order) can also be written as

$$2d \sin\left(\frac{\alpha+\beta}{2}\right) \cos\left(\frac{\alpha-\beta}{2}\right) = \lambda. \tag{2.1}$$

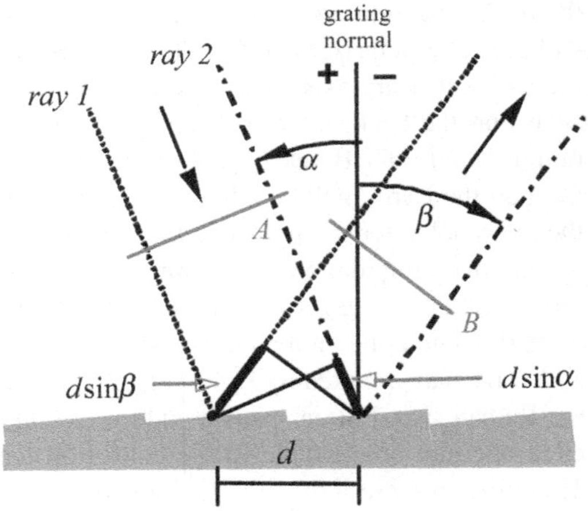

Fig. 2.3 Illustration of the light path for a particular wavelength, λ, by a diffractive planar grating, where the incident light is a parallel beam, and grating grooves are separated by *d*. Figure adapted from Ref. [4]

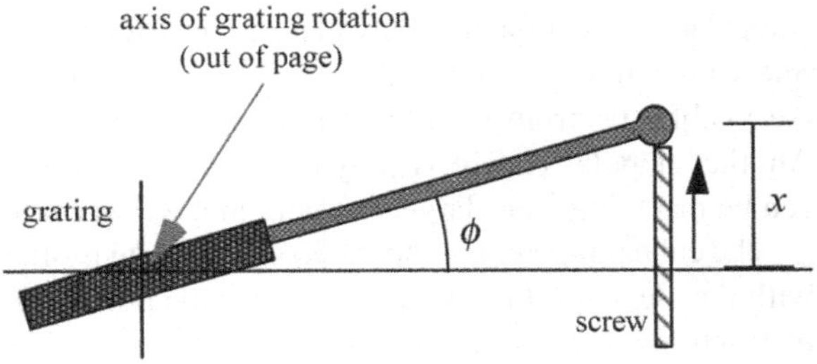

Fig. 2.4 Illustration of the mechanism of a sine bar grating mount. As the screw is extended linearly by the distance x, the grating rotates through an angle ϕ in such a way that $\sin\phi$ is proportional to x. Figure adapted from Ref. [4]

In most designs of spectrometers, the mirrors are fixed, meaning that the angle between incident and diffracted light is a constant, $\alpha - \beta = 2K$. Therefore, it is more convenient to project different parts of the spectrum onto detector by rotating grating about the central axis, i.e., scanning $\alpha + \beta = 2\phi$. The wavelength λ fulfilling Eq. (2.1) is then proportional to $\sin\phi$, which is again proportional to the linear displacement of a device called the "sine bar," shown in Fig. 2.4. Thus by rotating the lead screw connected to the sine bar, we *mechanically* project different parts of the spectrum onto CCD detector.

The angular dispersion of light, which ultimately determines the spectral resolution of the spectrometer, is a function of λ, α, and β. For incident white light at a particular angle α, the first order diffracted light within wavelength regime $\delta\lambda$ is dispersed into a small cone given by $\delta\beta$:

$$\frac{\delta\beta}{\delta\lambda} = \frac{1}{d\cos\beta}. \tag{2.2}$$

For small $\delta\beta$, the linear span of light projected on CCD, $\delta x = f\tan(\delta\beta) \approx f\delta\beta$, with f being the focal length of the spectrometer. Thus the linearly dispersion of light on the CCD, which is one of the most important factors for determining the spectral resolution, is

$$\frac{\delta x}{\delta \lambda} = \frac{f}{d \cos \beta}. \tag{2.3}$$

Thus, a spectrometer's spectral resolution is inversely proportional to the grating grooves distance or proportional to the grooves density. In our lab, we have gratings with three different groove densities: 150, 600, and 1800 grooves/mm. Another way of improving angular dispersion is by having larger β, which corresponds to longer wavelength [Eq. (2.1)].

Another important factor for determining the spectral resolution is the spectrometer's slit size. The entrance slit of the spectrometer, S1, is composed of a pair of horizontal and vertical precision slits. The image on S1 can be thought of a projection of the sample image (confocal image). Thus by controlling the opening of S1, we limit the area on the sample from which we collect the signal. The laser spot should be positioned in the middle of the slits to avoid aberrations and to maximize the signal. The width of the horizontal slit in S1 should match the exit slit, S4, which is also the confocal plane of S1 and the entrance slit of the last stage of spectrometer, and will ultimately determine the spectral resolution. In the design of our spectrometer, the width of the horizontal slits of S1 and S4 should match the pixel size of the CCD detector. In practical operations, we often bin the CCD pixels into superpixels to increase the signal intensity, with the compromise of decreasing spectral resolution. The width of S1 and S4 should be adjusted accordingly to maximize the efficiency.

One of the novel designs of this spectrometer is equipping a removable camera behind S1 (not shown in Fig. 2.2). A computer controlled flat mirror can be translated into the optical path behind S1, which sends the light into a high sensitivity color camera. Thus we can simultaneously see the images of the entrance slits and sample, which ensures the alignment of laser spot as well as collecting data from the sample area of interest.

The first two stages of the spectrometer form a coupled double spectrometer (ARC AM-505F) that usually operates in the "subtractive mode." In this operation, the double spectrometer acts as a spectral block filter of the input light, which is dispersed in the first stage by grating G1 and then recombined by inversely positioned identical grating G2 back into white light. The dispersion of the collected spectra is entirely achieved by the last stage (ARC AM-506), which can be viewed as a separate monochromator with a focal length of 660 mm, and the aperture ratio (defined as the focal length divided by the grating length) is $f/4.7$. The spectral resolution with 1800 g/mm gratings at 532 nm is about 0.9 cm^{-1}. In "additive mode," G1 and G2 are positioned in the same direction, and thus about 3 times better resolution can be achieved.

The advantage of operating in the subtractive mode is for better rejection of elastic line and stray light. Since Raman spectroscopy is only concerned with inelastically scattered light, the typically 8 orders of magnitude stronger elastically scattered laser line needs to be blocked before projecting the spectrum onto CCD camera. Otherwise, not only the CCD could suffer permanent damage, but also all spectral features would be overwhelmed by the "overflow" of the strong elastic line.

There are several ways of blocking the elastically scattered light [5], but triple stage spectrometer has the advantage of the highest resolution (cutoff wavelength being closest to the laser line) and best flexibility and therefore remains the most popular choice among Raman spectroscopist.

The basic working principle of the subtractive stages is that in the first stage, mirrors M1 and M2 focus the dispersed light on a horizontal precision slit, S2. The opening of S2 determines the cutoff wavelength of the spectral range allowed to pass, thus acting as a block filter. Mirrors M3 and M4 form the so-called relay stage, which focus the light onto another precision slit, S3, that is identical to S2. This additional slit allows significantly improved elastic line and stray light rejection. The efficiency of S2 and S3 mainly depends on the resolving power of the first two stages of spectrometer and the focusing power of the mirrors. Therefore, all of the mirrors in the spectrometer are off-axis parabolic mirrors to minimize the spherical aberrations.

The resolving power of a spectrometer is mainly determined by the grating grooves density, focus length, and effective grating area. In our lab, we have three sets of gratings: 150, 600, and 1800 g/mm. In order to optimize the dispersed light intensity, all of the gratings are "blazed," i.e., a specially designed groove profiles to maximize the intensity in the first order diffraction.

Detector

The invention and popularization of charge-coupled device (CCD) detectors are probably the single most important breakthrough for Raman spectroscopy in the past 30 years. The photon to electron conversion (quantum efficiency) is close to 100% in the visible range for the most advanced commercially available CCD detectors. The pixel size is as small as 20 μm while still possessing 200,000 electrons "well depth" (dynamical range), thus allowing both high resolution and high sensitivity.

We have several CCD cameras in our lab for different purposes. The most commonly used one is a back-illuminated silicon-based CCD from Princeton Instrument with 400 vertical pixels and 1340 horizontal pixels. Since a grating spectrometer does not use the vertical degrees of freedom, we always bin the illuminated vertical pixels (typically about 150) into a superpixel. If the signal is very small, we can further bin the horizontal pixels to increase the signal to noise ratio (S/N) at the cost of lowering the spectral resolution.

At liquid nitrogen temperature, the dark current of our CCD is about 0.3 electrons/px/hr, and the readout noise is about 3 electrons. To reduce the noise and maximize S/N, the CCD temperature is locked at $-120\,^{\circ}$C, and we usually use long exposure times (300 s) to minimize the number of readouts.

Cryostat

Many of the interesting physics and emergent behaviors, e.g., charge/spin/orbital density waves, composite quasiparticles and superconductivity, regretfully only happen at very low temperatures (typically below 50 K). The working principle of most of the cryostats is similar to a Dewar, where one or more "vacuum jackets" separate the sample space from the room temperature atmosphere [6]. The thermal insulation is mainly provided by the vacuum space and thermal radiation shielding. The shielding is connected to a heat reservoir below 100 K, such as a liquid nitrogen jacket, which reduces the thermal radiation on the sample by about 100 times due to Stefan–Boltzmann law. In our design, we utilize the back-flow of cold helium gas to cool down the shielding without additional coolant and thus also able to reach lower temperatures than using liquid nitrogen. In the sample space, the sample under inspection is thermally connected to a cooling source, thermometer and heating coil. These three components together provide a feedback loop for thermal stability at low temperatures.

Depending on the purposes and working temperatures of the cryostat, vastly different designs can be found on the market. In our lab, we use mainly a custom designed continuous flow liquid "helium bath cryostat" made by Oxford Instruments. This cryostat is connected to a small liquid reservoir through a needle valve, which is again connected to external liquid helium Dewars. Therefore, in principle, a temperature of 1.2 K can be reached by pumping on the sample space. In reality, mainly due to radiation heating, the base temperature is about 1.8 K. The main advantage of the bath cryostat is the large temperature adjustability and cooling efficiency. By adjusting the needle valve opening and helium supply rate into the reservoir, temperature between 2 and 300 K can be very easily stabilized with little help of heater. Also, since the sample is in direct contact with the cold helium gas, the cooling is particularly efficient on the sample surface, therefore ideal for Raman spectroscopy where the laser heating is focused on a small region on the sample surface. Since the heat is mainly conducted through the helium gas, the degree of laser heating on the sample is not sensitive to sample thermal conductivity. This is particularly important, because many of the physical phenomena we are interested in are second order phase transitions, across which the thermal conductivity experiences large change.

For most metallic samples, in optical regimes where most of the light is reflected, 1 mW of laser light typically results in about 0.5 K of heating in our cryostat at low temperatures (below 20 K).

2.2 Data Acquisition and Analysis

The double spectrometer and the cryostat temperature are controlled by two LabView programs, where the detailed spectrometer settings, response ratios, backgrounds, temperatures, and laser powers are automatically recorded for each experiment. The third stage monochromator is coupled to the CCD camera, and both controlled by Winspec (commercial program by Princeton Instruments), which records raw counts in each frame, spectral calibrations, grating positions, exposure times, etc. With all the information, we can convert the recorded counts per CCD pixel into Raman response function, after accounting for several calibrations and corrections listed below.

2.2.1 Instrumental Calibrations

The reliability of the data produced by a Raman spectroscopy lab depends a lot on how the system is being calibrated. In our lab at Rutgers, we calibrate our system in the following ways.

The temperature fluctuations of the room and cooling water may introduce drifting of laser power during experiments, which has to be accounted for before further data analysis. A thin quartz plate is placed behind the cube polarizer in the excitation light path (darker green in Fig. 2.2) to reflect about 5% of light intensity into a power meter. This reading is constantly recorded during the experiments, and all spectra are automatically normalized to the laser power to avoid drifting over long hours of experiments.

The grating positions of the spectrometer are controlled by step motors coupled with mechanical arms ("sine bars") [Fig. 2.4] [4], which needs to be related to the wavelength of light imaged by the CCD. This is done by using "spectral lamps" that produce narrow, intense lines of UV to NIR wavelengths due to the excitation of various gases, commonly Hg, Ar, and Ne. Since the emission lines of these gases are well documented by the National Institute of Standards and Technology (NIST), we can readily relate the wavelength of these lines to the grating positions and CCD pixels using the so-called spectrometer equations. This calibration procedure is done semi-automatically across the entire UV-NIR spectrum by the computer software provided by Princeton Instrument (Winspec manual, Appendix C), which controls the CCD detector and AM-506 monochromator. Such global calibration usually has larger error bars, which can be locally calibrated in the spectral window of interest, again by spectral lamps. The ultimate spectral resolution is limited by the spectrometer's mechanics and CCD's pixel size, which is better than 0.02 nm at 435 nm wavelength (for 1800 grooves/mm gratings).

The polarization cube, lens, spectrometer, and detector's throughput all have non-linear wavelength and polarization dependences, which, if not accounted correctly, could lead to spurious excitation and polarization profile. This can be calibrated by measuring the flux through all of the above optics from a known white light source (standard candle). Since the flux from the lamp is known, by measuring the recorded intensity on the CCD, we can derive the combined throughput ratio of linearly polarized light at each wavelength, which we call a spectrometer's "response ratio." In our lab at Rutgers, we use a NIST certified spectral lamp to calibrate the combined throughput of the polarization cube, L_2 lens, spectrometer, and CCD, for all individual gratings, grating positions, and cube orientations. A metallic flip mirror is placed in front of the cube polarizer [Fig. 2.2], which easily switches between the collimated light from the sample and from the spectral lamp.

Even when all lights and lasers are turned off, a small amount of background room light (usually from instrument and computer LED indicators) and CCD dark current can still contribute to the spectra. These contributions (about 0.3 counts/pixel/hour, or 3 counts/superpixel/min under 150×4 binning) are much smaller than the typical signal from most samples and are usually ignored. However, sometimes this is not entirely negligible compared to the sample signal and has to be carefully measured and subtracted from the signal before further analysis.

2.2.2 *Cosmic Ray Rejection*

In order to minimize readout noise in samples with weak signal, we have to take a long exposure time as possible. However, the earth surface is constantly under the bombardment of highly energized muons, typically created by the decay of pions in the upper atmosphere. These are very penetrating particles, thus very difficult to block by shielding. They deposit energy in the CCD detector via the Coulomb interaction as they traverse through the silicon sensor, which produce a usually large spike in the spectra [7]. After about 10–20 min, the spectra would be dominated by cosmic spikes, covering a large portion of the CCD detector. Once a pixel is covered by a cosmic spike, the information within that pixel is lost. Therefore, it is better to split the exposure time into several shorter consecutive "frames" and remove the spikes in each frame and then average the frames. Thus the number of spikes is still the same, but the amount of usable pixels is much more.

In practice, we take no longer than 5 min per frame, which typically contains less than 5 spikes in the spectra. There are many different strategies in removing the cosmic spikes, from simpler spacial or temporal filters to much more complicated wavelet transformation algorithms [8]. In our lab, we choose to use a temporal filter, requiring us to always take at least 3 consecutive frames per experiment. The removal of spikes is done automatically by a C++ code developed by Adrian Gozar [3], where we compare the data in each pixel across the frames and simply remove the data in a particular pixel in a particular frame, if we find that pixel to have a much higher photon counts ($> 5\sqrt{N}$, where N is the median counts of that pixel across the frames) than the same pixel in the other frames. In the end, we only average the data in the remaining pixels.

2.2.3 *Optical Corrections*

The signal recorded at the detector is in the unit of counts/second/pixel. This has to be converted into differential cross section and corrected for the optical constants of the sample, before quantitative analysis of the data can be done. The detailed conversion between measurement and theory was first done by Loudon in 1965 [9], although discussions of optical corrections have already begun as early as 1951 [10]. Here, we follow the simplified semi-quantitative approach as given in Ref.[11, 12], where we convert the measured photon counts into the dimensionless "scattering efficiency" per unit solid angle per unit energy (in units of cm^{-1}), $\frac{d^2S}{d\Omega d\omega}$, which is a dimensionless quantity.

Due to the construction and calibration of grating spectrometers, we first convert the measured data (after subtracting CCD dark current and environmental background) into the cross section per unit wavelength, $\frac{d^2S'}{d\Omega' d\lambda}$, by normalizing the CCD count rate to laser power and response ratio. Here $d\Omega'$ is the solid angle defined by the collection lens L_1. This can be easily converted into per unit energy by noting that $\frac{d}{d(-\omega')} \sim \lambda^2 \frac{d}{d\lambda}$. Since we are usually interested in the Stokes process, we usually write $\omega = -\omega'$.

However, this quantity depends on the details of optical properties of the sample. Most importantly, due to refraction at the sample/vacuum interface, the solid angle inside the crystal, $d\Omega$, could be very different than the collection solid angle fixed by the lens, $d\Omega'$. In fact, one can easily see that

$$\frac{\Omega'}{\Omega} = \frac{1 - \cos\theta'}{1 - \cos\theta}. \tag{2.4}$$

For small θ' as in most experimental setups (notice not the case in micro-Raman setups where the NA of the objective is easily larger than 0.5), $d\Omega'$ can be approximated as $n(\lambda)^2 d\Omega$ (assuming vacuum outside sample), where $n(\lambda)$ is the refractive index. This could be quite sizable in many materials, for example, $n(\lambda) \sim 4$ for germanium in the visible range.

In addition, only a portion of incident photon transmits across vacuum/sample interface and the same for the scattered photon. Therefore, the real scattering efficiency in the sample should be $\frac{1}{(1-R(\lambda_L))(1-R(\lambda))}$ times larger than we measured. Lastly, we should account for the scattering volume in different samples, which is roughly $V_{eff} \sim \frac{D^2}{\alpha(\lambda_L)+\alpha(\lambda)}$, where D^2 is the laser spot size, and α is the absorption coefficient which determines the penetration depth of light in the sample. More comprehensive corrections are given in Ref. [13].

Combining the above corrections, we can calculate the scattering efficiency in the sample:

$$\frac{d^2S}{d\Omega d\omega} \sim n^2(\lambda) \frac{d^2S'}{d\Omega' d\omega} \frac{\alpha(\lambda_L) + \alpha(\lambda)}{D^2(1 - R(\lambda_L))(1 - R(\lambda))}. \tag{2.5}$$

2.2.4 Data Analysis

The data analysis of this monograph is mainly done in Mathematica and then exported into OriginPro for plotting. Since the Raman signals in this study are mostly small, it is important to carefully address the issue of noise, signal to noise ratio (S/N), and statistical analysis. The signal noise in our spectra comes from several sources, with mainly the following ones.

Photon Shot Noise

This is due to the discrete nature of photons, where the probability of collecting N photons with a particular energy within a fixed time interval is described by the Poisson distribution [14]:

$$P_{\mu}(N) = e^{-\mu}\frac{\mu^N}{N!}, \tag{2.6}$$

where μ is just the average number of photon counts we would expect, $\overline{N}$, if we repeat the finite time interval experiment infinitely many times (or equivalently, in a single experiment with infinitely long collection time). In this limit, the standard deviation (SD) of a Poisson distribution can be shown to be proportional to $\sqrt{\mu}$ [14]. Therefore, using the principle of maximum likelihood, the SD of any finite time experiment that counted N photons is just simply $\sqrt{N}$, i.e., S/N= $N/\sqrt{N}$ is also proportional to $\sqrt{N}$. Since the photon events are assumed to occur randomly in time (white noise in time domain), in order to achieve higher S/N, the length of exposure time we need increases quadratically. For reasonably large signal $\mu \gtrsim 10$, the photon events approach a normal (Gaussian) distribution around its mean value, and thus the SD of two distributions becomes identical, i.e., $G_{X=\mu,\sigma=\sqrt{\mu}}(N)$. This approximation is basically fulfilled in all our experiments, and thus we usually consider the data to be following a Gaussian distribution. Sometimes, the signal is extremely low and the uncertainty is too large to make a claim, and then we will have to either increase the laser power (at the cost of laser heating) or increase the number of frames in each experiment.

Vibration and Focus Drift

Due to cosmic spikes issue, we prefer averaging signal from many consecutive frames than to take a single long exposure. This assumes that each frame was taken under the *exact* same conditions, i.e., the signal is perfectly homogeneous in time. This is obviously not possible due to realistic concerns, such as temperature fluctuations, laser power fluctuations, vibrations, etc. While the temperature and laser power are easier to monitor in real time and compensate through software, the focus and position of the laser spot drifting due to sample vibration are much more difficult to avert. The common practice is to setup all of the equipments on an optical table with pneumatic damper legs. However, the minute vibrations from pumping lines, liquid helium transfer line, and room air flow are unavoidable and will eventually lead to noticeable laser defocus and spot drift after roughly an hour. Therefore, we usually prefer to split a single long experiment into several 30 min shorter experiments, such that we can promptly adjust the drifting issues. This will of course introduce additional human errors, and therefore care has to be taken to check that the averaged signals between each 30 min experiments are consistent with each other within SD.

CCD Detector Noise

A charge-coupled device (CCD) is a photodetector composed of an array of top-gate capacitors on a semiconductor (usually silicon), where each capacitor works as a quantum well (pixels) that stores electrons proportional to the light intensity being illuminated [15]. The photons are absorbed from the back of the detector (in back illumination configuration) and are converted into electrons through photogeneration of electron–hole pairs across the semiconductor band gap. The photogenerated electrons are captured and integrated in the quantum well during light exposure and to be "readout" at a later time. There are two main sources of noise in a CCD device: (1) the "dark current," which is due to thermal generated charge carriers in each pixel at finite temperature. Thus the noise arising from the dark current can be reduced by lowering the operation temperature of the CCD detector and minimizing the readout pixels by using only the parts illuminated by the scattered light. Typically, the dark current can be reduced bellow 1 electron/pixel-hour in commercially available liquid nitrogen cooled silicon-based CCDs; (2) the readout noise, which comes from the uncertainties in the electrical measurement of a very small charge current, such as the charge stored in each pixel. This noise is contributed by many elements and can be somewhat reduced by better electronic design, for example, on-chip pre-amplifiers and more efficient analogue–digital converters, but it is impossible to entirely remove. In well-designed CCDs, the readout noise is as homogeneous in time as possible and therefore can be reduced by averaging many frames or increasing exposure time; (3) electronic interferences, such as power line voltage fluctuations, radio frequency interference, electronic noise from other power line sharing instruments, etc. The CCD readout signal is DC and usually within the microvolt range, which can easily couple to many sources of electronic interferences, making it very challenging for a precise measurement. In our lab, the CCD camera power is fed through a separate UPS system with line filter and connected to an isolated ground line. All of the cables are shielded by braided metal strands and using twisted pairs whenever possible; and (4) spikes due to energetic charged particles generated by cosmic rays. While intense cosmic spikes are easy to remove by taking multiple frames in every experiment as we discussed in the previous section, weak spikes comparable to the signal can be confusing and hard to distinguish. Luckily, this almost never happens and can be further reduced by using minimal area of CCD detector.

Statistical Analysis

From the above discussion, it is clear that noise and therefore uncertainty are unavoidable in the experiment. The Raman scattering efficiency is almost always very small, meaning that only a small fraction of incident photon is inelastically scattered. If the temperature regime of interest is relatively high, e.g., above 20 K, then we can increase the S/N by increasing laser power. Since we use continuous wave laser with a pretty large spot, laser damage of the sample is usually not a consideration for most of our studies. However, most of the phenomena we are interested in can only be observed below 20 K, and thus we are usually limited to a few milliwatts of laser power. Therefore, it is important to design the experimental apparatus to be as stable as possible, such that we can increase the S/N by averaging more experiments or increasing exposure times.

At the end of the day, we would like to know the uncertainty in our signal and the confidence level of our claims. The error of the intensity associated within $\delta\lambda$ or $\delta\omega$ can be easily found from statistical analysis of signal in repeated experiments or frames. In most practical cases, the probability function of measurement value can be considered as a Gaussian distribution, and thus the *best estimates* can be easily computed following text books [14].

For example, let the secondary emission intensity recorded by a certain CCD pixel (in unit of photon counts) to be ν. This is of course only a theoretical value that we do not know. But following the above discussion, if ν is a reasonably large number, then the probability function of the measurement at this pixel, n, will follow a Gaussian distribution centered at ν with SD of σ:

$$G_{\nu,\sigma}(n) = \frac{1}{\sigma\sqrt{2pi}} e^{\frac{-(n-\nu)^2}{2\sigma^2}}. \tag{2.7}$$

Regretfully, since we do not know ν, we do not know σ either. However, the properties of the Gaussian distribution help us to estimate and put constraints on the most probable values of ν and σ from N repeated measurements. Let the value of the i-th measurement be n_i, then the best estimates of ν and σ would be $\tilde{n} \pm \sigma_{\tilde{n}}$ and $\tilde{\sigma} \pm \sigma_{\tilde{\sigma}}$, respectively. From the properties of Gaussian distribution, we know that [14]

$$\tilde{n} = \frac{1}{N}\sum_{i=1}^{N} n_i \tag{2.8}$$

$$\tilde{\sigma} = \sqrt{\frac{1}{N-1}\sum_{i=1}^{N}(n_i - \tilde{n})^2} \tag{2.9}$$

$$\sigma_{\tilde{n}} = \tilde{\sigma}/\sqrt{N} \tag{2.10}$$

$$\sigma_{\tilde{\sigma}} = \tilde{\sigma}/\sqrt{2(N-1)}\,. \tag{2.11}$$

One has to be careful when plotting or interpreting the error bars of data points, which are usually the averages of many measurements. The error bars could reflect either $\tilde{\sigma}$ or $\sigma_{\tilde{n}}$, depending on the content of the plot. $\tilde{\sigma}$ reflects the standard deviation of the measurements, where one would expect the results of the next measurement would likely be, whereas $\sigma_{\tilde{n}}$ reflects the uncertainty in our estimate of $\tilde{n}$, which is important for determining the confidence interval of the spectra, and therefore the significance of any claims that follow from it.

References

1. C.V. Raman, A new radiation. Indian J. Phys. **2**, 387–398 (1928)
2. I.L. Fabelinskiĭ, Seventy years of Raman scattering. Physics-Uspekhi **41**(12), 1229 (1998)
3. A.M. Gozar, Inelastic Light Scattering in Low Dimensional Quantum Spin Systems. PhD thesis. University of Illinois at Urbana-Champaign, 2004
4. C. Palmer, E.G. Loewen, *Diffraction Grating Handbook* (Newport Corporation, New York, 2005)
5. W. Hayes, R. Loudon, *Scattering of Light by Crystals* (Courier Corporation, Chelmsford, 2012)
6. N.H. Balshaw, *Practical Cryogenics – An Introduction to Laboratory Cryogenics* (Oxford Instruments Superconductivity Limited, Abingdon, 2001)
7. M. Fisher-Levine, A. Nomerotski, Characterising CCDs with cosmic rays. J. Instr. **10**(08), C08006 (2015)
8. Y. Tian, K.S. Burch, Automatic spike removal algorithm for Raman spectra. Appl. Spectrosc. **70**(11), 1861–1871 (2016). eprint: http://dx.doi.org/10.1177/0003702816671065
9. R. Loudon, Theory of the resonance Raman effect in crystals. J. Phys. France **26**(11), 677–683 (1965). https://doi.org/10.1051/jphys:019650026011067700
10. L.A. Woodward, J.H.B. George, Refractive index correction in relative Raman intensity measurements. Nature **167**(4240), 193–193 (1951). https://doi.org/10.1038/167193a0
11. M. Cardona, Resonance phenomena, in *Light Scattering in Solids II*, ed. by M. Cardona, G. Güntherodt (Springer, Berlin, 1982), pp. 45–49
12. J. Menéndez, Characterization of bulk semiconductors using Raman spectroscopy, in *Raman Scattering in Materials Science*, ed. by W.H. Weber, R. Merlin. Springer Series in Materials Science (Springer, Berlin, 2000), pp. 55–103
13. W. Richter, C.R. Becker, A Raman and far-infrared investigation of phonons in the rhombohedral V_2–VI_3 compounds Bi_2Te_3, Bi_2Se_3, Sb_2Te_3 and $Bi_2(Te_{1-x}Se_x)_3$ $(0 < x < 1)$, $(Bi_{1-y}Sb_y)_2Te_3$ $(0 < y < 1)$. Phys. Status Solidi (B) **84**(2), 619–628 (1977). https://doi.org/10.1002/pssb.2220840226
14. J. Taylor, *Introduction to Error Analysis, the Study of Uncertainties in Physical Measurements*, 2nd edn. (University Science Books, Sausalito, 1997)
15. S.M. Sze, K.K. Ng, *Physics of Semiconductor Devices*, 3rd edn. (Wiley, Hoboken, 2006)

Chapter 3
Raman Scattering in URu_2Si_2

Abstract In this chapter, I will introduce our Raman scattering study of the "hidden order" (HO) and antiferromagnetic (AFM) phases in URu_2Si_2, a heavy fermion material of undiminished interest in the condensed matter community for over thirty years. I will first give a short overview on the heavy fermion physics and the exotic phases in this compound, where focus will be given on the experimental results [Sect. 3.1]. Then I will describe typical sample preparation and characterization procedures, which are particularly important for detecting the electronic phases in URu_2Si_2 [Sect. 3.2]. In Sect. 3.3, I will present our Raman spectroscopic results in the pristine and Fe substituted $URu_{2-x}Fe_xSi_2$ samples, which are fully consistent with a minimal model that explains the origin of the unusual A_{2g} susceptibility in URu_2Si_2 [Sect. 3.4]. We will also discuss the origin of the collective mode in the ordered phase at low temperature and the interrelation between the HO and AFM phases.

The work in this chapter is done in collaboration with E.D. Bauer (LANL), R.E. Baumbach (NHMFL and FSU), Kristjan Haule (Rutgers), M.B. Maple (UCSD), and J.A. Mydosh (Leiden U.). The results on the symmetry breaking in the HO phase of pristine URu_2Si_2 are published in Ref. [1]. The results on the interrelation between the HO and AFM phases in $URu_{2-x}Fe_xSi_2$ are published in Ref. [2].

3.1 Introduction

3.1.1 Kondo Effect and Heavy Fermion Systems

The heavy fermion metals have been a constant source of emergent behavior related to their unconventional superconductivity and novel magnetism for more than 30 years. At the heart of the heavy fermion physics lie strong electron correlations, leading to variety of emergent phenomena, e.g., heavy Fermi liquid [3], quantum critical transitions [4, 5], multipolar orderings [6], and unconventional supercon-

H.-H. Kung, *Collective Excitations in the Antisymmetric Channel of Raman Spectroscopy*, Springer Theses, https://doi.org/10.1007/978-3-030-89332-3_3

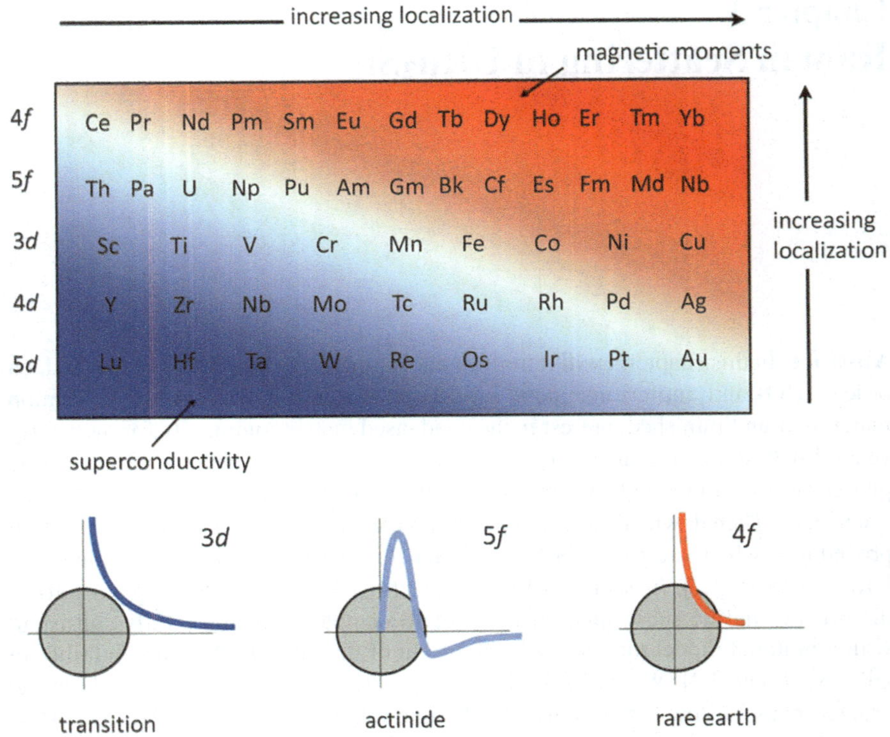

Fig. 3.1 The Kmetko–Smith diagram showing the degree of electron localization and itinerancy [9]. Figure adapted from Ref. [10]. The f-electrons are usually very localized, leading to magnetic ground states at low temperature, as opposed to d electrons which commonly form itinerant bands. However, along every row, there would be elements where the electrons behave dual character, leading to rich and unpredictable ground states at low temperature

ductivity [7, 8]. Figure 3.1 is a modified periodic table know as the Kmetko–Smith diagram, showing the degree of electron localization and itinerancy for d and f electrons. In general, f electrons are more localized than the d electrons due to the more positive nucleus. But higher principle quantum number also means more radial nodes in the electron wave function, thus increasing the orbital radius. The result is that the electrons in many elements (white region) possess a dual character of being partly localized and partly itinerant, leading to rich and unpredictable ground states at low temperature. Moreover, when strong interaction is introduced between the highly localized and partly filled f or d orbitals and the itinerant conduction bands from spd electrons, the result is a zoo of novel phenomena including the curious Kondo effect, non-Fermi liquid, unconventional superconductivity, and so on.

The Kondo Effect

Armed with the cryogenic techniques developed in Leiden University, experimentalists systematically studied temperature dependence of resistivity in metals. It was noted by W.J. de Haas in 1933 that good conductors, such as gold and copper, sometimes develop a strange resistivity minimum at very low temperature [11]. This phenomenon has been widely observed across many labs globally and is against the common belief that the resistivity of metal should decrease monotonically with decreasing temperature. One of the main complications to understanding the effect was due to various contradicting experimental reports on the role played by the impurities; namely, whether the impurity needs to be a particular element, or is it generic to any material possessing a certain (yet unknown) property? Physicists were clueless about this mysterious effect until Bernd Matthias and Myriam Sarachik at Bell Lab proved that the resistivity minima are associated with a trace amount of magnetic impurities in the metal [12–14]. More precisely, a nonmagnetic metal host mixed with trace amount of magnetic impurity would display the resistivity minimum if and only if the doped alloy is magnetic. The big question was when does local magnetic moment (1% Fe for example) in metallic alloy would induce global magnetism and when would not? How does a macroscopic electronic property such as the resistivity minimum has anything to do with the tiny local magnetic moments? Almost immediately, Philip W. Anderson also working in Bell Lab developed a model explaining the formation of local magnetic moments in metals [15], which is essential for solving this three decades-old problem.

Anderson's model elegantly combines the ideas of local Coulomb interaction and virtual bound states between free electron and local atomic orbitals. We use its simplest form, with one electron in the conduction band (*spd* orbital) and one electron from the f orbital as an example here:

$$H = \sum_{\sigma,k} \epsilon_k n_{\sigma,k} + \sum_{\sigma,k} (V_k c^{\dagger}_{\sigma,k} f_\sigma + V_k^* f^{\dagger}_\sigma c_{\sigma,k}) + \sum_{\sigma} \epsilon_f n_{f\sigma} + U n_{f\uparrow} n_{f\downarrow}. \quad (3.1)$$

The first three terms account for the virtual bound state with conduction electrons. Here, σ and k are, respectively, the spin and momenta quantum numbers, ϵ_k is the energy of a conduction band, and $n_{\sigma,k}$ is the conduction electron density. V_k is the hybridization matrix element between f orbitals and conduction band, $c^{\dagger}_{\sigma,k}$ and $f^{\dagger}_\sigma$ are creation operators for an electron in the conduction band and f orbital, respectively. The last two terms describe the local interaction in the atomic limit, where ϵ_f is the f orbital energy for single occupancy, $n_{f\sigma}$ is the density, and U is associated with on-site Coulomb energy for double occupancy. At the heart of Anderson's model is Coulomb's interaction of the f-orbitals, $U \sim \int_{r,r'} \rho_f(r)\rho_f(r')/|r - r'|$, where $\rho_f(r)$ is the electron density of the f-orbital at position r.

In the atomic limit, the energy cost of adding or removing one electron from the magnetic f^1 configuration is $\Delta E = U/2 \pm (\epsilon_f + U/2)$. That is, for larger and positive $U > |U + 2\epsilon_f|$, the ground state would prefer the doubly degenerate and magnetic f^1 configuration. On the other extreme where we put $U = 0$, the "semi-core" f-electron can tunnel out to hybridize with the surrounding electron sea. The hybridization broadens the localized f-states to a finite width, $w(\epsilon) \sim \pi\rho(\epsilon)|V_k|^2$, where $\rho(\epsilon)$ is the density of states. Therefore, we see that the Coulomb interaction and hybridization are competing with each other for the ground state. Anderson concluded that, within the mean field limit, the critical Coulomb interaction is $U_c = \pi w_0$, where w_0 is the f-orbital broadening averaged over the conduction band width around the Fermi level [10]. For metallic hosts with higher density of states, w_0 is typically larger, and therefore the system will favor hybridization over a magnetic ground state. That is, the Fermi level will cross the very broadened and half-filled f-orbital, resulting in a nonmagnetic (because spin up and down are degenerate) and metallic phase. However, if one reduces w_0 such that $U > U_c$, then the relatively sharp spin up and down orbitals separated by U are split across the Fermi level. Therefore, local moments form and the metallic alloy becomes magnetic.

One of the most interesting results proving the Anderson model is the appearance of a third peak crossing the Fermi energy in the photoemission (particle removal) or the inverse photoemission (particle addition) spectra. This peak is also known as the "Kondo resonance," which arises purely due to interaction, and therefore the peak area is equal to the quasiparticle weight Z. The peak width is roughly the Kondo temperature T_K, which sets the crossover between local impurity behavior where the spin is free and the low temperature Kondo regime where the spin and conduction electrons are highly entangled. This peak intensity is set by the scattering phase shift at the E_F, which when combined with Friedel sum rule represents the number of f-electrons bound inside the resonance. The most remarkable feature here is that the Kondo resonance is guaranteed to peak around E_F with width about 1 meV, while the Coulomb interaction U giving rise to it is typically on the order of 10 eV.

The key consequence of Anderson's model is that the local magnetic moments would develop *antiferromagnetic* coupling with the conduction electron sea at low temperature when charge fluctuations are neglected, instead of the more intuitive ferromagnetic coupling. If we start from an isolated atomic limit where $V_k = 0$, then we are looking at local atomic physics where the dominating features are due to valence fluctuations. But as we turn on hybridization V_k while keeping U much larger than the conduction band width D, the f^2 configuration is integrated out by the so-called Hubbard operators, $X_{\sigma 0} = |f^1_\sigma\rangle\langle f^0|$ and $X_{\sigma\sigma'} = |f^1_\sigma\rangle\langle f^1_{\sigma'}|$, which does not allow double occupancy of the f-electrons. The Anderson model now reduces to the Hubbard model with infinite U:

$$H = \sum_{\sigma,k} \epsilon_k n_{\sigma,k} + \sum_{\sigma,k} (V_k c^\dagger_{\sigma,k} X_{0\sigma} + V^*_k X_{\sigma 0} c_{\sigma,k}) + \sum_{\sigma} \epsilon_f X_{\sigma\sigma'}. \tag{3.2}$$

This is equivalent to the high temperature regime where the local impurity is free to do a spin flip, with finite probability of hopping into the conduction sea.

However, if we keep lowering the temperature, the high energy degrees of freedom are continuously integrated out, leaving the f-electron set in the f^1

configuration and the local moment has only spin degrees of freedom. Then the local spin can only flip through virtual charge fluctuations, where the spin of the f-electron is exchanged with the spin of a conduction band electron. In this case, the energy of the singlet bound state between the f and conduction electron is lowered by $J \sim V_k^2[\frac{1}{\epsilon_f+U} - \frac{1}{\epsilon_f}]$. The low energy Hamiltonian is now reduced to the famous (single impurity) Kondo model:

$$H = \sum_{\sigma,k} \epsilon_k n_{\sigma,k} + \sum_{k,k'} J c^{\dagger}_{k\alpha} \hat{\sigma}_{\alpha\beta} c_{k',\beta} \cdot \hat{S}_f. \tag{3.3}$$

Here, $\hat{\sigma}_{\alpha\beta}$ measures the electron spin at the impurity site, and $\hat{S}_f = \frac{1}{2} f^{\dagger}_{\alpha} \hat{\sigma}_{\alpha\beta} f_{\beta}$ measures the spin of the f-electron.

Jun Kondo soon found out that when using Anderson's model to calculate the electron scattering in metals with dilute magnetic impurities, the scattering rate of conduction electrons has a logarithmic divergence at low temperature, leading to a resistivity minimum. The resistivity upturn eventually saturates when all local electrons form singlet virtual bound states with conduction electrons. Thus, the local magnetic moments are "Kondo screened" at low temperature. This is as if the magnetic moments of the local impurities "*dissolve*" into the conduction electron sea, leaving behind the spinless "Kondo singlets" or "Kondo holes" at local impurity sites [10].

Heavy Fermion

After finding the exact solutions to the Kondo Hamiltonian in the entire temperature range by Natan Andrei and Paul Weigman in 1980, the Kondo problem for dilute magnetic impurities in metals is completely solved. However, the physics is much more complicated when the magnetic atoms form a lattice that interacts strongly with the conduction electron sea, known as the Kondo lattice. The first well-studied Kondo lattice is probably the Kondo insulator SmB_6 in 1969 [16]. But the field did catch much interest until superconductivity was observed at about 0.5 K in $CeCu_2Si_2$ by Frank Steglich in 1979 [17], where superconductivity seems to emerge from a Fermi liquid like ground state with unusually heavy electron mass, i.e., heavy fermion liquid. I will only qualitatively describe the phenomenology in heavy fermion materials, where a comprehensive description of the heavy fermion physics can be found in monographs such as Ref. [18] and in some manybody physics textbooks such as Ref. [10].

The heavy fermion compounds often display a giant Sommerfeld coefficient ($\gamma = \lim_{T\to 0} C(T)/T \sim$450–1600 mJ/mol K^2, compared to roughly unity in normal metals) and a giant magnetic susceptibility at low temperatures ($\chi \sim$8–50$\times 10^{-3}$ emu/mol, compared to roughly 10^{-6} emu/mol in normal metals). The ratio of the two is proportional to the dimensionless quantity known as the Wilson ratio,

$$R_w = \frac{\pi^2 k_B^2 \chi}{g^2 \mu_B^2 J(J+1)\gamma} \sim \frac{218.7\chi}{\mu_{eff}\gamma},$$

which remains the same order of magnitude as unity across most heavy fermion metals. Notably, $R_w = 1$ is expected in a free electron gas model such as Fermi liquid but is typically larger than 1 due to electron correlations. Here, the extremely large effective mass and magnetic susceptibility all point to strong electron correlations, which eventually renormalize to quasiparticles that are closely described by the Fermi liquid theory. A trend was observed that with increasing Wilson ratio, the low temperature ground state seems to shift from superconductivity, to nonmagnetic metal, to long range ordered magnetism [19]. The observation suggests the competitive roles played by magnetism and superconductivity. Therefore, much focus has been put in understanding the magnetism in heavy fermion systems, which interestingly share several similarities; Most notably, they all possess a large antiferromagnetic susceptibility with a temperature dependence that qualitatively follows the Curie–Weiss law near the room temperature, $\chi \sim (T + \Theta)^{-1}$. The fit to Curie's law typically yields a relatively large effective moment around 2–5 μ_B at high temperatures, close to the free ion limit. Moreover, the separation between f-atoms is typically larger than 4 Å, above the "Hill limit" of about 3.25–3.50 Å, where the f-orbitals cease to overlap and long range ordered magnetism is predicted to prevail. This is quite surprising because the low temperature ground state of heavy fermions is not all long range ordered and sometimes not even magnetic. The origin of this contradiction is usually explained similar to the Kondo effect, where the local magnetic moment is dissolved into the conduction sea.

Within the Kondo model, the local magnetic moment at x_i induces Friedel oscillations in the nonlocal spin density, which again couples to a magnetic moment at x_j, thus giving rise to a long ranged conduction electron mediated magnetic interaction, known as the RKKY interaction. The typical energy scale for RKKY interaction in solid is $T_{RKKY} \sim J^2\rho$, where J is the effective single impurity Kondo coupling between local f orbital electrons and conduction band, and ρ is the conduction electron density of states. However, there is actually another energy scale in a Kondo lattice competing with T_{RKKY}, that is the Kondo energy, $T_K \sim D\exp[\frac{-1}{2J\rho}]$, where D is the conduction band width. If $J\rho$ is small, then RKKY interaction dominates the low temperature phase and typically favors an antiferromagnetic ground state [Fig. 3.2]. For large $J\rho$, however, Kondo interaction stabilizes a ground state where each magnetic atom resonantly scatters conduction electrons. Therefore, one can expect a quantum phase transition when $J\rho$ is at a critical value where RKKY and Kondo energy scales are comparable.

Due to Bloch's theorem, the *coherent* elastic scattering in the heavy fermion ground state should produce a renormalized quasiparticle band, with band width roughly equal to T_K. In a Kondo lattice, this new band can be understood as a hybridization of the local f states with the conduction band, as illustrated in Fig. 3.3 [21]. The hybridized bands have a narrow indirect gap, $\Delta_h \sim T_K$, and a larger direct gap $2v \sim \sqrt{DT_K}$. Due to the flatness of the bands near

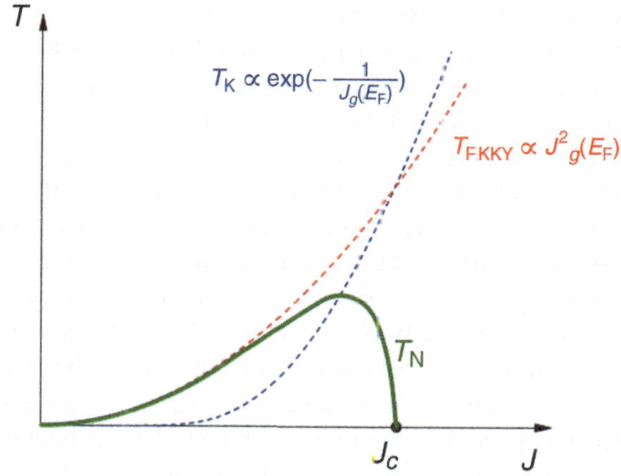

Fig. 3.2 The Doniach phase diagram illustrating the competition between the RKKY and Kondo energy scales in a Kondo lattice system [20]. With increasing J, the ground state typically shifts from an antiferromagnet to paramagnetic heavy Fermi liquid, resulting in a quantum critical point at $J = J_c$. Figure adapted from Ref. [20]

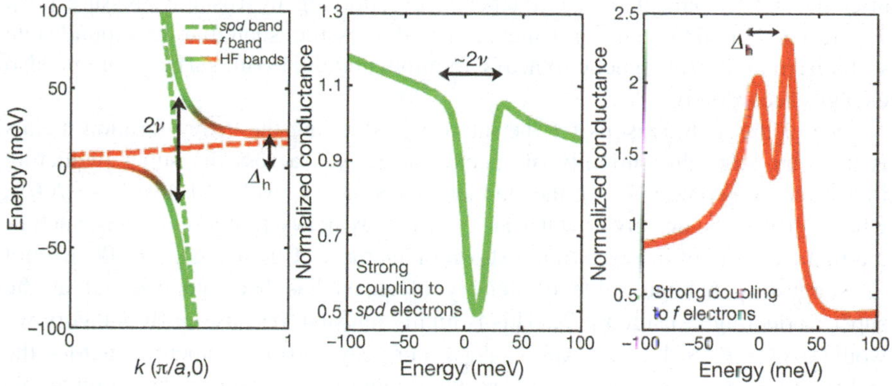

Fig. 3.3 Illustrations showing band dispersion and tunneling spectra of a Kondo lattice, adapted from Ref. [21]. (**a**) Schematic drawing of heavy quasiparticle dispersion showing the hybridization between local f states with the itinerant *spd* conduction band in a heavy fermion material below the coherence temperature, displaying direct (2ν) and indirect Δ_h hybridization gaps. (**b**, **c**) Computed tunneling spectra into the itinerant and localized electron bands. The interference between both tunneling channels (cotunneling) results in a Fano lineshape [24]. Figure adapted from Ref. [21]

Fermi surface, the Sommerfeld coefficient γ in heavy fermion systems is typically hundreds of times larger than in normal metals [3, 19], suggesting enhancement in the conduction electron effective mass and hence the name "heavy fermion." The drastic mass enhancement and direct band gap opening below the coherent temperature were confirmed by infrared spectroscopy measurements, where the mass enhancement $m^*/m \sim D/T_K$ was verified in many materials [3]. The low temperature resistivity possesses T^2 dependence as expected from Fermi liquid, and optical conductivity shows a renormalized Drude peak.

A remarkable feature predicted by the above picture is the Fermi surface volume expansion [22]. Without any hybridization, the Fermi surface is composed entirely of the conduction *spd*-electrons, which would enclose a volume not counting the f-electrons. However, according to the Luttinger sum rule, the Fermi surface volume is determined by the number of valence electrons (including both local and conduction), independent of the interaction or local impurities involved. This is achieved through the Friedel sum rule, which claims that the number of valence electrons must be increased by the amount proportional to the interaction induced phase shift at the Fermi surface. Combining the two sum rules, one concludes that the hybridization must renormalize the *spd*-band in such a way that the Fermi surface encloses a larger volume consistent with both local and conduction electron counting. This is basically just another version of the Kondo resonance peak in the single impurity Kondo model. Here, the resonance peak at the Fermi energy originates from the flat part of the renormalized quasiparticle bands in Fig. 3.3, which results in the "double peak" structure in the density of states, as opposed to the single peak in the impurity model. Experimentally, it has been demonstrated by quantum oscillations that the Fermi surface in $CeRhIn_5$ expands to a larger volume after the antiferromagnetism is quenched by applying hydrostatic pressure [23]. The Kondo coupling constant J increases with pressure, such that the ground state shifts from antiferromagnetic to heavy fermion, and the Fermi surface volume also changes accordingly.

One of the main reasons for the interest in studying the heavy fermion metals is the very large discontinuity of specific heat, ΔC, across the superconducting transition temperature, T_c. In the standard BCS theory, $\Delta C = 1.43\gamma T_c \sim N(0)$, where $N(0)$ is the density of states at E_F. In many heavy fermion systems, such as the first example of heavy fermion superconductor, $CeCu_2Si_2$, $\Delta C \sim 500$ mJ/mol K, suggesting a huge value of density of states has been gapped out at the superconducting transition [17]. This is highly unusual, because in BCS theory we would expect $T_c \sim 1.13\omega_D \exp[-\frac{1}{VN(0)}]$. For heavy fermion superconductors, the Debye frequency ω_D and electron–phonon coupling potential V are comparable to many standard BCS superconductors. That is, the 3 orders of magnitude larger $N(0)$ implies that we should have a room temperature superconductor. However, the transition temperatures of heavy fermion superconductors are typically a disappointingly low number, regularly below 1 K. Therefore, it is highly desirable to understand the mechanisms for such "suppression" of T_c in these materials, in the hope that we might one day be able to drastically raise the transition temperature.

Although the thermodynamic and transport properties of many heavy fermion materials have been thoroughly studied in the 80s [19], and the electrodynamic response was more or less understood after the 90s from infrared spectroscopic studies [3]. Many details of the band hybridization and dynamical properties of the heavy electrons are still not fully understood even to date and remain an interesting and progressing field of study. Recent development of new experimental techniques, in particular the high resolution and dilution fridge temperature scanning tunneling spectroscopy (STS), has helped us understand the "composite" nature of the heavy fermions, which produces a characteristic Fano lineshape in the tunneling spectra [Fig. 3.3] [21, 25–27]. On the theoretical side, the development of dynamical mean field theory and relevant computational techniques has greatly helped our understanding and even predictive powers in the strongly correlated electron systems [28–30]. Very recently, the topological properties in heavy electron systems rekindled physicists' interests in the field, leading to emergence of new concepts such as topological Kondo insulators [31].

URu_2Si_2

Among the many known exotic heavy fermion materials, URu_2Si_2 has been particularly interesting due to a mysterious nonmagnetic and non-structural phase transition at $T_{HO} = 17.5K$, also known as the hidden order (HO) transition [32]. URu_2Si_2 has body centered tetragonal structure characterized by the D_{4h} point group symmetry ($I4/mmm$ space group), as illustrated in Fig. 3.4. It can be seen from Fig. 3.1 that uranium sits at the border of locality and itinerancy in the Kmetko–Smith diagram, thus showing a dual character of localization and itinerancy. While it is commonly accepted that the uranium ions in URu_2Si_2 are in a mixed valence state between tetravalent $5f^2$ and trivalent $5f^3$ configurations [33], recent resonant X-ray emission spectroscopic studies strongly suggest a predominant $5f^2$ configuration [34]. This is also consistent with the results of DMFT+DFT calculation [35].

Notice, however, that the valence configuration does not directly imply the locality or itinerancy of the *5f* electrons at low temperature. As we have introduced in the previous section, the local f orbitals can be delocalized through Kondo hybridization, which conventionally assumes Krammers doublet ground state as in the $5f^3$ configuration. However, other less trivial versions of Kondo effect have also been proposed in materials with even number of electrons, such as two channel Kondo effects [37, 38]. URu_2Si_2 is believed to be one of these cases, where the most probable atomic configuration is two electrons from the $j = 5/2$ spin–orbit subshell, forming a multiplet of total angular momentum $J = 4$. This multiplet further splits into 5 singlet states $2A_{1g} \oplus A_{2g} \oplus B_{1g} \oplus B_{2g}$ and 2 doublet states $2E_g$

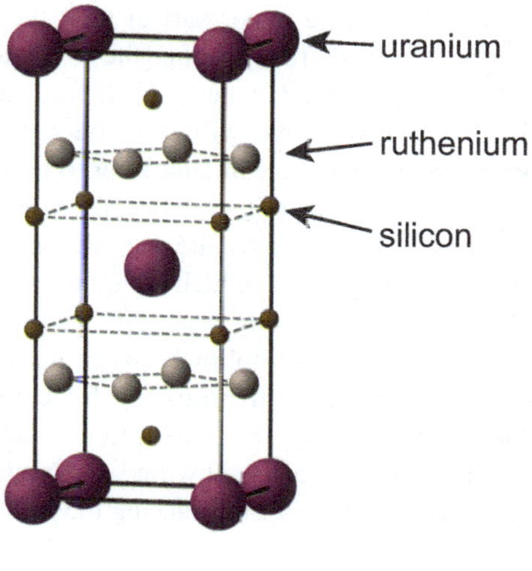

Fig. 3.4 Crystal structure of body centered tetragonal URu_2Si_2 in the paramagnetic state. Lattice constants are $a = 4.124$ Åand $c = 9.582$ Å at low temperature [36]

Table 3.1 The crystal field states of U-5*f* electrons with two electrons in the angular momentum $j = 5/2$ subshell, categorized by irreducible representation (IRR) of the D_{4h} group. These states have total angular moment $J = 4$ and magnetic quantum number M_j runs from $-4 \cdots 4$. The states on the right hand side are labeled by their M_j value, e.g., $|0\rangle \equiv |J = 4, M_j = 0\rangle$

IRR	state
$A_{2g}\ (\Gamma_2)$	$\|A2g\rangle = \frac{i}{\sqrt{2}}(\|4\rangle - \|-4\rangle)$
$A_{1g}\ (\Gamma_1^{(1)})$	$\|A1g\rangle = \frac{\cos\theta}{\sqrt{2}}(\|4\rangle + \|-4\rangle)$ $- \sin\theta\,\|0\rangle$
$A_{1g}\ (\Gamma_1^{(2)})$	$\|\nvDash\rangle = \frac{\sin\theta}{\sqrt{2}}(\|4\rangle + \|-4\rangle)$ $+ \cos\theta\,\|0\rangle$
$E_g\ (\Gamma_{5,1}^{(1)})$	$\|\nVDash\rangle = \cos\phi\,\|-3\rangle + \sin\phi\,\|1\rangle$
$E_g\ (\Gamma_{5,2}^{(1)})$	$\|\ntrianglerighteq\rangle = \cos\phi\,\|3\rangle + \sin\phi\,\|-1\rangle$
$E_g\ (\Gamma_{5,1}^{(2)})$	$\|\ntrianglelefteq\rangle = \sin\phi\,\|-3\rangle - \cos\phi\,\|1\rangle$
$E_g\ (\Gamma_{5,2}^{(2)})$	$\|\ntriangleleft\rangle = \sin\phi\,\|3\rangle - \cos\phi\,\|-1\rangle$
$B_{1g}\ (\Gamma_3)$	$\|\ntriangleright\rangle = \frac{1}{\sqrt{2}}(\|2\rangle + \|-2\rangle)$
$B_{2g}\ (\Gamma_4)$	$\|\nleftarrow\rangle = \frac{i}{\sqrt{2}}(\|2\rangle - \|-2\rangle)$

under the tetragonal crystal electric field, as listed in Table 3.1. Note that fluctuations into other atomic configurations such as different valence or even into $j = 7/2$ subshell remain finite at all temperatures, which greatly increases the complexity and difficulty for a detailed theoretical description. Various versions of two channel Kondo effect have been proposed within this crystal field scheme [35, 39, 40].

3.1.2 *The Phase Diagrams of URu_2Si_2 and the Enigmatic "Hidden Order"*

Several reviews focused on the low temperature ordered phases in URu_2Si_2 have already been available [32, 41–50]. More general review of "122 type" uranium based heavy fermion systems is given in Ref. [51]. Ever since the first discovery of this material, many PhD students have either focused on this compound or included this as a major part of their thesis dissertation. The authors have particularly benefited from reading the dissertations of Collin Broholm [52], Jonathan Buhot [53], Lance Cooper [54], Elena Hassinger [55], and Maxwell Shapiro [56]. In the following, we will limit ourselves to the introduction of properties and experiments that we will later refer to in our Raman scattering study.

At room temperature, URu_2Si_2 is considered a bad metal, with conduction band (CB) electrons scattering off uranium ions incoherently as suggested by transport properties, which is quite conventional for a heavy fermion system [Fig. 3.5] [57, 58]. The resistivity shows a slight upturn upon cooling in the high temperature regime, followed by a drastic drop as the lattice coherence develops and the 5*f* electrons hybridize with the conduction band, forming a heavy fermion band [36]. The magnetic susceptibility along *c*-axis follows Curie–Weiss law down to about 150 K, also showing a maximum around the same temperature as in resistivity at about 70 K, marking the coherence temperature T_{coh} [36].

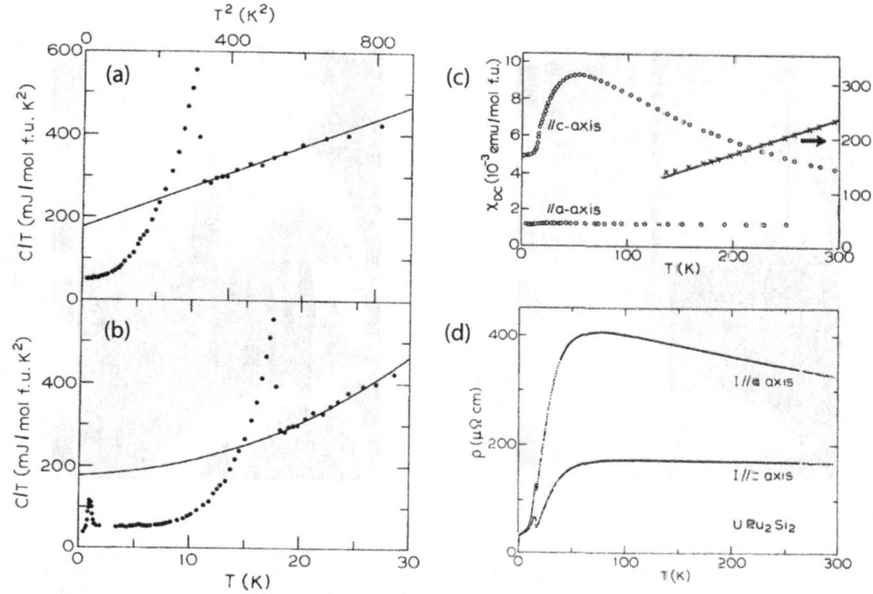

Fig. 3.5 (**a** and **b**) The specific heat data from Ref. [36] plotted as C/T vs. T^2. The λ-like anomaly at 17.5 K is characteristic of a second order phase transition. (**c**) Magnetic susceptibility from Ref. [36]. (**d**) Resistivity data from Ref. [57]

The Kondo hybridization and the formation of heavy bands are most clearly seen in the optical conductivity measurements [59–61]. Figure 3.6 shows temperature and frequency dependent optical conductivity data from Ref. [61]. Clear development of a narrow Drude peak below 5 meV is observed below T_{coh} (marked by the red arrows) in the a-axis, followed by a suppression between 10 and 50 meV due to the opening of hybridization gap.

Below a second order phase transition, URu_2Si_2 orders into two density wave-like phases involving long range ordering of the uranium-5f electrons when a critical parameter x is tuned [62], where x can be chemical substituent concentration [63, 64], pressure [65, 66], or magnetic field [67, 68]. We will denote the second order transition temperature to be $T_{DW}(x) \approx 18$ K. We note that the transition temperature for the special case of $x = 0$ is $T_{DW}(0) = T_{HO} = 17.5$ K.

At $x < x_c$, the system settles in the enigmatic "hidden order" (HO) phase [Fig. 3.5] [36, 69, 70], which transforms into an unconventional large moment antiferromagnetic (AFM) phase through a first order transition for $x > x_c$ [Fig. 3.7], in which uranium atoms carry a magnetic moment of about $0.4\,\mu_B/U$ [66, 71]. The moments are staggered along c-axis and are arranged ferromagnetically in-plane, as determined by neutron scattering [72, 73] and resonant elastic X-ray scattering (REXS) [74] [Fig. 3.8b].

It is deduced from C/T data that about 40% of the Fermi surface is gapped and the entropy change is about $0.2R\log 2$ per uranium, which is an order of magnitude

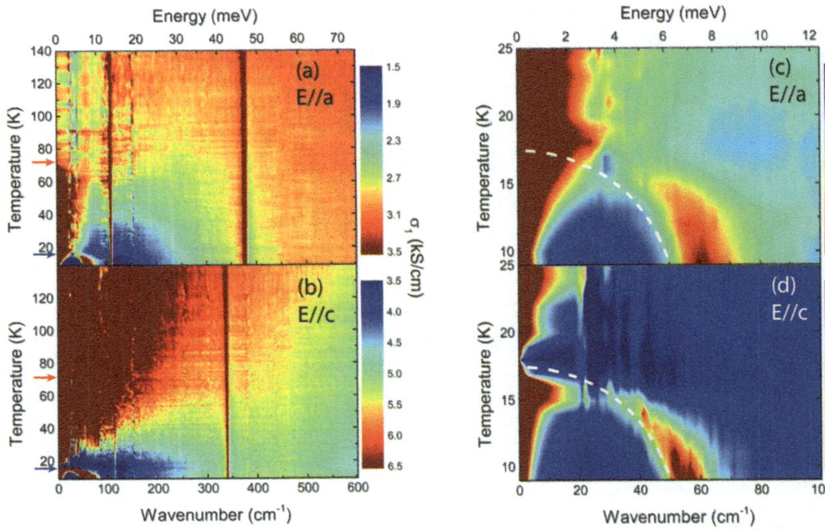

Fig. 3.6 (**a**) a-axis and (**b**) c-axis, optical conductivity data from Ref. [61]. The red and blue arrows indicate the coherence temperature T_{coh} and HO temperature T_{HO}, respectively. Enlarged low temperature optical conductivity along (**c**) a-axis and (**d**) c-axis. The white dashed line shows a BCS gap function with $T_c = 17.5$ K. Figure adapted from Ref. [61]

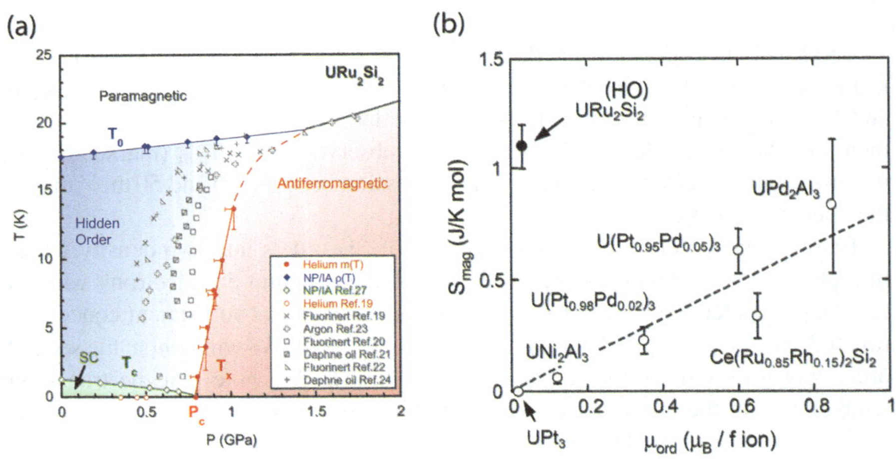

Fig. 3.7 (**a**) The pressure–temperature phase diagram for URu_2Si_2, adapted from Ref. [65]. The symbol "SC" stands for superconductivity, which is fully suppressed for $\geq P_c$ where antiferromagnetism appears. (**b**) Magnetic entropy S_{mag} versus ordered moment m_{ord} for several uranium compounds, figure adapted from Ref. [41]. The broken line is a guide to the eye

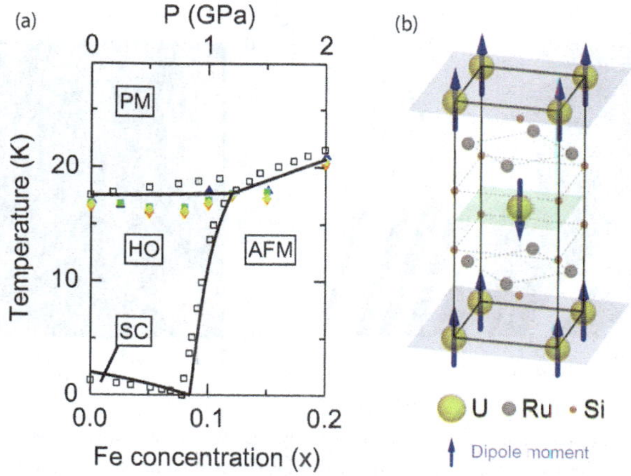

Fig. 3.8 The phase diagram of URu_2Si_2 system, where the black lines show the phase boundaries. (**a**) The measurements on the iron substituted $URu_{2-x}Fe_xSi_2$ crystals from neutron diffraction [80] (blue triangle), electrical resistivity [63] (green square), magnetic susceptibility [63] (purple triangle), and heat capacity [64] (yellow diamond) are overlaid with the neutron diffraction results for URu_2Si_2 under hydrostatic pressure [65] (open square) to show the similarity between the two tuning parameters. (**b**) The magnetic lattice structure in the ordered phases in URu_2Si_2, determined by neutron scattering [73] and resonant elastic X-ray scattering (REXS) [74]. In the HO phase, the local moment was determined to be less than 0.04 μ_B/U and in the pressurized AFM phase about 0.4 μ_B/U. But the magnetic structure is determined to be the same as illustrated here

too large to be explained by the small magnetic moment of about 0.04 μ_B per uranium. Hall effect shows that carrier concentration decreased by about an order of magnitude across the transition, while in-plane resistivity increased less than 10% [75].

Below 1.5 K, a superconducting state, which likely breaks time reversal symmetry [76], emerges from the HO phase and is absent in AFM phase [Fig. 3.7] [65]. The HO and AFM phases are known to exhibit "adiabatic continuity" [77]; that is, both phases possess similar electronic properties [63, 78], and the Fermi surface practically shows no change across the phase boundary [77]. Furthermore, inelastic neutron scattering observed a dispersive collective excitation in the HO phase [66, 73] and recently in the AFM phase of pressurized URu_2Si_2 [79]. This raises the intriguing question of the symmetry relation between the two phases. However, experimental progress is hindered due to inherent constraints of low temperature pressurized experiments.

The availability of $URu_{2-x}Fe_xSi_2$ crystals [63, 64] made it possible to perform high resolution spectroscopic experiments at low temperature and ambient pressure in both the HO and AFM phases. With increasing iron concentration, the a-axis lattice constant decreases linearly while leaving the c-axis roughly unchanged [63]. Thus, iron substitution mimics the effect of applying small pressure or in-plane stress on the URu_2Si_2 lattice, and the iron (Fe) concentration, x, can be approximately treated as an effective "chemical pressure" [63]. Recently, the phase diagram

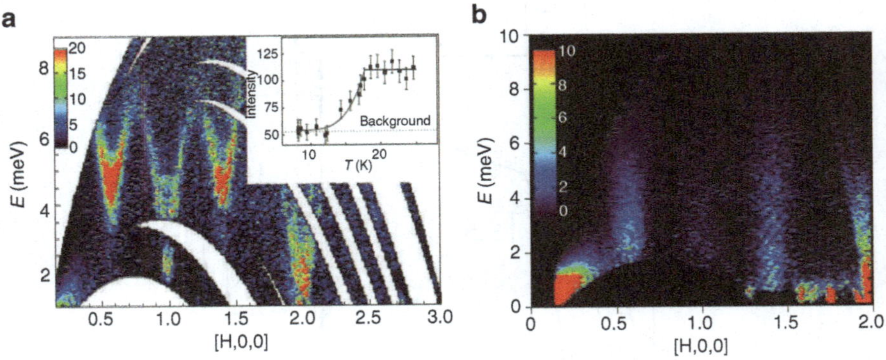

Fig. 3.9 Inelastic neutron scattering results at temperatures above and below the HO transition, from Ref. [84]. (**a**) The excitation spectrum in the HO phase (data measured at 1.5 K, just above the superconductivity) and (**b**) above the HO phase at 20 K. Notice the ungapped continuum of excitations at incommensurate wave vectors at 20 K

of $URu_{2-x}Fe_xSi_2$ single crystals has been determined [62, 64, 80–82], which resembles the low pressure phase diagram of pristine URu_2Si_2 [32, 65] [Fig. 3.8]. The inelastic neutron scattering measurements again illustrate the analogies of the AFM phase to the HO phase [82, 83], albeit differences remain relating to the existence of the resonance in the AFM state of pressurized [79, 83] or Fe substituted crystals [82].

In the paramagnetic phase above T_{HO}, inelastic neutron scattering (INS) detected a continuous spectrum at incommensurate wave vector $\mathbf{Q}_1 = (\pm 0.6, 0, 0)$ [84] [Fig. 3.9]. Below T_{HO}, this continuum is gapped at $\Delta_1 = 4$ meV, and another excitation appears at commensurate wave vector $\mathbf{Q}_0 = (1, 0, 0)$ with energy gap $\Delta_0 = 2$ meV [72, 84, 85] [Fig. 3.9]. Application of hydrostatic pressure at low temperature showed that the commensurate mode is fully suppressed in AFM state, while the incommensurate mode hardens with increasing pressure [86]. Consistent results were obtained from Fe substituted crystals [82, 83].

Particular caution should be taken here with the conventions of the ordering vectors. Historically, the wave vectors of the *high temperature* BCT phase (the structure lattice) were given in the units of the *low temperature* ST phase (the density wave lattice) parameters, $a = 4.13$ Åand $c = 9.58$ Å. This was because of the observation of a small magnetic moment staggered along c-axis, with ordering vector $(0, 0, \frac{\pi}{c})$ Å^{-1}, as shown in Fig. 3.8b. In the reciprocal lattice units of the ST phase, the ordering vector is written as (0, 0, 1). However, the structural Brillouin zone (BCT) is twice the size of density wave Brillouin zone (ST), as clearly seen in Fig. 3.10. That is, the ordering vector $\mathbf{Q}_0 = (0, 0, 1)$ given in the ST reciprocal lattice units would have been (0, 0, 1/2) if given in the structural lattice (BCT) units.

With the above clarification in mind, the "incommensurate" wave vector, $\mathbf{Q}_1$, is in fact a high symmetry point of the BCT Brillouin zone. In the reciprocal lattice units of the ST phase, $\mathbf{\Sigma}_{BCT} \equiv (\pm\frac{1}{2}[1 + \frac{a^2}{c^2}] \approx \pm 0.6, 0, 0)$, where a and c are lattice

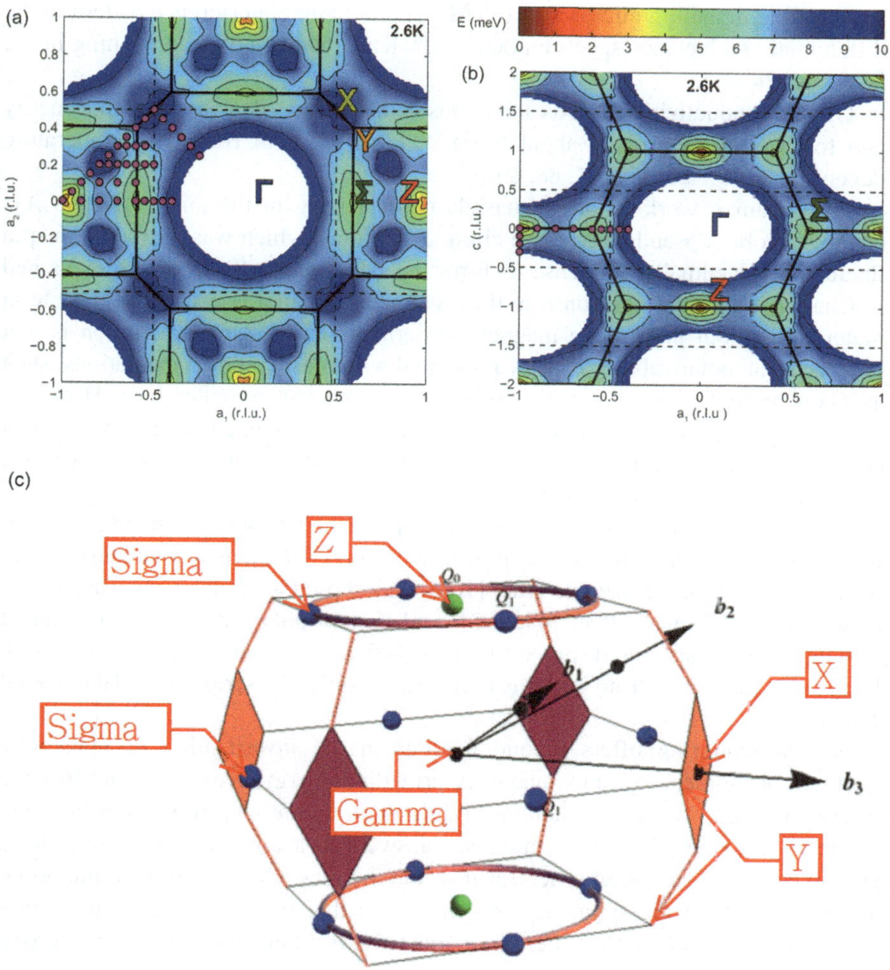

Fig. 3.10 Magnetic dispersion in HO phase (2.6 K) in (**a**) the *ab*-plane and (**b**) the *ac*-plane, measured by inelastic neutron scattering (INS) [87]. The heavy solid lines show the Brillouin zone boundaries of the high temperature paramagnetic body centered tetragonal (BCT) phase, where as the dotted lines show the low temperature simple tetragonal (ST) phase. The reciprocal lattice unit (r.l.u.) is given in the ST lattice parameters. (**c**) The BCT Brillouin zone, high symmetry points and where the commensurate (green balls) and incommensurate (blue balls) excitations were observed

parameters of the ST phase. As evident from Fig. 3.10, $(\pm0.6, 0, 0)$ would also be equivalent to wave vectors such as $(\pm0.4, 0, 1)$ in the second zone. Moreover, in the BCT Brillouin zone, the high symmetry point $\mathbf{Z} \equiv (0, 0, \pm1)$ (given in the reciprocal lattice units of the ST phase as before) is equivalent to $(\pm1, 0, 0)$ [Fig. 3.10].

Scanning tunneling microscopy (STM) reported Fano interference pattern in the differential conductance spectrum below 120 K and a 4–5 meV gap opening in the HO state [26, 27].

Optical conductivity in a-axis and c-axis both showed a density wave-like energy gap followed by a peak at about 65 cm^{-1} (8 meV), with BCS like temperature dependence [Fig. 3.6] [59–61, 88, 89].

Early Raman work reported quasielastic scattering in the antisymmetric A_{2g} channel, probing pseudovector like chiral excitations, which was attributed to spin fluctuations eventually condense to form the AFM phase [90]. As we explained in Chap. 1, the Raman response in this symmetry channel is usually negligible in materials without long range magnetic ordering. The fundamental reason is that the electronic polarizability tensors associated with most common excitations, such as phonons and plasmons, are symmetric with respect to reflections. However, it has previously been experimentally identified in nonmagnetic heavy element compounds due to crystal field excitations [91–93] and in some systems with exotic electronic ground states [94–96].

Several theories have been suggested to explain the HO phenomena [32], including the recently proposed hexadecapole order [35, 97], dynamical symmetry breaking [98, 99], E-type quadrupole order [100], rank-5 pseudo-spin density wave [101], under-screened Kondo effect [49, 102], rank-5 multipole order [103], modulated spin liquid [104], staggered spin–orbit order [105], and hastatic order [40, 106, 107]. Until now, there is still no consensus for the complete description of HO ground state [43].

Raman scattering offers unique features in the investigation of manybody quantum phenomena by simultaneously providing energy, symmetry, and lifetime information of collective excitations. While the ultimate goal in solving the "HO mystery" is to find the order parameter, answers to a few questions would have greatly constrain the phase space for possible theories. One of the most important questions is about the symmetry or symmetries that were removed at the phase transition. A typical second order phase transition is related to a reduction of group symmetry, which usually occurs through removing a combination of translation, rotation, and time reversal symmetry operators.

3.2 Material and Methods

Preparation of high quality single crystals with controlled chemical compositions has always been a critical factor for the understanding of physical phenomena and reproducibility of experiments in condensed matter physics. This is particularly important in the quest of the HO mystery due to the proximity of several electronic phases at low temperature. Small amount of residual stress in URu_2Si_2 results in puddles of antiferromagnetic regions in the crystal [71], leading to confusing observations in the early years [32].

At low temperature, the residual resistivity is dominated by the impurity or defect scatterings in URu_2Si_2. Therefore the resistivity ratio between the room temperature and the base temperature, ρ_{RT}/ρ_{base-T}, also known as the residual resistivity ratio (RRR), is often used as a characterization of crystal purity for metallic compounds. It is known that the superconducting transition temperature T_c in URu_2Si_2 is strongly affected by RRR [108, 109], not uncommon for nodal superconductors [110, 111]. On the other hand, the dependence of RRR with the HO transition temperature and the sharpness of the phase transition is much less obvious [108, 109]. Recently, C_4 symmetry breaking features were reported in the HO phase only in the samples with highest RRR values [112, 113]. However, the subtlety of these signatures seems inconsistent with the large entropy change and the robustness of the HO phase transition. Therefore, the C_4 symmetry breaking may have been due to secondary effect of an (unknown) primary order parameter.

Traditionally, large single crystals were grown by the Czochralski pulling method in the arc furnaces [36, 114]. In order to minimize the contribution from extrinsic impurity phases, several improvements have been made. First, the starting uranium has purity of only about 99.9% and is further purified by the so-called electro-refinement method, i.e., solid state electrotransport method under ultrahigh vacuum to remove trace amount of transitional metals such as iron and copper [115]. The as-grown crystals using refined uranium by the Czochralski pulling method have typical RRR over 10 and untraceable impurity levels [45]. The superconducting transition temperature T_{sc} in these as-grown samples is typically less than 1 K. The crystals are then followed by annealing process using the electro-refinement method in high vacuum for over 10 days, which was shown to significantly increase the RRR up to 1000 in some cases [116], and T_{sc} is increased to 1.5 K [109, 116].

3.2.1 Raman Studies

The $URu_{2-x}Fe_xSi_2$ single crystals used in this monograph are grown by the Czochralski method in a tetra-arc furnace in argon environment. The pristine crystals were further post-processed by electro-refinement as described in Ref. [45, 108, 116] and were provided by Dr. Ryan E. Baumbach and Dr. Eric D. Bauer from Los Alamos National Laboratory. The crystals with Fe substitution were provided by Dr. Sheng Ran, Dr. Noravee Kanchanavatee, and Dr. M. Brian Maple from University of California, San Diego. The quality of the synthesized single crystals was confirmed by X-ray diffraction measurements.

The Raman scattering data shown in the following are all collected from the *ab*-surface of $URu_{2-x}Fe_xSi_2$ single crystals. Unless otherwise given, the Raman data were acquired using 752.5 nm line of a Kr^+ laser for excitation. The scattered light was analyzed by a custom triple-grating spectrometer equipped with a liquid nitrogen cooled CCD detector. The scattered light intensities were corrected for the spectral response of the spectrometer and CCD. The laser spot size on the sample is roughly $50 \times 100\,\mu m^2$. All temperatures shown in the following are corrected for

	E	$2C_4$ (z)	C_2	$2C'_2$	$2C''_2$	i	$2S_4$	σ_h	$2\sigma_v$	$2\sigma_d$	linears, rotations	quadratic
A_{1g} (Γ_1^+)	1	1	1	1	1	1	1	1	1	1		x^2+y^2, z^2
A_{2g} (Γ_2^+)	1	1	1	-1	-1	1	1	1	-1	-1	R_z	
B_{1g} (Γ_3^+)	1	-1	1	1	-1	1	-1	1	1	-1		x^2-y^2
B_{2g} (Γ_4^+)	1	-1	1	-1	1	1	-1	1	-1	1		xy
E_g (Γ_5^+)	2	0	-2	0	0	2	0	-2	0	0	(R_x, R_y)	(xz, yz)
A_{1u} (Γ_1^-)	1	1	1	1	1	-1	-1	-1	-1	-1		
A_{2u} (Γ_2^-)	1	1	1	-1	-1	-1	-1	-1	1	1	z	
B_{1u} (Γ_3^-)	1	-1	1	1	-1	-1	1	-1	-1	1		
B_{2u} (Γ_4^-)	1	-1	1	-1	1	-1	1	-1	1	-1		
E_u (Γ_5^-)	2	0	-2	0	0	-2	0	2	0	0	(x, y)	

Fig. 3.11 The character table for the point group D_{4h}. The linear basis functions (x,y) and z denote vectors in and out of the *ab*-plane, respectively. Similarly, (R_x,R_y) and R_z are pseudovectors in and out of the *ab*-plane, respectively. The quadratic functions are also shown

laser heating. The power on the sample is about 10 mW for most temperatures and kept below 7 mW for the lowest temperatures to reduce laser heating.

$URu_{2-x}Fe_xSi_2$ single crystals have body centered tetragonal structure characterized by the D_{4h} point group symmetry ($I4/mmm$ space group), as illustrated in Fig. 3.4. In our measurements, we align the incident and scattered light along the *c*-axis of the crystal. Thus, the electric field of light is in-plane and transforms as the E_u irreducible representation of the D_{4h} point group 3.11. Therefore, the inelastically scattered light couples to excitations that transform as

$$E_u \otimes E_u = A_{1g} \oplus A_{2g} \oplus B_{1g} \oplus B_{2g}$$

irreducible representations. The basis functions of these representations are shown in Fig. 3.11. The A_{2g} symmetry channel is special in the sense that it only couples to the pseudovector like excitations, forbidden for phonons.

The symmetry of excitations is selected through varying the experimentally controlled combination of incident and scattered light directional vector and polarization, called the "scattering geometry" [details in Sect. 1.3]. We adopt Porto's notation to indicate the scattering geometries, $\mathbf{k}_i(\mathbf{e}_i\mathbf{e}_s)\mathbf{k}_s$, where the directional vector and polarization of incident (scattered) light are denoted by $\mathbf{k}_i$ and $\mathbf{e}_i$ ($\mathbf{k}_s$ and $\mathbf{e}_s$), respectively. Since we are only using quasi-backscattering geometry, we omit the directional vector in the following. We use X, Y, X′, Y′, R, and L for shorthand notations of $\mathbf{e}_i$ and $\mathbf{e}_s$. X=[100] and Y=[010] are aligned along crystallographic a-axis, X′=[110] and Y′=[1$\bar{1}$0] are aligned 45° to the a-axis, and R=(X+iY)/$\sqrt{2}$ and L=(X-iY)/$\sqrt{2}$ are right and left circularly polarized, respectively.

The measured spectral intensity in the $\mathbf{e}_i\mathbf{e}_s$ scattering geometry, $I_{\mathbf{e}_i\mathbf{e}_s}(\omega, T)$, is related to the Raman response function, $\chi''_{\mathbf{e}_i\mathbf{e}_s}(\omega, T) \equiv \text{Im}[\chi_{\mathbf{e}_i\mathbf{e}_s}(\omega, T)]$, by the following: $I_{\mathbf{e}_i\mathbf{e}_s}(\omega, T) = [1 + n(\omega, T)]\chi''_{\mathbf{e}_i\mathbf{e}_s}(\omega, T) + L(\omega, T)$, where $\chi_{\mathbf{e}_i\mathbf{e}_s}(\omega, T)$ is the Raman susceptibility, $n(\omega, T)$ is the Bose factor, and $L(\omega, T)$ is a background mainly resulting from luminescence of the sample.

We notice that other than the phononic contribution, the response in B_{1g} and B_{2g} symmetry channels is the same and weakly temperature dependent. This suggests that electronic Raman contributions in the B_{1g} and B_{2g} channels are negligibly small, and the measured spectral intensity in RL scattering geometry, $I_{RL}(\omega, T) = [1+n(\omega, T)][\chi''_{B1g}(\omega, T)+\chi''_{B2g}(\omega, T)]+L(\omega, T)$, is predominantly contributed by luminescent background. This background signal is very small in the near infrared excitation and can be approximated as a small constant.

3.2.2 Characterization of Strain Free Areas

Due to the proximity of the HO and AFM phases [Fig. 3.7], any small strain field in the crystal can induce small ~50 nm size AFM puddles in the HO phase, which blurs fine spectroscopic features coming from the HO phase. Therefore, it is important not only to measure from a flat and clean surface as in most Raman scattering experiments but also to measure from a strain free area. In our experiment, the ab-plane residual resistivity ratio (RRR) for crystals from the same batch as used for this Raman scattering study is over 300 [109]. The cleaved samples were examined under a reflective Nomarski differential interference contrast (DIC) microscope to find a stress free area for Raman study. Our spectrometer is equipped with a "behind entrance slits" periscope as described in Sect. 2.1, which allows us to measure from the exact same spot predetermined by the Nomarski microscope.

The reflective Nomarski DIC microscope is commonly used by the semiconductor industry to optically image the surface roughness and defects [117]. In comparison to the more common bright field microscopy, the microscopic lattice deformation can be much more clearly distinguished in Nomarski DIC image [118] and hence a sensitive tool for finding an optically flat and strain free area on the crystal for Raman scattering studies. Figure 3.12 is a white light illuminated Nomarski microscope image of URu_2Si_2 single crystal, glued on a copper sample holder by epoxy and silver paint. The color gradients are proportional to the strain gradients on the sample surface induced by defects, steps, and impurity centers. The white arrows contrast two areas, both seem flat and featureless under bright field microscopy, while Normarski microscope clearly shows detailed differences due to minor crystal imperfections. The black rectangle is 100 by 200 μm, about twice the laser spot size of our experiment. The magnetic inclusions are known to exist in

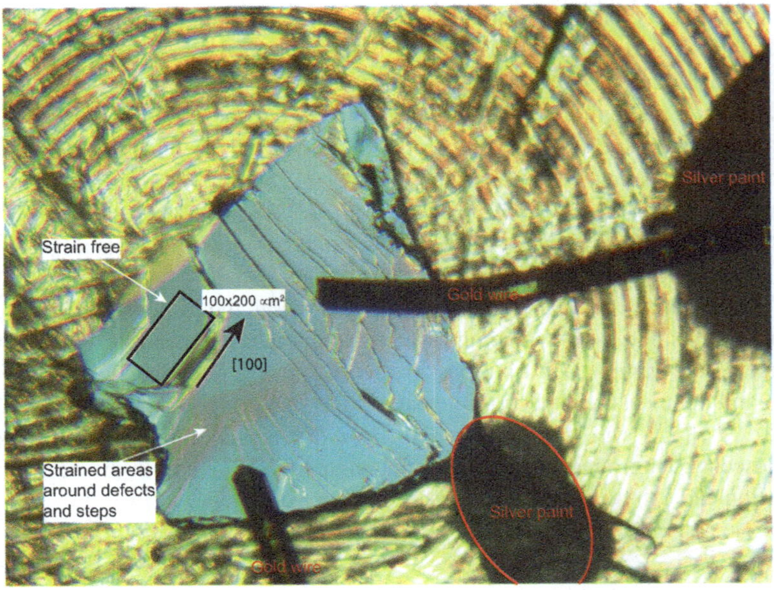

Fig. 3.12 Nomarski microscope image of URu_2Si_2 single crystal glued on a copper sample holder by epoxy and silver paint. The color gradients are proportional to the strain gradients on the sample surface induced by defects, step, and impurity centers. The white arrows contrast two areas, both seem flat and featureless under bright field microscopy. The gold wires are attached to the copper holder by conductive silver paint

bulk URu_2Si_2 under very small pressure or stress [65, 66], and thus it is crucial to study the intrinsic properties of the HO phase from a strain free area. The attached gold wires are glued to the copper holder by conductive silver paint to improve the thermal conductivity between the sample and copper holder.

3.3 Experiment and Results

Phase transitions are emergent phenomena in condensed matter materials that arise due to interactions between large quantity of particles. Despite the astronomical number of materials in this world, the "general behavior" of phase transitions is surprisingly similar. Experimentally, phase transitions are associated with divergence of some measurable quantity, such as heat capacitance, particle density, pressure, or magnetization. The phenomena are so common in nature, yet the concept remains one of the most puzzling fields in physics, and scientists are only starting to understand them in the recent hundred years. Theoretically, phase transitions occur as singularities in the partition function after taking the thermal dynamic limit [119]. Depending on the details of the singularity (or the way the measurable diverges), phase transitions can be sorted into a few categories [120].

In particular, a second order phase transition, e.g., ferromagnetic or superconducting transition, is associated with the emergence of an "order parameter" and spontaneous symmetry breaking. Above the phase transition temperature, both temporal and spatial short range fluctuations occur across the sample. Approaching the transition temperature, the fluctuations experience "critical slow down" as both the coherence time and length approach the thermal dynamic limit. Below the transition temperature, the fluctuations are quenched as some long range ordering form simultaneously and break the ground state symmetry of the high temperature phase. In the language of group theory, at least one symmetry operator is removed from the space group of the high temperature phase across a second order phase transition. To quantify the phase transition, one defines a measurable quantity, i.e., the order parameter ψ, that is only nonzero below the transition temperature. The Ginzburg–Landau free energy is then written as a function of ψ, and the thermal dynamic quantities can be derived from the free energy [121].

While the ultimate goal is to write down the partition function (microscopic theory) of a system, it is typical to use the Ginzburg–Landau theory for predicting the measurable properties of the system. This is because the partition functions defer from material to material, but the Ginzburg–Landau theory for a certain type of transition, e.g., superconductivity, can be very general and applies to many similar systems. In principle, infinitely many possible order parameters can arise due to a certain symmetry breaking. However, real materials embody many constraints, such as the lattice parameters and band structures, and therefore only allow a handful of imaginable order parameters. Thus, knowing the broken symmetry in a second order phase transition significantly reduces the complexity of the problem by constraining the allowed order parameters.

In our experiment, we will focus on the determination of symmetry breaking in the HO phase. Raman scattering is unique for simultaneously measuring the energy and symmetry of collective modes at the Brillouin zone center. Therefore, we study the *Raman selection rules* of two collective modes for determining the point group symmetry in the HO phase. (1) The 163 cm^{-1} B_{1g} phonon mode, corresponding to the lattice vibration of Ru atoms along c-axis and (2) a new mode of electronic origin with A_{2g} symmetry, which emerges only below the second order phase transition. The B_{1g} phonon is sensitive to the in-plane lattice C_4 rotational symmetry breaking. Furthermore, the Raman selection rule is capable of distinguishing the new Bravais lattice being orthorhombic (D_{2h}) or monoclinic (C_{2h}). We will show our low temperature data below and discuss in relation to the recent reports of C_4 symmetry breaking in the HO phase. The A_{2g} collective mode is typically associated with the magnetic Raman scattering [122]. Here, this mode is a new type of electronic Raman excitation, for which we will explain its origin and significance. Most importantly, by studying the Raman selection rules of this mode, we determine that the reflection symmetry is broken in the HO phase.

3.3.1 Phonon Modes

URu_2Si_2 crystallizes in the body centered tetragonal $ThCu_2Si_2$ structure type, with space group $I4/mmm$ (point group D_{4h}). The material contains 5 atoms in the primitive cell and therefore has a total of 15 phonon modes, with 3 acoustic (A_{2u}+E_u) and 12 optical bands. The U, Ru, and Si atoms occupy the 2a, 4d, and 4e Wyckoff positions, respectively, and it can be determined from the Bilbao crystallographic database that the optical phonons are half Raman active (A_{1g}+B_{1g}+2E_g) and half IR active (2A_{2u}+2E_u) [123, 124] (Fig. 3.13). The atomic displacements of each optical phonons are shown in Fig. 3.14, reproduced from Ref. [125]. We can see that the A_{1g} and B_{1g} phonons only involve the vibration of Si and Ru atoms, respectively. The role played by the lattice degrees of freedom has been studied by Raman scattering in the very early days of URu_2Si_2 [90]. It was concluded that the A_{1g} phonon intensity shows anomalous increase at low temperature, which may suggest strong electron–phonon coupling.

Recently, a detailed study of the lattice degrees of freedom from 2 to 300 K has been done by Raman scattering, IR spectroscopy, inelastic neutron scattering, and *ab initio* DFT calculations [125]. The study reproduced the anomalous temperature dependence of the A_{1g} phonon found in Ref. [90] and also reported a small energy softening of the B_{1g} phonon below about 100 K. The authors of Ref. [125] suggested a tendency toward lattice instability with orthorhombic distortion below 100 K, which is possibly related to the observation of a "pseudogap" in the E_g channel [126]. However, Ref. [125] found no phononic anomaly anywhere in the BZ across the HO transition. This study thereby put a strong constraint on the theoretical models explaining the HO phase transition, such that the order parameter must not couple to the lattice or at least couple in a manner that the phonons remain

Fig. 3.13 A table of the IR and Raman active phonons (acoustic modes not included), given the Wyckoff positions of the crystal structure of URu_2Si_2, determined from the Bilbao crystallographic database [123, 124]

IR Ative Modes

WP	A_{1g}	A_{1u}	A_{2g}	A_{2u}	B_{1g}	B_{1u}	B_{2g}	B_{2u}	E_u	E_g
4e	-	-	-	1	-	-	-	-	1	-
4d	-	-	-	1	-	-	-	-	1	-
2a	-	-	-	1	-	-	-	-	1	-

Infrared active Modes: $2A_{2u}$ + E_u (Acoustic modes not included)

Raman Active Modes

WP	A_{1g}	A_{1u}	A_{2g}	A_{2u}	B_{1g}	B_{1u}	B_{2g}	B_{2u}	E_u	E_g
4e	1	-	-		-	-	-	-		1
4d	-	-	-		1	-	-	-		1
2a	-	-	-		-	-	-	-		-

Raman active Modes: A_{1g} + B_{1g} + $2E_g$

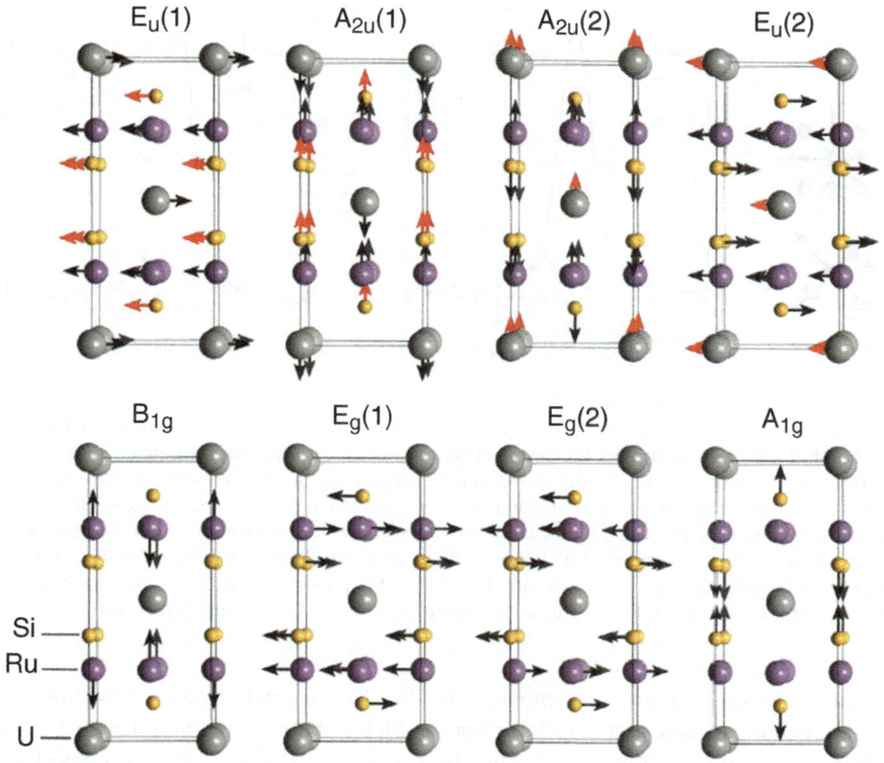

Fig. 3.14 The atomic displacement of the IR and Raman active phonons, where the arrows indicate the relative motions in a frozen time frame. Figure adapted from Ref. [125]

unaffected (energy, lifetime, and intensity) even with the strong electron–phonon couplings.

Figure 3.15 plots the Raman response function in the XX scattering geometry at 7 K, containing the B_{1g} phonon centered at 20 meV. In general, the phonon width is composed of its decay rate and inhomogeneous broadening due to stress field [125, 127]. Hence, the phonon width at low temperature is an indication of crystal quality. It has been demonstrated that stress field on the sample induces local AFM domains[32, 128]. Here, the deconvoluted full-width-at-half-maximum of the B_{1g} phonon is about 0.1 meV, smaller than the values reported elsewhere for this material[125, 127], indicating that the measured surface is single domain and stress free as it is required to study the symmetry of the order parameter in the HO phase.

Recently, a small lattice distortion was discovered by X-ray diffraction [129], where the observed orthorhombicity is 6.2×10^{-5}. In general, the broken fourfold rotational symmetry allows the B_{1g} phonon at 20.0 meV to leak into other channels. As illustrated in Fig. 3.15, the B_{2g} type lattice distortion reported in Ref. [129] would leave X and Y no longer aligned with crystallographic axes, leading to

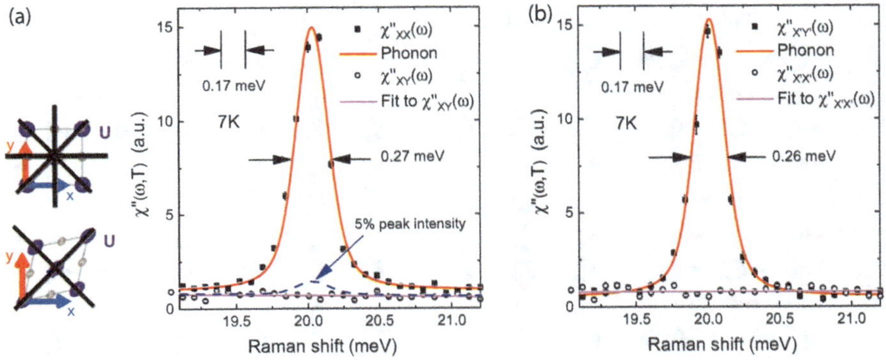

Fig. 3.15 (**a**) The Raman active B_{1g} phonon measured at 7 K in XX and XY scattering geometries and (b) in $X'X'$ and $X'Y'$ scattering geometries. The data were taken from the same surface as the rest of data presented in this report. The red curve is a fit of the Raman response function (**a**) in XX and (**b**) in $X'Y'$, scattering geometry (solid squares) containing the B_{1g} phonon, whereas the pink line is a fit to the Raman response function (**a**) in XY and (**b**) in $X'X'$, scattering geometry (open circles) containing no phonon. The deconvoluted full-width-at-half-maximum of the B_{1g} phonon is about 0.1 meV, close to resolution limitation of the instrument. There is no observable B_{1g} phonon leakage into other symmetry channels at 7 K, where a 5% phonon intensity is shown as guide to the eye (blue dashed line). The instrumental resolution is shown in the upper-left corner

intensity leakage of the B_{1g} phonon into the XY channel. However, within the experimental accuracy, no such orthorhombicity induced phonon leakage was observed in our data [Fig. 3.15a]. The absence of the phonon leakage is consistent with the tininess of the orthorhombicity observed by X-ray diffraction.

Similar argument can be extended to B_{1g} type of lattice distortion, which would allow finite intensity in the otherwise forbidden geometry $X'X'$. We also did *not* observe such feature experimentally [Fig. 3.15b]. The s/n of the data allows us to confidently rule out any C_4 breaking lattice distortion (orthorhombicity) in the *ab*-plane larger than 5%.

3.3.2 Raman Response in the A_{2g} Symmetry Channel

In Fig. 3.16, we plot the Raman response in the chiral A_{2g} channel where the most significant temperature dependence was observed. We show that $\chi''_{A2g}(\omega, T)$ displays enhancement of quasi-elastic scattering upon cooling.

At high temperatures, the Raman spectra exhibits a Drude line shape, $\chi''_{A2g}(\omega, T) \sim \frac{\omega\Gamma(T)}{\omega^2+\Gamma^2(T)}$, which in Ref. [90] was attributed to quasi-elastic scattering. The Drude line shape is due to incoherent scattering from the uranium crystal field states, and the maxima energy $\Gamma(T)$ reflects the scattering rate or the inverse quasiparticle lifetime. As temperature decreases, the coherent scattering leads to longer quasiparticle lifetime, thus decreasing $\Gamma(T)$. Below 70 K, the line

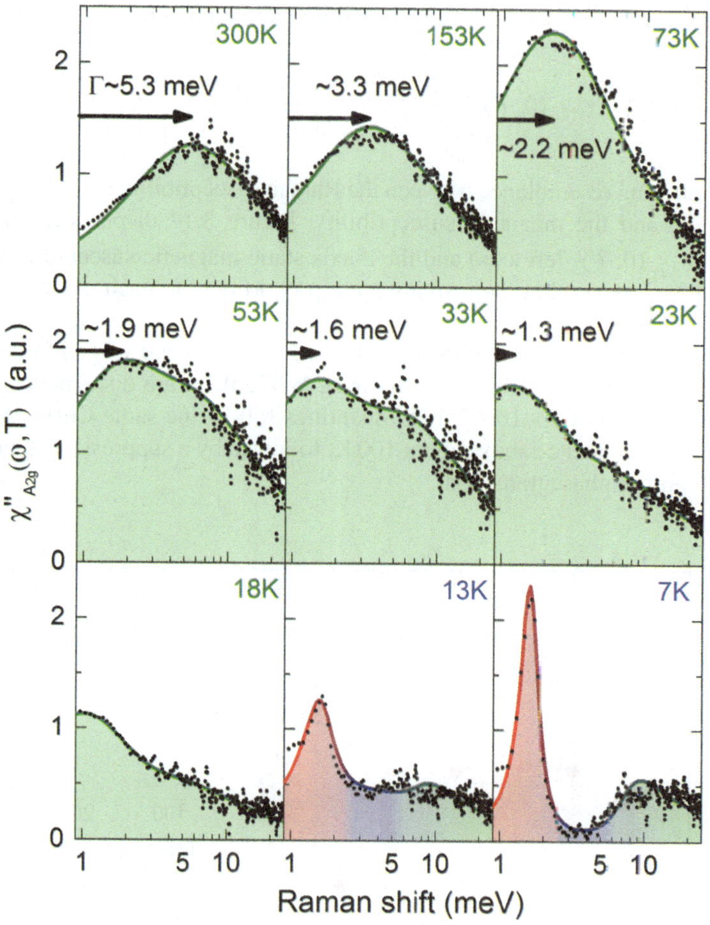

Fig. 3.16 The A_{2g} Raman response function decomposed from the spectra measured in the *XY*, $X'Y'$, and *RL* scattering geometries. The solid lines are guide to the eyes illustrating the narrowing of the Drude function[90, 130]: $\chi''_{A2g}(\omega, T) \sim \text{Im}\left[\frac{\omega}{\omega+i\Gamma(T)}\right]$, where $\Gamma(T)$ is the Drude scattering rate (indicated by the arrows), which decreases on cooling. Below 70 K, the Raman response deviates from the Drude function. Below T_{HO}, the Raman response shows a spectral weight suppression below 6 meV and the appearance of an in-gap mode at 1.6 meV (7 and 13 K)

shape deviates slightly from the Drude function, tracking the formation of the heavy fermion states by the hybridization of the itinerant conduction band and the U-5*f* states. In Fig. 3.16, the maximum in the Raman response function decreases from 5 meV at room temperature to 1 meV just above T_{HO}, reflecting the dramatic increase of the quasiparticle lifetime.

Further analysis of data in Fig. 3.16 by doing Krammers–Kronig transformation,

$$\chi'_{A2g}(0, T) = \frac{2}{\pi} \int_0^{25\,\mathrm{meV}} \left[\frac{\chi''_{A2g}(\omega, T)}{\omega} \right] d\omega, \tag{3.4}$$

reveals a striking resemblance between the Raman susceptibility in the A_{2g} channel, $\chi'_{A2g}(0, T)$, and the magnetic susceptibility. Figure 3.17 displays a comparison between $\chi'_{A2g}(0, T)$ (left axis) and the c-axis static magnetic susceptibility $\chi_c^m(T)$ (right axis), showing that the responses are proportional to each other at temperatures above T_{HO}.

Figure 3.18 shows the comparison between $\chi'_{A2g}(0, T)$ and $\chi_c^m(T)$ for a different iron concentration x in $URu_{2-x}Fe_xSi_2$ [64]. While there are discrepancies around the maxima at about 50–100 K, both quantities follow the same Curie–Weiss-like temperature dependence above about 100 K, followed by a suppression approaching the second order phase transition.

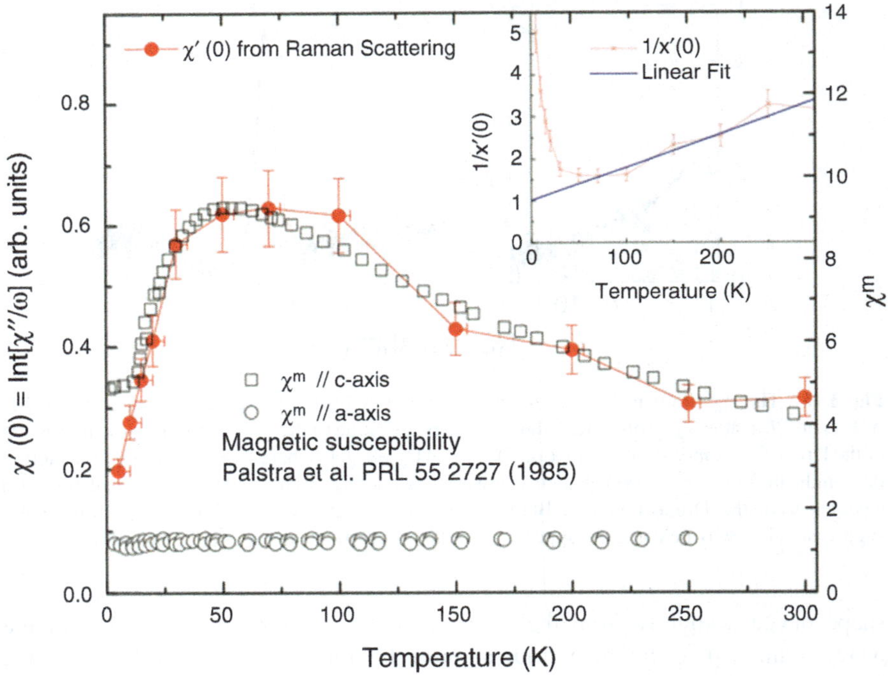

Fig. 3.17 Temperature dependence of the static Raman susceptibility in A_{2g} channel: $\chi'_{A2g}(0, T) = \frac{2}{\pi} \int_0^{25\,\mathrm{meV}} \left[\frac{\chi''_{A2g}(\omega, T)}{\omega} \right] d\omega$ (red dots), and the static magnetic susceptibility along c-axis and a-axis from Ref.[36] is plotted as blue squares and black circles, respectively. T_{HO} is marked by the dashed line

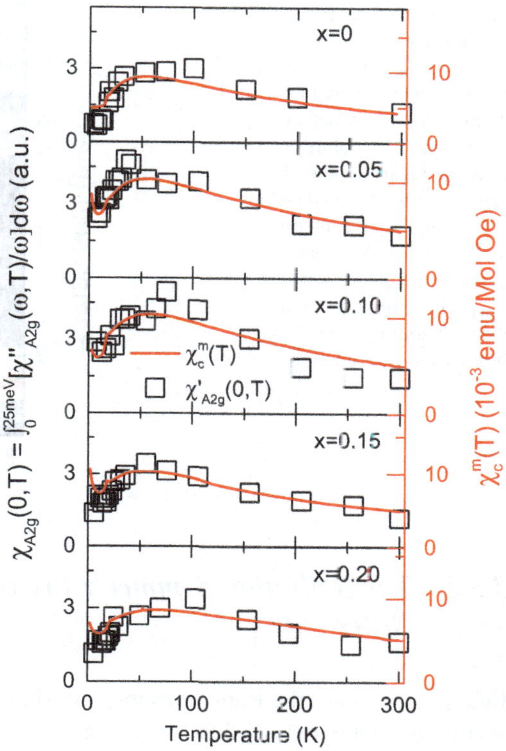

Fig. 3.18 The static Raman susceptibility in the A_{2g} symmetry channel (open squares) $\chi_{A2g}(0, T)$, compared with the magnetic susceptibility with field applied along the c-axis [64] (solid line)

The Curie–Weiss-like behavior above 100 K in static magnetic [64, 131] and Raman susceptibilities [1, 90, 132] suggests A_{2g} pseudovector-like instabilities at low temperature. Below about 50 K, the Kondo screening begins setting in [32, 59, 89, 131, 133] and the correlation length of the HO [134] or AFM [65, 128] phase builds at the ordering vector $Q_0 = (0, 0, 1)$; therefore, both the magnetic and Raman uniform susceptibilities start to decrease [Fig. 3.18]. Close to the transition temperature, both the HO and AFM order parameters fluctuate regardless of the low temperature ordering. However, the static magnetic susceptibility at Q_0 diverges only across the PM–AFM phase transition [65, 80], whereas it becomes "near critical" from the PM-HO phase [134]. Thus, HO is a nonmagnetic transition, but there is the "ghost" of AFM present. Here, we find that the temperature dependencies of the static A_{2g} Raman susceptibility $\chi_{A2g}(0, T)$ are similar and track $\chi_c^m(T)$ in all measured samples.

Below $T_{HO} = 17.5$ K, the A_{2g} Raman response function shows suppression of low energy spectral weight below 55 cm^{-1} (6.8meV) and the emergence of a sharp in-gap mode centered at 13 cm^{-1} (1.6 meV) [Fig. 3.16]. The temperature evolution of the gap and in-gap mode is most clearly seen in the colored intensity plot, where a mode at 1.6 meV appears sharply below T_{HO} without much temperature dependence [Fig. 3.19]. The low energy spectral weight gap suppression roughly follows Bardeen–Cooper–Schrieffer (BCS) gap function (white line in Fig. 3.19).

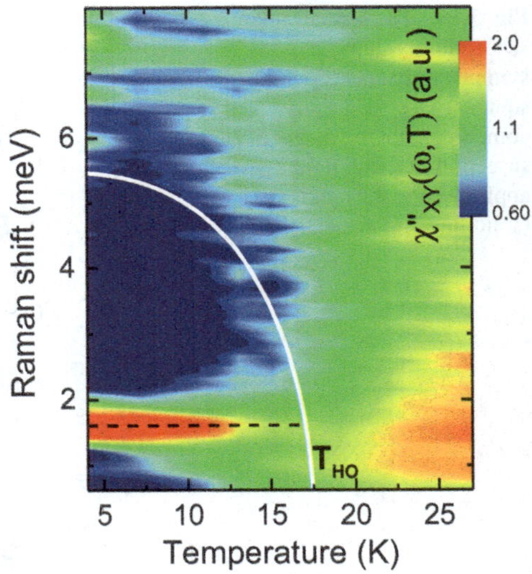

Fig. 3.19 Temperature dependence of the low frequency Raman response in the *XY* scattering geometry, dominantly comprised of A_{2g} excitations. The colors are coded in logarithmic scale. A gap-like suppression develops on cooling, and an in-gap mode at 1.6 meV (black dashed line) emerges below T_{HO}. The white line shows the temperature dependence of the BCS gap function

3.3.3 The Reflection Symmetry Breaking in the "Hidden Order" State

Figure 3.20 shows the Raman response in six scattering geometries at 7 K. The six scattering geometries are denoted as $\mathbf{e}_s\mathbf{e}_i$ =XX, XY, X'X', X'Y', RR, and RL, with $\mathbf{e}_i$ being the direction vector for incident light polarization and $\mathbf{e}_s$ being the scattered light polarization. X=[100] and Y=[010] are aligned along crystallographic axes, X'=[110] and Y'=[1$\bar{1}$0] are at 45° to the *a*-axis, and R=(X+iY)/$\sqrt{2}$ and L=(X-iY)/$\sqrt{2}$ are right and left circularly polarized light, respectively.

The intense in-gap mode is observed in all scattering geometries containing A_{2g} symmetry. In the most naive sense, the mode can be interpreted as a $\omega_0 = 1.6$ meV resonance between two quasi-localized crystal field states, which have symmetry difference of A_{2g}, e.g., between $|A_{1g}\rangle$ and $|A_{2g}\rangle$ states [132]. However, crystal field excitations in heavy element compounds appear gradually with decreasing temperatures [92, 93]. The abrupt appearance of the mode below the phase transition temperature is highly unusual for a crystal field excitation. We will explain the origin of this mode in the next section.

The most interesting and important feature in Fig. 3.20 is a much weaker intensity observed at the same energy of the A_{2g} mode in *XX* and $X'X'$ geometries, containing the excitations of the A_{1g} symmetry.

The data are fitted to a Lorentzian centered at 1.6 meV (convoluted by instrumental resolution). The intensities can be separated into distinct symmetry channels, $\mathrm{I}_{A_{2g}} = (\mathrm{I}_{XY}+\mathrm{I}_{RR}-\mathrm{I}_{X'X'})/2 = 2.6\pm0.1$, $\mathrm{I}_{A_{1g}} = (\mathrm{I}_{X'X'}+\mathrm{I}_{RR}-\mathrm{I}_{XY})/2 = 0.7\pm0.1$, $\mathrm{I}_{B_{1g}} = (\mathrm{I}_{X'Y'}+\mathrm{I}_{XX}-\mathrm{I}_{RR})/2 = 0.3\pm0.1$, and $\mathrm{I}_{B_{2g}} = (\mathrm{I}_{X'X'}+\mathrm{I}_{XY}-\mathrm{I}_{RR})/2 = 0.1\pm0.1$. Hence, the intensity ratio of A_{1g} to A_{2g} channels is $25\pm3\%$, B_{1g} to A_{2g} channels is $11\pm3\%$, and B_{2g} to A_{2g} channels is $3\pm3\%$.

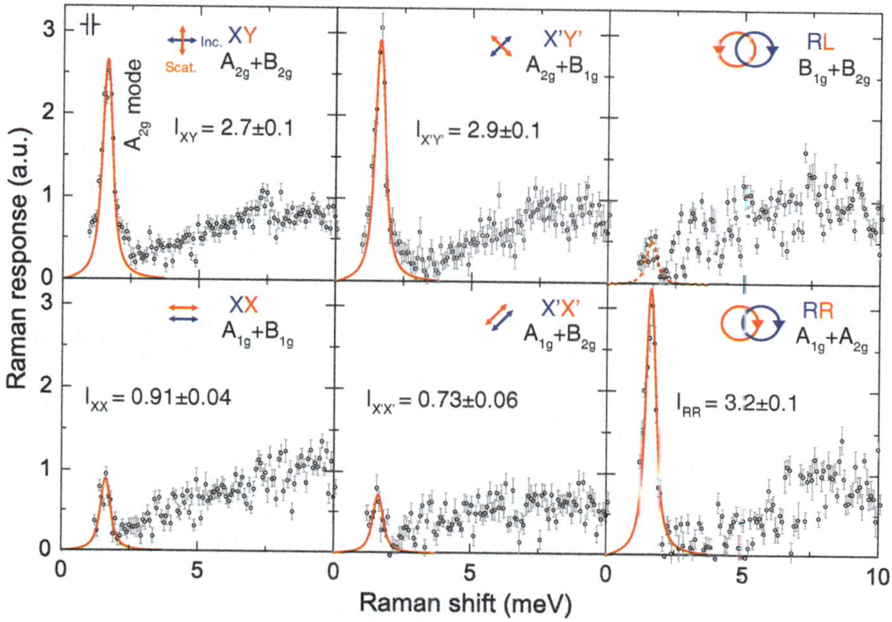

Fig. 3.20 The Raman response function in six scattering geometries at 7 K. The arrows in each panel show the linear or circular polarizations for incident (blue) and scattered (red) light. The irreducible representations for each scattering geometry are shown within the D_{4h} point group. The data are shown in black circles, where the error bars show one standard deviation. The red solid lines are fits of the in-gap mode to a Lorentzian, and the fitted intensity using the method of maximum likelihood is noted in each panel. The full width at half maximum of the in-gap mode is about 0.3 meV at 7 K (instrumental resolution of 0.17 meV is shown in the XY panel)

Therefore, we claim the observation of the 1.6 meV mode dominantly presenting in the A_{2g} symmetry channel with a weaker intensity "leakage" into the A_{1g} symmetry channel in the HO phase. Similarly, the small intensity in the B_{1g} channel could also be interpreted as small "leakage" of the collective mode from A_{2g} channel. However, the intensity borders on the noise level, and we do not intend to claim any observation in B_{1g} or B_{2g} channel.

The selection rules between scattering geometries and symmetries of Raman response are governed by the Raman tensors [Sect. 1.3] and summarized in Table 3.2. At the Γ point, collective modes have definite symmetries and therefore do not appear in more than one symmetry channel. However, due to spontaneous symmetry breaking, the point group symmetry of the system is reduced, and it is possible that one collective mode "leaks" into more than one symmetry channel. For example, upon the reduction of symmetry from point group D_{4h} to C_{4h}, the A_{1g} and A_{2g} irreducible representations merge into A_g, B_{1g} and B_{2g} merge into B_g, and E_g remains doubly degenerate [137]. Thus, a collective mode with "pure" A_{2g} symmetry of D_{4h} group should only appear in XY, $X'Y'$, and RR scattering geometries. But in the lower symmetry C_{4h} point group, there is no A_{2g} irreducible

Table 3.2 The Raman selection rules of URu_2Si_2 [135, 136]. Upon the reduction of symmetry from point group D_{4h} to C_{4h}, the A_{1g} and A_{2g} irreducible representations merge into A_g, B_{1g} and B_{2g} merge into B_g, and E_g remains doubly degenerate [137]

Scattering geometry	Irreducible representations (D_{4h})	(C_{4h})
RR	$A_{1g} + A_{2g}$	A_g
RL	$B_{1g} + B_{2g}$	B_g
XX	$A_{1g} + B_{1g}$	$A_g + B_g$
YX	$A_{2g} + B_{2g}$	$A_g + B_g$
$X'X'$	$A_{1g} + B_{2g}$	$A_g + B_g$
$Y'X'$	$A_{2g} + B_{1g}$	$A_g + B_g$

Character table

$T > T_{HO}$ (D_{4h} group)

	E	$2C_4$	C_2	$2C'_2$	$2C''_2$	i	$2S_4$	σ_h	$2\sigma_v$	$2\sigma_d$	linears, rotation	quadratic
A_{1g}	1	1	1	1	1	1	1	1	1	1		x^2+y^2, z^2
A_{2g}	1	1	1	-1	-1	1	1	1	-1	-1	R_z	
B_{1g}	1	-1	1	1	-1	1	-1	1	1	-1		x^2-y^2
B_{2g}	1	-1	1	-1	1	1	-1	1	-1	1		xy
E_g	2	0	-2	0	0	2	0	-2	0	0	(R_x, R_y)	(xy, yz)

$T < T_{HO}$ (C_{4h} group)

	E	$2C_4$	C_2	i	$2S_4$	σ_h	linears, rotation	quadratic	D_{4h} group
A_g	1	1	1	1	1	1	R_z	x^2+y^2, z^2	$A_{1g}+A_{2g}$
B_g	1	-1	1	1	-1	1		x^2-y^2, xy	$B_{1g}+B_{2g}$
E_g	2	0	-2	2	0	-2	(R_x, R_y)	(xy, yz)	E_g

Fig. 3.21 The character tables for the point group D_{4h} and C_{4h}, showing only parity even irreducible representations

representation any more. The previous A_{2g} of D_{4h} group is now reduced to A_g in C_{4h} group, which appears in all but RL scattering geometry in Table 3.2. Comparing the selection rules of C_{4h} group to Fig. 3.20, one can easily see that the symmetry properties of the collective mode are entirely consistent with C_{4h} point group. In fact, C_{4h} is the only point group consistent with our data.

Comparing the character tables (showing only parity even irreducible representations) for the point group D_{4h} and C_{4h} in Fig. 3.21, we see that the reflection symmetry operators, σ_d and σ_v (equivalently the out-of-plane rotation operators C_2' and C_2''), have to be removed from the D_{4h} group. In other words, the *local* vertical and diagonal reflection symmetry operators at the uranium sites have to be removed. Therefore, the conclusion from the data shown in Fig. 3.20 is that the reflection symmetries are broken in the HO phase.

The tiny intensity leakage into the *RL* scattering geometry measures the strength of orthorhombic distortion caused by broken fourfold rotational symmetry. However, the signal in the RL geometry borders on the noise level and therefore is at most a subdominant order parameter induced by the broken reflection symmetries.

3.3.4 Iron Substitution Dependence of the A_{2g} Collective Mode

Albeit the distinct discrete symmetries are broken above and below the critical concentration x_c, we find that the A_{2g} collective mode continuously evolves with parameter x. In the HO phase, the mode energy decreases as x is increased, disappearing at the critical Fe concentration x_c. In the AFM phase, the collective mode again emerges in the same A_{2g} symmetry channel with the energy increasing with x. The continues transformation of this collective excitation, a photoinduced transition between the HO and AFM electronic phases, provides direct experimental evidence for a unified order parameter for both nonmagnetic and magnetic phases arising from the orbital degree of freedom of the uranium-5f electrons

Figure 3.22 shows the temperature dependence of the Raman response in the eminent A_{2g} symmetry channel of the D_{4h} group. The upper panels of Fig. 3.22 show the intensity plots of the low energy Raman response $\chi''_{A2g}(\omega, T)$ below 30 K. Above $T_{DW}(x)$ at about 18–20 K, a quasi-elastic peak (QEP) comprises most of the spectral weight for all samples, narrowing toward the transition. The observed QEP originates from overdamped excitations between quasidegenerate crystal field states [1, 35], and the narrowing of the QEP with cooling is due to the increase of excitation lifetime, related to the development of a hybridization gap and formation of a heavy Fermi liquid [59–61]. Our data show that this process is similar for both pristine and iron doped samples, regardless of the low temperature ordered phase

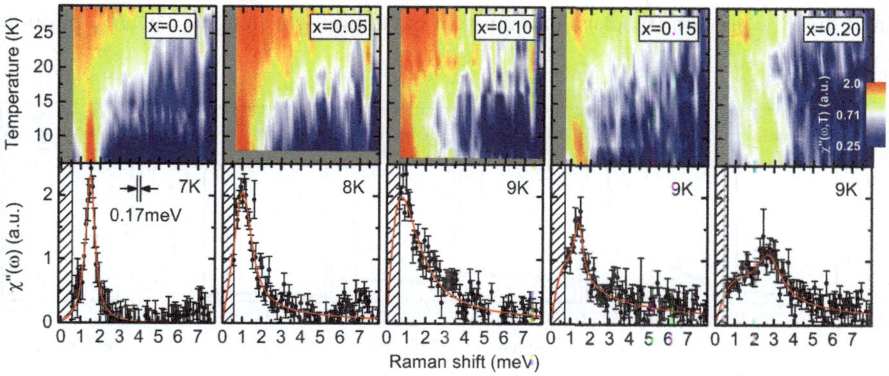

Fig. 3.22 Raman response in the A_{2g} symmetry channel, $\chi''_{A2g}(\omega, T)$, for $URu_{2-x}Fe_xSi_2$ below the second order phase transition temperature. The upper panels show intensity plots, where the intensities are color coded in logarithmic scale. The lower panels show the spectra at about half the transition temperature to emphasize the collective mode, where the error bars represent one standard deviation, and the red solid lines are guides to the eye

being HO or AFM phase. This indirectly shows that the Kondo hybridization cannot be a critical ingredient for distinguishing HO from AFM phase.

Below $T_{DW}(x)$, the most significant feature in the A_{2g} channel is a sharp collective mode. The sharpness of this resonance suggests the lack of relaxation channels due to the opening of an energy gap [59, 61, 62, 89]. In order to see the mode's lineshape more clearly, we plot $\chi''_{A2g}(\omega, T)$ for each Fe concentration x in the lower panels, with $T \approx T_{DW}(x)/2$. The lineshapes broaden with increasing x owing to the inhomogeneity of the local stress field, or unsuppressed relaxation channels introduced by doping that interact with the collective mode, which may also be related to the increasing continuum in the $x = 0.15$ and 0.2 spectra. In contrast to the monotonic broadening of the lineshape width, the collective mode frequency shows nonmonotonic behavior as function of x. The energy decreases with increasing x in the HO phase, until it vanishes below the instrumental resolution at $x = 0.10$, which is close to the HO and AFM phase boundary determined by elastic neutron scattering [80] and thermal expansion measurements [64]. The mode reappears in the AFM phase, where the energy increases with increasing x, and also appears in the same A_{2g} symmetry channel as in the HO phase.

3.3.5 *Excitation and Sample Dependence*

Figure 3.23a shows Raman response in the $X'Y'$ scattering geometry in the HO phase, measured with different excitation energies at about 8 K. All spectra were normalized by laser power, spectrometer response, and corrected for optical constants obtained from Ref. [61] [Chap. 2]. We demonstrate that the collective mode and the background have very different excitation profile as the conventional phonon. Figure 3.23b plots the integrated intensity of the A_{2g} collective mode (red) and the B_{1g} phonon (black) as function of excitation wavelength. The B_{1g} phonon

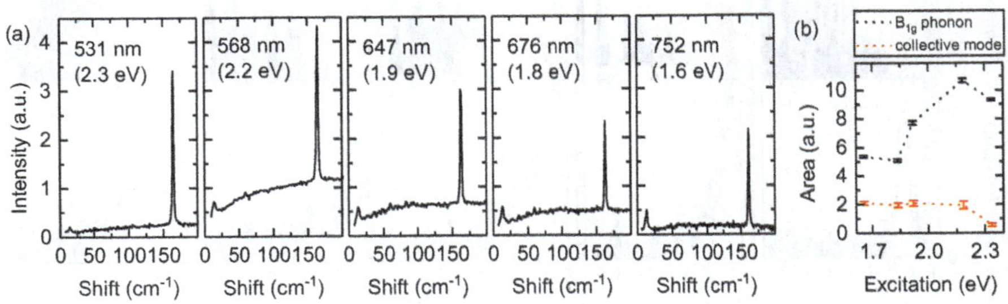

Fig. 3.23 (**a**) Low frequency secondary intensity in the $X'Y'$ scattering geometry, measured with different excitation energies in the HO phase (about 8 K). This includes contribution from both Raman response and photoluminescence background. All spectra were normalized by laser power, spectrometer response, and corrected for refractive index. (**b**) Integrated intensity of the A_{2g} collective mode (red) and the B_{1g} phonon (black) as function of excitation wavelength. Error bars are shown for one standard deviation

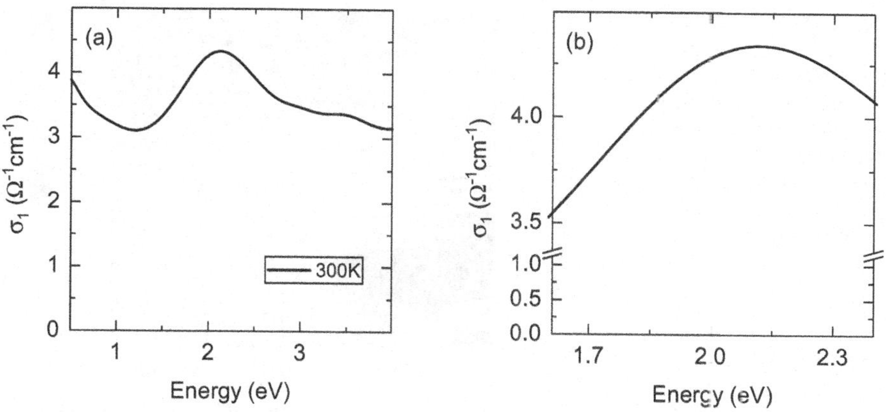

Fig. 3.24 (**a**) The *a*-axis optical conductivity of URu_2Si_2 from infrared to ultraviolet, measured at 300 K, from Ref. [61]. (**b**) Enlarged from panel (**a**) in the visible range between 1.6 and 2.4 eV

intensity seems to peak at around 2.2 eV, as reported in the dissertation of Dr. Buhot [Fig. 4.19 in Ref. [53]]. This may be due to the resonance enhancement from an interband transition, as suggested by the optical conductivity in Fig. 3.24 [61].

The A_{2g} collective mode has however a different profile, where the intensity is peaked at slightly longer wavelength than 676 nm. Notice that the spectrometer and CCD response is about 5 times larger around 550 nm than in 750 nm. Therefore despite the slightly larger signal in the NIR regime, the required spectra collection time is actually longer than 500–600 nm excitations. In most Raman experiments, one is usually tempted to pick the excitation where the signal (before normalizing to spectrometer and detector response) is the strongest. However, for the data analysis we employ here for URu_2Si_2, it is also important to minimize the photoluminescent contribution to the *scattering cross section*. For example, the Krammers–Kronig analysis applies to linear susceptibility functions, and therefore the photoluminescent background needs to be subtracted *before* the transformation. However, any subtraction of background introduces uncertainties, rendering the Krammers–Kronig analysis pointless if signal becomes comparable to the background. In fact, even by adjusting a simple linear background, one can get almost any temperature dependence for the "static susceptibility." Moreover, the analysis of "gap suppression" is meaningless unless either the luminescence background is known, or the background is negligibly small. In reality, the luminescent background has no reason to be linear or even any simple analytic function.

Unfortunately, it can be seen in Fig. 3.23a that the broad continuum has yet another excitation dependence different from the A_{2g} signal, peaking at around 568 nm. This broad background extends to much higher energies and is therefore assigned to be *mainly* due to photoluminescence, and it should be subtracted from the intensity before calculating Raman response. However, it must be emphasized here that there is no way to distinguish Raman response from photoluminescence

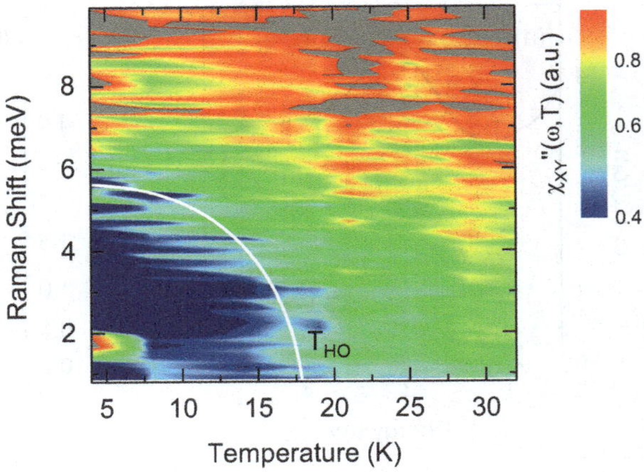

Fig. 3.25 Temperature dependence of the low frequency Raman response in the *XY* scattering geometry, taken from a different sample than in Fig. 3.19 and measured with 647 nm excitation. The colors are coded in logarithmic scale

for such broad continuum. Therefore, in our study [1, 138], we do the experiment with 752 nm excitation from Kr^+ ion laser, which contains minimal background.

Another challenging aspect of this study is to find a good sample *and* a good cleave. The residual resistivity ratio (RRR) from different batch of URu_2Si_2 crystals can differ from 20 to 700, likely due to varying degree of impurity concentrations and crystalline homogeneity [109]. Although the varying RRR does not greatly modify the HO temperature, it will likely introduce different inhomogeneous broadening in all spectroscopic features, such as phonons and collective modes. Figure 3.25 shows the temperature dependence of the low frequency Raman response in the *XY* scattering geometry, measured from a batch of samples with lower RRR. The RRR in this batch of samples ranges between 50 and 200, which is consistently lower compared to another batch of samples with RRR ≈200–600, where Fig. 3.19 was measured from. In all measured sample batches, T_{HO} is characterized by transport measurements and consistently found at 17.5 K.

Compared to Fig. 3.19, we first notice that the A_{2g} collective mode in Fig. 3.25 appears at significantly lower temperature, about 8 K. Surprisingly, the temperature dependence of the gap suppression is still consistent with T_{HO} characterized by transport measurements, indicating that the laser heating has been correctly accounted for in both figures. Also, the FWHM (deconvoluted for the instrumental resolution) of the A_{2g} peak is about 0.7 meV at 7 K, which is much broader than about 0.4 meV found in samples with higher RRR. Therefore, the reason for the seemingly weaker A_{2g} peak is likely just due to the inhomogeneous broadening. In fact, we found that the integrated intensity of the A_{2g} peak is roughly the same in all samples, resulting the *peak height* in low RRR samples to be much lower, hindering us to distinguish it from noise background. Another possible aspect is that Fig. 3.25 was measured with 647 nm, which has slightly higher background than in 752 nm, introducing higher uncertainty in background subtraction.

In all measured samples from three different sources, grown by the Czochralski method with and without electrorefinement, and also samples grown by molten indium flux method, we consistently find the A_{2g} peak to saturate at 1.6–1.7 meV in low temperature. On the other hand, we find that the FWHM of the peak fluctuates between samples and even slightly varies for different spots on the same cleave. On average, we find the FWHM roughly between 0.4 and 0.5 meV in high RRR samples and about 0.7 meV in low RRR samples. In all measured samples, the phonon width is close to instrumental resolution limited and therefore cannot serve as a reliable standard for sample quality.

3.4 Discussion

We now discuss the origin and the observed doping dependence of the collective mode in the ordered phases within a phenomenological Ginzburg–Landau approach, based on a minimal model where the two order parameters for both HO and AFM phases can be constructed from $|A_{2g}\rangle$ and $|A_{1g}\rangle$ [35, 97]. This model has recently been directly confirmed by experiments using inelastic X-ray scattering [139–141] and resonant ultrasound spectroscopy [142].

3.4.1 *The Minimal Model and the Hexadecapole Order Parameter*

For low energy spectroscopy, it is sufficient to introduce a minimal model containing the two lowest energy states in Table 3.1, $|A_{2g}\rangle$ and $|A_{1g}\rangle$, separated by a small energy $\approx 30\,\text{cm}^{-1}$. From this minimal model, Haule and Kotliar proposed an A_{2g} type local interaction operator, $\Psi(r) \equiv V\,\langle |A_{1g}\rangle\,\langle A_{2g}|\rangle_r$, with V being the interaction strength, and r is the lattice site [35]. This is a complex order parameter, where the real part is responsible for the condensation in the HO phase, and is proportional to the electric hexadecapolar order parameter: $\Psi_{HO} \sim \langle (J_x J_y + J_y J_x)(J_x^2 - J_y^2)\rangle \sim \langle J_+^4 - J_-^4\rangle$. A hexadecapolar order parameter breaks the reflection symmetries but does not couple to the lattice directly (in the first order), and thus no lattice distortion (such as C_4 symmetry breaking) is expected. It was pointed out in Ref. [143] that the A_{2g} local order parameter and the antiferro-staggering along c-axis with ordering vector $Q_{DW} = (0, 0, \frac{c}{2\pi})$ lower the group symmetry from the high temperature $I4/mmm$ (No. 139) to $P4/mnc$ (No. 128).

This A_{2g} hexadecapolar order parameter alone cannot explain the C_4 rotational symmetry breaking observed in magnetic torque [112], X-ray diffraction [129], and elastoresistivity [113] experiments. However, notice that external magnetic field or strain can induce small C_4 symmetry breaking as explained in Ref. [143] from symmetry arguments or in Supplementary Materials in Ref. [1] from direct calculations by involving a small subdominant order parameter of different symmetry (B_{1g} was included for example).

Consider now the minimal model introduced above and the low energy sector of the purely real hexadecapolar order parameter, which is appropriate for the low energy spectroscopy. We can write an effective Hamiltonian as

$$\begin{array}{c|cc} H_0 & |A_{2g}\rangle & |A_{1g}\rangle \\ \hline |A_{2g}\rangle & 0 & V\sin\theta \\ |A_{1g}\rangle & V\sin\theta & \omega_0 \end{array} \tag{3.5}$$

where $|A_{2g}\rangle = \frac{i}{\sqrt{2}}(|4\rangle - |-4\rangle)$ with A_{2g} symmetry and $|A_{1g}\rangle = \frac{\cos\theta}{\sqrt{2}}(|4\rangle + |-4\rangle) - \sin\theta\,|0\rangle$ with A_{1g} symmetry are the singlet crystal field states of the U-5f electrons. Notice that we have absorbed the $\sin\theta$ factor into V in the main text.

From perturbation theory, the perturbed states up to second order in V are

$$\begin{aligned} |\mathrm{HO}\rangle &= \left(1 - \frac{V^2\sin^2\theta}{2\omega_0^2}\right)|A_{2g}\rangle + \frac{V\sin\theta}{\omega_0}|A_{1g}\rangle + \mathcal{O}(V^3) \\ |\beth\rangle &= \frac{-V\sin\theta}{\omega_0}|A_{2g}\rangle + \left(1 - \frac{V^2\sin^2\theta}{2\omega_0^2}\right)|A_{1g}\rangle + \mathcal{O}(V^3), \end{aligned} \tag{3.6}$$

where $|\mathrm{HO}\rangle$ is the HO ground state, and $|\beth\rangle$ (reads "veht") is the first excited state.

Notice that V can take up positive or negative values. In the following, we will use the notation:

$$\begin{aligned} |\mathrm{HO}^+\rangle &= \left(1 - \frac{V^2\sin^2\theta}{2\omega_0^2}\right)|A_{2g}\rangle + \frac{|V|\sin\theta}{\omega_0}|A_{1g}\rangle + \mathcal{O}(V^3) \\ |\mathrm{HO}^-\rangle &= \left(1 - \frac{V^2\sin^2\theta}{2\omega_0^2}\right)|A_{2g}\rangle - \frac{|V|\sin\theta}{\omega_0}|A_{1g}\rangle + \mathcal{O}(V^3) \end{aligned} \tag{3.7}$$

to distinguish the right- and left-handed states.

The mixing between $|A_{2g}\rangle$ and $|A_{1g}\rangle$ states by the hexadecapolar order parameter in HO phase is illustrated in Fig. 3.26. The resulted two states, $|\mathrm{HO}^+\rangle$ and $|\mathrm{HO}^-\rangle$,

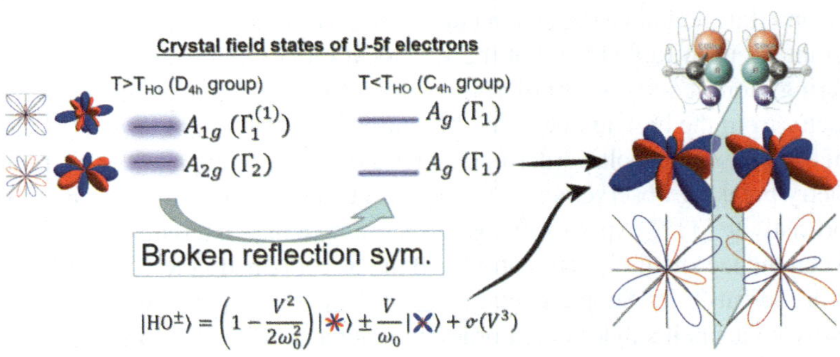

Fig. 3.26 The formation local chiral states, $|\mathrm{HO}^+\rangle$ and $|\mathrm{HO}^-\rangle$, in the HO phase due to reflection symmetry breaking and an A_{2g} type local interaction. The illustrations are meant to emphasize the symmetry of the wave function, where red and blue are positive and negative phases

cannot be transformed into one another by any remaining C_{4h} group operators. In particular, the two states are related to each other by a mirror reflection with an additional sign change: a property known as chirality (or handedness). The choice of either the right-handed or the left-handed state on a given uranium site, $|HO^+\rangle$ or $|HO^-\rangle$, defines the local chirality in the HO phase. Notice that these two degenerate states preserve the time reversal symmetry, carry no spin, and contain the same total charge but differ only in charge distribution, thus "hidden" from most experimental probes for decades.

3.4.2 The Ordering Vector

As we explained in Sect. 3.1, the order parameter was historically "believed" to be staggered in HO phase along $\mathbf{Q}_0 = (0, 0, 1)$ due to observation of a weak magnetic Bragg peak [73, 74]. However, since the magnetic feature is now attributed to extrinsic reasons, the question of whether translational symmetry being broken or not is back on the table. In particular, if the ordering is truly magnetic and along $\mathbf{Q}_0$ as shown in Fig. 3.8b, then one would expect certain periodicity of the magnetic dispersion. That is, the Z point of the BCT zone $(\pm 1, 0, 0)$ will become the Γ point $(0, 0, 0)$ in the HO phase [Fig. 3.10]. Also, the Σ point of the BCT zone $(\pm 0.6, 0, 0)$ should be equal to $(\pm 0.4, 0, 0)$ in the ordered phase. However, the inelastic neutron scattering results show that both magnetic and lattice excitations (phonons) obey the structural lattice periodicity (BCT zone) in the HO phase [87].

The Brillouin zone folding predicted for an ordering vector $\mathbf{Q}_0$ was recently observed in the HO phase by angle resolved photoemission spectroscopy (ARPES) [144]. The electronic band periodicity changes from BCT in the paramagnetic phase into ST in the HO phase, suggesting the formation of some density wave. The observed density wave gap $\sim$ 7 meV is consistent with the optical conductivity results [88], and the estimated gapping of density of states at Fermi surface could account for the entropy lost measured by heat capacitance. Thus, it seems clear that the ordering is electronic origin and does not break the lattice periodicity.

Besides the ARPES results [144], we show in the following that the Raman spectroscopy results also suggest ordering along $\mathbf{Q}_0$ (Fig. 3.27).

Detailed studies of the temperature dependence of the energy and FWHM of the 1.6 meV sharp A_{2g} mode by Buhot et al. [132], and its comparison to the collective mode observed by inelastic neutron scattering at Q_0 [85], revealed that the two modes measured at very different crystal momentum is likely the same mode. This is a "smoking gun" evidence of unit cell doubling along c-axis, i.e., the ordering vector is indeed $Q_0 = (0, 0, 1)$ as in the AFM phase. Thus the HO ground state is the staggering of local symmetry breaking order parameters, forming "some kind" density wave as suggested by infrared spectroscopy [88] and ARPES measurements [144].

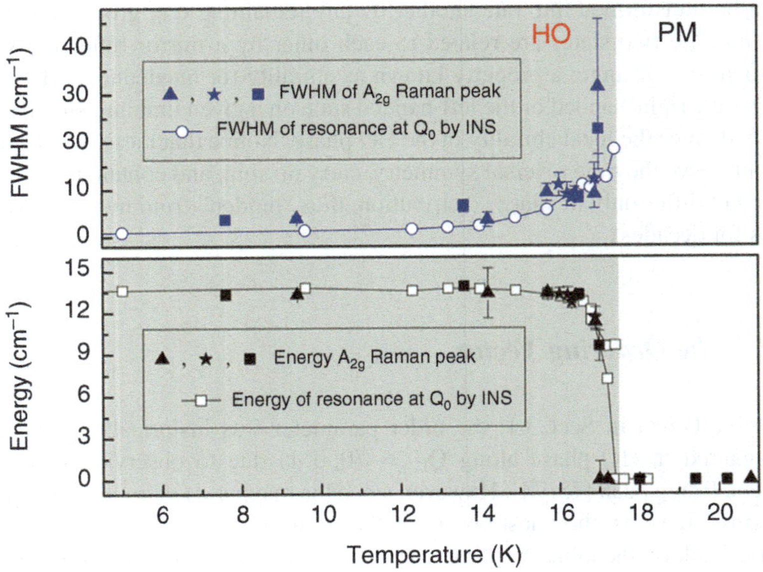

Fig. 3.27 The comparison between the A_{2g} collective mode in Raman scattering and Q_0 mode in inelastic neutron scattering, adapted from Ref. [132]. The figure compares the temperature dependence of the energy and FWHM of the A_{2g} Raman peak (full symbols) and the Q_0 mode measured by inelastic neutron scattering (open symbols) [85]. The full symbols (stars, triangle, and square) are extracted from various measurements on different samples from the same batch. The gray area corresponds to the temperature range of the HO phase

However, detailed studying of phonon branches by inelastic neutron scattering showed that within experimental sensitivity no evidence of translational symmetry breaking in any direction is observed [87]. This means that the long range ordered HO ground state does not induce any periodic lattice distortion, which is a very strange case since the phononic thermal conductivity [145] and linear thermal expansion coefficients [64] all show drastic change across the HO phase transition. In conclusion, the HO ground state must be an electronic ordering, which although couples strongly to lattice, and does not induce lattice distortion.

The most intuitive explanation to reconcile the above experiments would be to have *c*-axis staggering of an electric multipolar order parameter, which breaks the reflection symmetries on uranium sites, but does not induce any lattice distortion. Within the crystal group of URu_2Si_2, a hexadecapole (rank 4 electric multipole) is the choice in the lowest order. Notice that the interaction between rank 4 multipoles is not necessarily weak. For example, dotriacontapole (rank 5) is the dominant interaction in PrO_2, even stronger than dipoles [6].

3.4.3 The $\chi''_{A2g}(\omega, T)$ Continuum

The Raman response above T_{HO} can be described within a low energy minimal model suggested in Ref. [35] that contains two singlet states of A_{2g} and A_{1g} symmetries, split by an energy ω_0. In the following, we denote the singlet states of A_{2g} and A_{1g} symmetries by $|A_{2g}\rangle$ and $|A_{1g}\rangle$; the conduction band is labeled $|CB\rangle$. At high temperatures, the crystal field states are quasidegenerate in energy and localized at the uranium f shells in space. The incoherent excitation between the two states thus leads to a broad Drude like quasi-elastic peak as in Fig. 3.16, where the inverse excitation lifetime determines the peak maxima $\Gamma(T)$.

The proportionality between $\chi'_{A2g}(0, T)$ and $\chi_c^m(0, T)$ in Fig. 3.17 can be understood by noting that both susceptibilities probe A_{2g} excitations [Fig. 3.11]. This is because the Raman susceptibility in A_{2g} channel can be expressed by

$$\chi_{A_{2g}}(\omega) \propto \int \langle O_{A_{2g}}(\tau) O^{\dagger}_{A_{2g}}(0)\rangle e^{i\omega\tau} d\tau, \tag{3.8}$$

where $O_{A_{2g}}$ is the Raman operator in the A_{2g} channel. The magnetic susceptibility along z-axis is

$$\chi_z^m(\omega) \propto \int \langle J_z(\tau) J_z(0)\rangle e^{i\omega\tau} d\tau. \tag{3.9}$$

Here, J_z can be approximated by the orbital angular momentum operator L_z. Notice that $O_{A_{2g}}$ and L_z both couple to wave functions that transform as A_{2g} representation in D_{4h} group and thus differ only in the structure of the bare vertex. Therefore, we have the following relations:

$$\begin{aligned}
\langle A2g|O_{A_{2g}}|A2g\rangle &= \langle A2g|J_z|A2g\rangle = 0 \\
\langle A1g|O_{A_{2g}}|A1g\rangle &= \langle A1g|J_z|A1g\rangle = 0 \\
\langle A1g|O_{A_{2g}}|A2g\rangle &= \beta \langle A1g|J_z|A2g\rangle = i\alpha \\
\langle CB|O_{A_{2g}}|A2g\rangle &= \beta' \langle CB|J_z|A2g\rangle = i\alpha',
\end{aligned} \tag{3.10}$$

where α, β, α', and β' are real numbers. We see that the matrix elements of $O_{A_{2g}}$ and J_z operators are proportional to each other within the minimal model. This explains the proportionality between $\chi'_{A2g}(0, T)$ and $\chi_c^m(0, T)$ in Fig. 3.17.

The extreme anisotropy between a- and c-axis magnetic susceptibility also follows from this minimal model. The in-plane magnetic susceptibility is probed by a E_g symmetry operator in D_{4h} group [137], which does not couple to excitations between the $|A_{2g}\rangle$ and $|A_{1g}\rangle$ states. Thus, the predicted in-plane magnetic susceptibility χ_a^m is zero.

3.4.4 The Chirality Density Wave

The result of antiferro-staggering of the chiral states, $|HO^+\rangle$ or $|HO^-\rangle$, along c-axis is a chirality density wave [Fig. 3.28]. The staggered condensate can be approximated by a form

$$|\psi_{HO}\rangle = \prod_{r=A\ site} |HO_r^+\rangle \times \prod_{r=B\ site} |HO_r^-\rangle . \tag{3.11}$$

Note that $|A_{2g}\rangle$ is the lowest energy crystal field state in the HO phase, and thus $|HO_r^\pm\rangle$ at uranium site r is dominantly $|A_{2g}\rangle$, with a small admixture of $|A_{1g}\rangle$, i.e., $|HO^\pm\rangle = \cos\theta\,|A_{2g}\rangle \pm \sin\theta\,|A_{1g}\rangle$. $\theta \equiv \arcsin(V/\omega_0)$, where ω_0 is the splitting between the lowest lying crystal field states in the minimal model, and V is the order parameter strength in the HO phase.

If we assume that in the AFM phase, the ordering of the crystal field states switches, i.e., the $|A_{1g}\rangle$ is the lowest energy crystal field state. Moreover, we let the AFM order parameter be the imaginary part of the complex local interaction, $\Psi(r) \equiv V\,\langle |A_{1g}\rangle\,\langle A_{2g}|\rangle_r$, which is proportional to the magnetic dipole operator: $\Psi_{AFM} \sim \langle J_z\rangle$. The new local wave functions in the AFM phase are $|AFM^\pm\rangle \approx \cos\theta'\,|A_{1g}\rangle \pm i\sin\theta'\,|A_{2g}\rangle$. Here, $\theta' \equiv \arcsin(V'/\omega_0)$, and V' is the order parameter strength in the AFM phase. Since the mixing is purely imaginary, the charge distribution on the uranium site does not break any spatial symmetry; instead, it acquires nonzero out-of-plane magnetic moments and thereby breaks time reversal

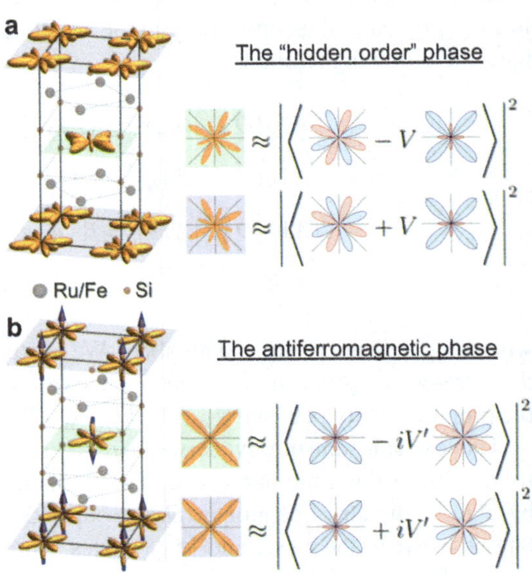

Fig. 3.28 The crystal structure of $URu_{2-x}Fe_xSi_2$ in (**a**) the HO and (**b**) the AFM phases. Illustrations capturing the symmetries of the charge distributions of the ground state wave functions are placed at the uranium atomic sites. On the right are illustrations showing the in-plane structures of the wave functions

symmetry. The staggering of these local states along Q_0 forms an antiferromagnetic lattice, as illustrated in Fig. 3.28. The Néel-type condensate takes the form

$$|\psi_{\text{AFM}}\rangle = \prod_{r=A\ site} |\text{AFM}_r^+\rangle \times \prod_{r=B\ site} |\text{AFM}_r^-\rangle\,. \tag{3.12}$$

The two apparently competing orders, the chirality density wave and the antiferromagnetic state, are constructed by mixing the two orbital wave functions on uranium sites with a real or an imaginary phase factor, $\sin\theta$ or $i\sin\theta'$, thus unifying the two order parameters.

3.4.5 Ginzburg–Landau Theory

Based on the A_{2g} complex order parameter, Haule and Kotliar built a phenomenological Ginzburg–Landau theory to explain the phase diagram of URu_2Si_2 [97]. The Ginzburg–Landau theory is later slightly modified by Kusunose and Harima to improve the agreement with experiments [143] and recently expanded by Boyer and Yakovenko to account for "low field training" experiments [76, 145].

The Ginzburg–Landau free energy takes the form

$$F[\Psi] = \Psi^T \hat{A}\Psi + \beta\left(\Psi^T\Psi\right)^2 + \gamma\left(\Psi^T\hat{\sigma}_1\Psi\right)^2, \tag{3.13}$$

where $\Psi^T \equiv (\psi_{\text{HO}}\ \ \psi_{\text{AFM}})$, $\hat{A} \equiv \begin{pmatrix}\alpha_{HO} & 0\\ 0 & \alpha_{AFM}\end{pmatrix}$, with α_{HO} and α_{AFM} vanish at the critical temperature, and $\hat{\sigma}_1 \equiv \begin{pmatrix}0 & 1\\ 1 & 0\end{pmatrix}$ is the Pauli matrix. γ ensures a finite energy barrier between the two phases and hence the phase separation between the HO and AFM phases [128].

A schematic of the Ginzburg–Landau free energy at various special points in the phase diagram of URu_2Si_2 is shown in Fig. 3.29. Below the second order phase transition T_{DW}, two global and two local minima develop on ψ_{HO} and ψ_{AFM} axes due to spontaneous discrete symmetry breaking, where the minima characterize the ground states in the HO and AFM phases, respectively. At the critical doping in the ordered phase [Fig. 3.29f], the four minima are degenerate, but the barrier between the minima remains finite due to the γ term in Ginzburg–Landau functional [Eq. (3.13)]. Therefore the transition between the HO and LMAF phases is of the first order, and the coexistence of both phases is allowed, explaining the LMAF puddles that have been observed in the HO phase [147, 148].

Here, the free energy parameters are introduced following the recipes given Refs. [97, 143, 146] with adjustments to match the phase diagram in Fig. 3.29. All energies and temperatures are given in the units of Kelvin, and all pressure units are in GPa. The quadratic couplings are defined as

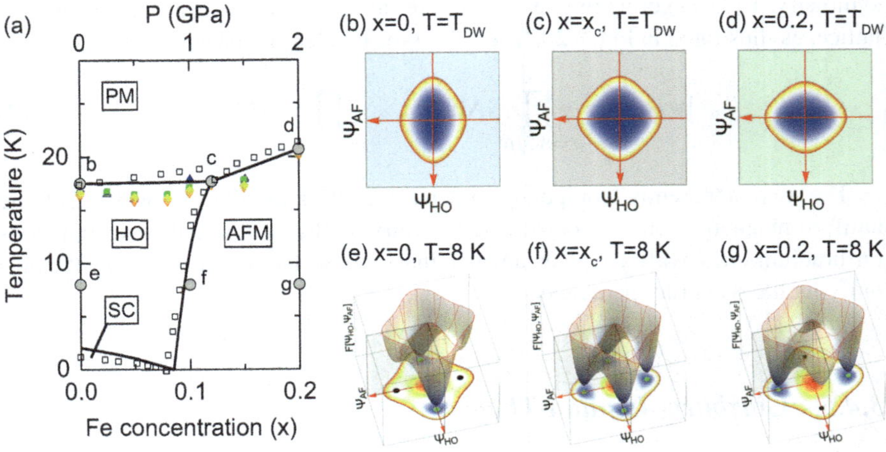

Fig. 3.29 (**a**) The phase diagram of URu_2Si_2 system, as shown in Fig. 3.8. (**b**)–(**g**) Schematics of the Ginzburg–Landau free energy at various special points in the phase diagram [solid gray circles in (**a**)]. ψ_{HO} and ψ_{AFM} are the real and imaginary part of the hexadecapole order parameter, respectively [35, 97, 143]

$$\alpha_{HO}(\omega_0, T, P) \equiv -\frac{1}{2}J_{HO}(\omega_0, P) + \tilde{\alpha}_{HO}(\omega_0, T, P)$$

and

$$\alpha_{AFM}(\omega_0, T, P) \equiv -\frac{1}{2}J_{AFM}(\omega_0, P) + \tilde{\alpha}_{AFM}(\omega_0, T, P),$$

where $\omega_0 = 35$ K is the effective energy separation between the singlet states in the minimal model, T is temperature, and P can be either hydrostatic pressure or effective chemical pressure.

$$J_{HO}(\omega_0, P) = \frac{\omega_0(1 + a_1 P)}{\tanh\left(\frac{\omega_0}{2T_{HO}}\right)}$$

and

$$J_{AFM}(\omega_0, P) = \frac{\omega_0(1 + a_2 P)}{\tanh\left(\frac{\omega_0}{2T_N}\right)}$$

are the effective nearest neighbor exchange constants, with $T_{HO} = 17.5$ K and $T_N = 15.5$ K.

$$\tilde{\alpha}_{HO}(\omega_0, T, P) = \frac{1}{2}\omega_0 \coth\left(\frac{\omega_0}{2T}\right)(1 - a_3 P)$$

and

$$\tilde{\alpha}_{AFM}(\omega_0, T, P) = \frac{1}{2}\omega_0 \coth\left(\frac{\omega_0}{2T}\right)(1 + a_3 P)$$

are the effective on-site couplings.

$$\beta(\omega_0, P) = \frac{\omega_0\left[\sinh\left(\frac{\omega_0}{T}\right) - \frac{\omega_0}{T}\right]\cosh^2\left(\frac{\omega_0}{2T}\right)}{8\sinh^4\left(\frac{\omega_0}{2T}\right)}$$

is the quartic coupling.

3.4.6 Origin of the A_{2g} Collective Modes in URu_2Si_2

Having established the x dependence of the Ginzburg–Landau free energy landscape, we are finally at a place to discuss the origin of the A_{2g} collective modes in the HO and AFM phases. As we mentioned in Sect. 3.3, the local picture of explaining the A_{2g} mode as crystal field excitation between $|A_{2g}\rangle$ and $|A_{1g}\rangle$ is too naive, completely ignoring the effect of phase transition that mixes these two states. Moreover, the condensate of these local orbital states implies off-site coupling, which results in a dispersive collective mode as observed by inelastic neutron scattering [72].

Therefore, the more accurate description of the A_{2g} collective mode should be an exciton carrying quadrapolar moment, which has the symmetry of the subdominant phase (e.g., $|\psi_{\text{AFM}}\rangle$ when the system orders in the HO phase). This mode propagates through the crystal that orders with the dominant symmetry (e.g., $|\psi_{\text{HO}}\rangle$). Likewise, when the ground state is $|\psi_{\text{AFM}}\rangle$, the propagating exciton is of $|\psi_{\text{HO}}\rangle$ symmetry. However, we will see below that at the Γ point, the local description actually captures the basic symmetry properties of the excitation and is appropriate for describing the symmetry breaking as we have done in Sect. 3.3. The symmetry difference between the two condensates is mainly A_{2g}, with a small admixture of A_{1g}; hence, such exciton can be detected by Raman in the A_{2g} channel and explains the sharp resonance shown in Fig. 3.19.

The energy of the collective mode at the Γ point then simply corresponds to the energy separation between the dominant long range order (e.g., $|\psi_{\text{HO}}\rangle$ in the HO phase) and the sub-dominant order (e.g., $|\psi_{\text{AFM}}\rangle$). Or equivalently, the energy difference between the Ginzburg–Landau free energy minima along ψ_{HO} and ψ_{AFM} axes in Fig. 3.29, which is vanishingly small at the critical Fe concentration and even away from this point, is much smaller than the size of the gap [62, 82]. Comparison of this energy difference with the A_{2g} mode is shown in Fig. 3.30. A very good agreement can be seen without any parameter adjustments.

The intensities of the collective modes in each symmetry channel can also be calculated from the matrix elements, $\langle\psi_{\text{AFM}}|\, O_\mu \,|\psi_{\text{HO}}\rangle$, where O_μ is the Raman

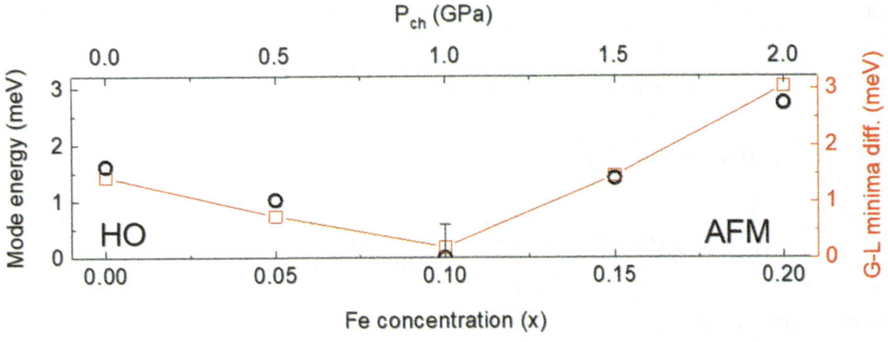

Fig. 3.30 The A_{2g} mode energy is plot against iron concentration x and compared to the energy difference between the Ginzburg–Landau free energy minima along ψ_{HO} and ψ_{AFM} axes in Fig. 3.29

operator in the μ symmetry channel. At the Γ point (long wavelength limit), without the loss of generality, we can consider the wave functions at site r,

$$|\psi_{\text{HO}}\rangle_r = |\text{HO}\rangle = \left(1 - \frac{V^2 \sin^2\theta}{2\omega_0^2}\right)|A_{2g}\rangle + \left(\frac{V\sin\theta}{\omega_0}\right)|A_{1g}\rangle + \mathcal{O}(V^2) \tag{3.14}$$

$$|\psi_{\text{AFM}}\rangle_r = |\text{AFM}\rangle = \left(1 - \frac{V'^2 \sin^2\theta'}{2\omega_0^2}\right)|A_{1g}\rangle + \left(\frac{iV'\sin\theta'}{\omega_0}\right)|A_{2g}\rangle + \mathcal{O}(V'^2). \tag{3.15}$$

To illustrate the symmetry selection rule in Raman scattering, we consider the simplest case where $V = V'$ and $\theta = \theta'$. In the HO phase, the matrix element in the A_{2g} channel is then

$$\begin{aligned}
&\langle\text{AFM}|\, O_{A2g}\, |\text{HO}\rangle \\
&= \left(1 - \frac{V^2\sin^2\theta}{\omega_0^2}\right)\langle A_{1g}|\, O_{A2g}\, |A_{2g}\rangle + \left(\frac{iV^2\sin^2\theta}{\omega_0^2}\right)\langle A_{2g}|\, O_{A2g}\, |A_{1g}\rangle + \mathcal{O}(V^4) \\
&= \langle A_{1g}|\, O_{A2g}\, |A_{2g}\rangle - \frac{V^2\sin^2\theta}{\omega_0^2}\left(\langle A_{1g}|\, O_{A2g}\, |A_{2g}\rangle - i\,\langle A_{2g}|\, O_{A2g}\, |A_{1g}\rangle\right) + \mathcal{O}(V^4) \\
&= \left(1 - \frac{(1+i)V^2\sin^2\theta}{\omega_0^2}\right)\langle A_{1g}|\, O_{A2g}\, |A_{2g}\rangle + \mathcal{O}(V^4).
\end{aligned} \tag{3.16}$$

We have used in the above $\langle A_{2g}|\, O_{A2g}\, |A_{2g}\rangle = \langle A_{1g}|\, O_{A2g}\, |A_{1g}\rangle = 0$ by symmetry selection rules and $\langle A_{1g}|\, O_{A2g}\, |A_{2g}\rangle = -\langle A_{2g}|\, O_{A2g}\, |A_{1g}\rangle$ by the fact that $\langle A_{1g}|\, O_{A2g}\, |A_{2g}\rangle$ is purely imaginary. Therefore, the intensity of collective mode in the A_{2g} channel is

$$I_{A2g} \sim |\langle \text{AFM}| O_{A2g} |\text{HO}\rangle|^2 = \left(1 - \frac{2V^2 \sin^2\theta}{\omega_0^2}\right) |\langle A_{1g}| O_{A2g} |A_{2g}\rangle|^2 + \mathcal{O}(V^4). \tag{3.17}$$

Similarly, the matrix element in the A_{1g} channel is

$$\begin{aligned}&\langle \text{AFM}| O_{A1g} |\text{HO}\rangle \\ &= \frac{iV\sin\theta}{\omega_0} \langle A_{2g}| O_{A1g} |A_{2g}\rangle + \frac{V\sin\theta}{\omega_0} \langle A_{1g}| O_{A1g} |A_{1g}\rangle + \mathcal{O}(V^3).\end{aligned} \tag{3.18}$$

We have used in the above $\langle A_{1g}| O_{A1g} |A_{2g}\rangle = \langle A_{2g}| O_{A1g} |A_{1g}\rangle = 0$ by symmetry selection rules. Therefore, the "leakage" intensity of the collective mode in the A_{1g} channel is

$$\begin{aligned}I_{A1g} &\sim |\langle \text{AFM}| O_{A1g} |\text{HO}\rangle|^2 \\ &= \left(\frac{V^2 \sin^2\theta}{\omega_0^2}\right) |\langle A_{1g}| O_{A1g} |A_{1g}\rangle + i\langle A_{2g}| O_{A1g} |A_{2g}\rangle|^2 + \mathcal{O}(V^4).\end{aligned} \tag{3.19}$$

Thus, the intensity in the A_{1g} channel is proportional to the square of the order parameter, $V^2 \sin^2\theta$, providing an indication of the symmetry breaking in the HO phase. Comparison of I_{A2g} to I_{A1g} is however difficult, because the evaluation of the expectation values depends on the details of the intermediate states and the laser excitation energy.

3.5 Summary

Summarizing this chapter, we used polarization resolved low temperature Raman spectroscopy to study the broken symmetry and collective modes in the long range ordered phases of $URu_{2-x}Fe_xSi_2$. By studying the temperature dependent Raman response, we observed a collective mode appearing in the A_{2g} symmetry channel below the second order phase transition, in both HO and AFM phases. The mode energy continuously evolves with increasing x, decreasing from 1.7 meV in the HO phase to almost zero at the HO/AFM phase boundary, and reappears with increasing energy in the AFM phase. The mode's evolution provides direct evidence for unified order parameter for both HO and AFM phases arising from the orbital degrees of freedom of the uranium-5f electrons.

While the long range ordering and broken time reversal symmetry in the pressurized and chemically substituted phases are undoubtedly due to AFM, the broken symmetry in the HO phase has been haunting the condensed matter physics community for many years. By carefully analyzing the Raman response in all available scattering geometries in the HO phase, we conclude that the reflection symmetries σ_v and σ_d must be broken. Therefore, the HO must have a ground state breaking the local chirality at uranium sites. Moreover, the A_{2g} mode is also observed by inelastic neutron scattering at finite momentum, Q_0, suggesting the lattice translational symmetry is also broken along c-axis, resulting in a chirality density wave.

In our study, we find that the uranium-5f orbitals in URu_2Si_2 can arrange into orders with broken chirality or time reversal symmetry. While such orders are competing for the same phase space at low temperature, they are also subtly connected and were here unified into a common order parameter, which are almost degenerate in energy. Here, we propose that the "unrealized" phase is also the first excited state and can be probed through inelastic light scattering. The low energy excitations in the Ginzburg–Landau theory are usually Goldstone modes, but here we detected a new type of collective mode which corresponds to the oscillation between the HO and AFM phases.

This is by no means the end to the story of the HO in URu_2Si_2. A direct experimental verification to the chiral wave function, $|HO\rangle$, is still yet to be done. Whether the Raman active collective mode observed in the AFM phase is indeed related to the HO phase, and why is it absent in the inelastic neutron scattering data? How does the Raman collective mode in the HO phase soften toward the AFM phase? Is it a first or second order phase boundary? Is there a quantum critical point at the phase boundary? Most importantly, how generic are the phenomena we observed in the URu_2Si_2 system? Can we reproduce the HO in other heavy fermion systems? Are we able to describe or even predict the HO by *ab initio* calculations? Knowing the symmetry of the order parameter is just a small step forward, and there are still many unanswered questions lying in front of us.

References

1. H.-H. Kung et al., Chirality density wave of the "hidden order" phase in URu_2Si_2. Science **347**(6228), 1339–1342 (2015). https://doi.org/10.1126/science.1259729
2. H.-H. Kung et al., Analogy between the "Hidden Order" and the orbital antiferromagnetism in $URu_{2-x}Fe_xSi_2$. Phys. Rev. Lett. **117**, 227601 (2016). https://doi.org/10.1103/PhysRevLett.117.227601
3. L. Degiorgi, The electrodynamic response of heavy-electron compounds. Rev. Mod. Phys. **71**, 687–734 (1999). https://doi.org/10.1103/RevModPhys.71.687
4. J.D. Thompson, Z. Fisk, Progress in heavy-fermion superconductivity: Ce115 and related materials. J. Phys. Soc. Jpn. **81**(1), 011002 (2012). https://doi.org/10.1143/JPSJ.81.011002
5. J.L. Sarrao, J.D. Thompson, Superconductivity in cerium- and plutonium-based '115' materials. J. Phys. Soc. Jpn. **76**(5), 051013 (2007). https://doi.org/10.1143/JPSJ.76.051013
6. P. Santini et al., Multipolar interactions in f-electron systems: the paradigm of actinide dioxides. Rev. Mod. Phys. **81**, 807–863 (2009). https://doi.org/10.1103/RevModPhys.81.807
7. R. Joynt, L. Taillefer, The superconducting phases of UPt$ 3$. Rev. Mod. Phys. **74**, 235–294 (2002). https://doi.org/10.1103/RevModPhys.74.235
8. C. Pfleiderer, Superconducting phases of f -electron compounds. Rev. Mod. Phys. **81**(4), 1551–1624 (2009). https://doi.org/10.1103/RevModPhys.81.1551

9. J.L. Smith, E.A. Kmetko, Magnetism or bonding: a nearly periodic table of transition elements. J. Less Common Met. **90**(1), 83–88 (1983)
10. P. Coleman, *Introduction to Many-Body Physics* (Cambridge University Press, Cambridge, 2015). https://doi.org/10.1017/CBO9781139020916
11. W.J. De Haas, J. De Boer, G.J. Van den Berg, The electrical resistance of gold, copper and lead at low temperatures. Physica **1**(7–12), 1115–1124 (1934). https://doi.org/10.1016/S0031-8914(34)80310-2
12. M.P. Sarachik, E. Corenzwit, L.D. Longinotti, Resistivity of Mo-Nb and Mo-Re alloys containing 1% Fe. Phys. Rev. **135**, A1041–A1045 (1964). https://doi.org/10.1103/PhysRev.135.A1041
13. B.T. Matthias et al., Magnetic moment of transition metal atoms in dilute solution and their effect on superconducting transition temperature. Phys. Rev. Lett. **5**, 542–544 (1960). https://doi.org/10.1103/PhysRevLett.5.542
14. A.M. Clogston et al., Local magnetic moment associated with an iron atom dissolved in various transition metal alloys. Phys. Rev. **125**, 541–552 (1962). https://doi.org/10.1103/PhysRev.125.541
15. P.W. Anderson, Localized magnetic states in metals. Phys. Rev. **124**, 1–53 (1961). https://doi.org/10.1103/PhysRev.124.41
16. A. Menth, E. Buehler, T.H. Geballe, Magnetic and semiconducting properties of SmB_6. Phys. Rev. Lett. **22**, 295–297 (1969). https://doi.org/10.1103/PhysRevLett.22.295
17. F. Steglich et al., Superconductivity in the presence of strong Pauli paramagnetism: $CeCu_2Si_2$. Phys. Rev. Lett. **43**, 1892–1896 (1979). https://doi.org/10.1103/PhysRevLett.43.1892
18. A.C. Hewson, *The Kondo Problem to Heavy Fermions*. Cambridge Studies in Magnetism (Cambridge University Press, Cambridge, 1997)
19. G.R. Stewart, Heavy-fermion systems. Rev. Mod. Phys. **56**(4), 755–787 (1984). https://doi.org/10.1103/RevModPhys.56.755
20. F. Steglich et al., Magnetism, f-electron localization and superconductivity in 122-type heavy-fermion metals. J. Phys. Condensed Matter **24**(29), 294201 (2012). https://doi.org/10.1088/0953-8984/24/29/294201
21. P. Aynajian et al., Visualizing heavy fermions emerging in a quantum critical Kondo lattice. Nature **486**(7402), 201–206 (2012). https://doi.org/10.1038/nature11204
22. R.M. Martin, Fermi-surface sum rule and its consequences for periodic Kondo and mixed-valence systems. Phys. Rev. Lett. **48**, 362–365 (1982). https://doi.org/10.1103/PhysRevLett.48.362
23. H. Shishido et al., A drastic change of the Fermi surface at a critical pressure in "CeRhIn5: dHvA" study under pressure. J. Phys. Soc. Jpn. **74**(4), 1103–1106 (2005). https://doi.org/10.1143/JPSJ.74.1103
24. M. Maltseva, M. Dzero, P. Coleman, Electron cotunneling into a Kondo lattice. Phys. Rev. Lett. **103**, 206402 (2009). 10.1103/PhysRevLett.103.206402
25. M.H. Hamidian et al., How Kondo-holes create intense nanoscale heavy-fermion hybridization disorder. Proc. Natl. Acad. Sci. **108**(45), 18233–18237 (2011). https://doi.org/10.1073/pnas.1115027108. eprint. http://www.pnas.org/content/108/45/18233.full.pdf
26. A.R. Schmidt et al., Imaging the Fano lattice to 'hidden order' transition in URu_2Si_2". Nature (London) **465**(7298), 570–576 (2010). https://doi.org/10.1038/nature09073
27. P. Aynajian et al., Visualizing the formation of the Kondo lattice and the hidden order in URu_2Si_2. Proc. Natl. Acad. Sci. U. S. A. **107**(23), 10383–10388 (2010). https://doi.org/10.1073/pnas.1005892107
28. A. Georges et al., Dynamical mean-field theory of strongly correlated fermion systems and the limit of infinite dimensions. Rev. Mod. Phys. **68**, 13–125 (1996). https://doi.org/10.1103/RevModPhys.68.13
29. G. Kotliar et al., Electronic structure calculations with dynamical mean-field theory. Rev. Mod. Phys. **78**, 865–951 (2006). https://doi.org/10.1103/RevModPhys.78.865
30. D.N. Basov et al., Electrodynamics of correlated electron materials. Rev. Mod. Phys. **83**, 471–541 (2011). https://doi.org/10.1103/RevModPhys.83.471

31. M. Dzero et al., Topological Kondo insulators. Annu. Rev. Condens. Matter Phys. **7**(1), 249–280 (2016). https://doi.org/10.1146/annurev-conmatphys-031214-014749
32. J.A. Mydosh, P.M. Oppeneer, Colloquium: hidden order, superconductivity, and magnetism: the unsolved case of URu_2Si_2. Rev. Mod. Phys. **83**(4), pp. 1301–1322 (2011). And references therein. https://doi.org/10.1103/RevModPhys.83.1301
33. J.R. Jeffries et al., Degree of 5f electron localization in URu_2Si_2: electron energy-loss spectroscopy and spin-orbit sum rule analysis. Phys. Rev. B **82**(3), 033103 (2010). https://doi.org/10.1103/PhysRevB.82.033103
34. K.O. Kvashnina et al., Resonant X-ray spectroscopy of uranium intermetallics at the $M_{4,5}$ edges of uranium. Phys. Rev. B **95**, 245103 (2017). https://doi.org/10.1103/PhysRevB.95.245103
35. K. Haule, G. Kotliar, Arrested Kondo effect and hidden order in URu_2Si_2. Nat. Phys. **5**(11), 796–799 (2009). https://doi.org/10.1038/nphys1392
36. T.T.M. Palstra et al., Superconducting and magnetic transitions in the heavy-fermion system URu_2Si_2. Phys. Rev. Lett. **55**, 2727–2730 (1985). https://doi.org/10.1103/PhysRevLett.55.2727
37. D.L. Cox, Quadrupolar Kondo effect in uranium heavy-electron materials? Phys. Rev. Lett. **59**, 1240–1243 (1987). 10.1103/PhysRevLett.59.1240
38. D.L. Cox, A. Zawadowski, Exotic Kondo effects in metals: magnetic ions in a crystalline electric field and tunneling centres. Adv. Phys. **47**(5), 599–942 (1998). https://doi.org/10.1080/000187398243500
39. A.I. Tóth, G. Kotliar, Hexadecapolar Kondo effect in $\mathbf{URu_2Si_2}$? Phys. Rev. Lett. **107**, 266405 (2011). https://doi.org/10.1103/PhysRevLett.107.266405
40. P. Chandra, P. Coleman, R. Flint, Hastatic order in the heavy-fermion compound URu_2Si_2". Nature (London) **493**, 621–626 (2013). https://doi.org/10.1038/nature11820
41. H. Amitsuka et al., Hidden order and weak antiferromagnetism in URu_2Si_2. Phys. B Condensed Matter **312**, 390 (2002). https://doi.org/10.1016/S0921-4526(01)01343-6
42. N. Harrison et al., Metamagnetism, quantum criticality, hidden order and crystal electric fields in URu_2Si_2. Physica B: Condensed Matter 346, 92 (2004). https://doi.org/10.1016/j.physb.2004.01.027
43. J.A. Mydosh, P.M. Oppeneer, Hidden order behaviour in URu_2Si_2 (A critical review of the status of hidden order in 2014). Philos. Mag. **94**(32–33), 3642–3662 (2014). And references therein. https://doi.org/10.1080/14786435.2014.916428
44. N. Kanchanavatee et al., Enhancement of the hidden order/large moment antiferromagnetic transition temperature in the $URu_{2-x}Os_xSi_2$ system. Philos. Mag. **94**(32–33), 3681–3690 (2014). https://doi.org/10.1080/14786435.2014.886022
45. Y. Haga et al., Single crystal growth and characterization of URu_2Si_2. Philos. Mag. **94**(32–33), 3672–3680 (2014). https://doi.org/10.1080/14786435.2014.974711
46. C. Tabata et al., X-ray backscattering study of crystal lattice distortion in hidden order of URu_2Si_2. Philos. Mag. **94**(32–33), 3691–3701 (2014). https://doi.org/10.1080/14786435.2014.952701
47. F. Bourdarot, S. Raymond, L.-P. Regnault, Neutron scattering studies on URu_2Si_2. Philos. Mag. **94**(32–33), 3702–3722 (2014). https://doi.org/10.1080/14786435.2014.935513
48. T. Durakiewicz, Photoemission investigations of URu_2Si_2. Philos. Mag. **94**(32–33), 723–3736 (2014). https://doi.org/10.1080/14786435.2014.937783
49. P.S. Riseborough, S.G. Magalhães, E.J. Calegari, Signatures of broken spin-rotational invariance in the "Hidden Ordered" compound URu_2Si_2? Philos. Mag. **94**(32–33), 3820–3837 (2014). https://doi.org/10.1080/14786435.2014.952256
50. P. Thalmeier, T. Takimoto, H. Ikeda, Itinerant multipolar order in URu_2Si_2 and its signature in magnetic and lattice properties. Philos. Mag. **94**(32–33), 3863–3876 (2014). https://doi.org/10.1080/14786435.2013.861615
51. O. Stockert et al., Superconductivity in Ce- and U-based "122" heavy-fermion compounds. J. Phys. Soc. Jpn. **81**(1), 011001 (2012). https://doi.org/10.1143/JPSJ.81.011001. eprint

52. C.L. Broholm, Magnetic fluctuations in heavy Fermion systems: a neutron scattering study of UPt_3, U_2Zn_{17} and URu_2Si_2. PhD thesis. The University of Copenhagen, 1989
53. J. Buhot, Ordres cachés et magnétisme étudiés par spectroscopie Raman en conditions extrêmes. PhD thesis. Université Paris Diderot-Paris 7, 2015
54. S.L. Cooper, Light Scattering Studies of Heavy Fermion and High Tc Superconductors. PhD thesis. University of Illinois at Urbana-Champaign, 1988
55. E. Hassinger, Pressure-temperature phase diagram of URu_2Si_2 by accalorimetry and resistivity measurements. PhD thesis. Heidelberg University, 2007
56. M.C. Shapiro, Elastoresistivity as a Probe of Electronically Driven Rotational Symmetry Breaking and Its Application to the Hidden Order State in URu_2Si_2. PhD thesis. Stanford University, 2016
57. T.T.M. Palstra, A.A. Menovsky, J.A. Mydosh, Anisotropic electrical resistivity of the magnetic heavy-fermion superconductor URu_2Si_2. Phys. Rev. B **33**, 6527–6530 (1986). https://doi.org/10.1103/PhysRevB.33.6527
58. M.W. McElfresh et al., Effect of pressure on competing electronic correlations in the heavy-electron system URu_2Si_2. Phys. Rev. B **35**, 43–47 (1987). https://doi.org/10.1103/PhysRevB.35.43
59. W.T. Guo et al., Hybridization gap versus hidden-order gap in URu_2Si_2 as revealed by optical spectroscopy. Phys. Rev. B **85**, 195105 (2012). https://doi.org/10.1103/PhysRevB.85.195105
60. R.P.S.M. Lobo et al., Optical conductivity of URu_2Si_2 in the Kondo liquid and hidden-order phases. Phys. Rev. B **92**, 045129 (2015). https://doi.org/10.1103/PhysRevB.92.045129
61. N. Bachar et al., Detailed optical spectroscopy of hybridization gap and hidden order transition in high-quality URu_2Si_2 single crystals. Phys. Rev. B **94**, 235101 (2016). https://doi.org/10.1103/PhysRevB.94.235101
62. J.S. Hall et al., Electrodynamics of the antiferromagnetic phase in URu_2Si_2. Phys. Rev. B **92**, 195111 (2015). https://doi.org/10.1103/PhysRevB.92.195111
63. N. Kanchanavatee et al., Twofold enhancement of the hidden-order/large moment antiferromagnetic phase boundary in the $URu_{2-x}Fe_xSi_2$ system. Phys. Rev. B **84**, 245122 (2011). https://doi.org/10.1103/PhysRevB.84.245122
64. S. Ran et al., Phase diagram and thermal expansion measurements on the system $URu_{2-x}Fe_xSi_2$. Proc. Natl. Acad. Sci. U. S. A. **113**(47), 13348–13353 (2016). https://doi.org/10.1073/pnas.1616542113
65. N.P. Butch et al., Antiferromagnetic critical pressure in URu_2Si_2 under hydrostatic conditions. Phys. Rev. B **82**, 060408 (2010). https://doi.org/10.1103/PhysRevB.82.060408
66. F. Bourdarot et al., Magnetic properties of URu_2Si_2 under uniaxial stress by neutron scattering. Phys. Rev. B **84**(18), 184430 (2011). https://doi.org/10.1103/PhysRevB.84.184430
67. M. Jaime et al., High magnetic field studies of the hidden order transition in URu_2Si_2". Phys. Rev. Lett. **89**, 287201 (2002). https://doi.org/10.1103/PhysRevLett.89.287201
68. D. Aoki et al., Field reentrance of the hidden order state of URu_2Si_2 under pressure. J. Phys. Soc. Jpn. **78**(5), 053701 (2009). https://doi.org/10.1143/JPSJ.78.053701
69. M.B. Maple et al., Partially gapped Fermi surface in the heavy-electron superconductor URu_2Si_2. Phys. Rev. Lett. **56**, 185–188 (1986). https://doi.org/10.1103/PhysRevLett.56.185
70. W. Schlabitz et al., Superconductivity and magnetic order in a strongly interacting Fermi-system: URu_2Si_2. Z. Phys. B **62**, 171–177 (1986). https://doi.org/10.1007/BF01323427
71. H. Amitsuka et al., Pressure–temperature phase diagram of the heavy-electron superconductor. J. Magn. Magn. Mater. **310**(2), Part 1, 214–220 (2007). https://doi.org/10.1016/j.jmmm.2006.10.008
72. C. Broholm et al., Magnetic excitations in the heavy-fermion superconductor URu_2Si_2. Phys. Rev. B **43**, 12809–12822 (1991). https://doi.org/10.1103/PhysRevB.43.12809
73. C. Broholm et al., Magnetic excitations and ordering in the heavy-electron superconductor URu_2Si_2. Phys. Rev. Lett. **58**, 1467–1470 (1987). https://doi.org/10.1103/PhysRevLett.58.1467
74. E.D. Isaacs et al., X-ray magnetic scattering in antiferromagnetic URu_2Si_2. Phys. Rev. Lett. **65**, 3185–3188 (1990). https://doi.org/10.1103/PhysRevLett.65.3185

75. J. Schoenes et al., Hall-effect and resistivity study of the heavy-fermion system URu_2Si_2. Phys. Rev. B **35**, 5375–5378 (1987). https://doi.org/10.1103/PhysRevB.35.5375
76. E.R. Schemm et al., Evidence for broken time-reversal symmetry in the superconducting phase of URu_2Si_2. Phys. Rev. B **91**, 140506 (2015). https://doi.org/10.1103/PhysRevB.91.140506
77. Y.J. Jo et al., Field-induced Fermi surface reconstruction and adiabatic continuity between antiferromagnetism and the hidden-order state in URu_2Si_2. Phys. Rev. Lett. **98**, 166404 (2007). https://doi.org/10.1103/PhysRevLett.98.166404
78. E. Hassinger et al., Temperature-pressure phase diagram of URu_2Si_2 from resistivity measurements and ac calorimetry: hidden order and Fermi-surface nesting. Phys. Rev. B **77**, 115117 (2008). https://doi.org/10.1103/PhysRevB.77.115117
79. T.J. Williams et al., Gapped excitations in the high-pressure antiferromagnetic phase of URu_2Si_2. Phys. Rev. B **95**, 195171 (2017). https://doi.org/10.1103/PhysRevB.95.195171
80. P. Das et al., Chemical pressure tuning of URu_2Si_2 via isoelectronic substitution of Ru with Fe. Phys. Rev. B **91**, 085122 (2015). https://doi.org/10.1103/PhysRevB.91.085122
81. M.N. Wilson et al., Antiferromagnetism and hidden order in isoelectronic doping of URu_2Si_2. Phys. Rev. B **93**, 064402 (2016). https://doi.org/10.1103/PhysRevB.93.064402
82. N.P. Butch et al., Distinct magnetic spectra in the hidden order and antiferromagnetic phases in $URu_{2-x}Fe_xSi_2$. Phys. Rev. B **94**, 201102 (2016). https://doi.org/10.1103/PhysRevB.94.201102
83. T.J. Williams et al., Hidden order signatures in the antiferromagnetic phase of $U(Ru_{1-x}Fe_x)_2Si_2$. Phys. Rev. B **95**, 104440 (2017). https://doi.org/10.1103/PhysRevB.95.104440
84. C.R. Wiebe et al., Gapped itinerant spin excitations account for missing entropy in the hidden-order state of URu_2Si_2. Nat. Phys. **3**(2), 96–99 (2007). https://doi.org/10.1038/nphys522
85. F. Bourdarot et al., Precise study of the resonance at $Q_0 = (1, 0, 0)$ in URu_2Si_2. J. Phys. Soc. Jpn. **79**(6), 064719 (2010). https://doi.org/10.1143/JPSJ.79.064719
86. A. Villaume et al., Signature of hidden order in heavy fermion superconductor URu_2Si_2: resonance at the wave vector $Q_0 = (1; 0; 0)$. Phys. Rev. B **78**, 012504 (2008). https://doi.org/10.1103/PhysRevB.78.012504
87. N.P. Butch et al., Symmetry and correlations underlying hidden order in URu_2Si_2. Phys. Rev. B **91**, 035128 (2015). https://doi.org/10.1103/PhysRevB.91.035128
88. D.A. Bonn, J.D. Garrett, T. Timusk, Far-infrared properties of URu_2Si_2. Phys. Rev. Lett. **61**, 1305–1308 (1988). https://doi.org/10.1103/PhysRevLett.61.1305
89. J.S. Hall et al., Observation of multiple-gap structure in hidden order state of URu_2Si_2 from optical conductivity. Phys. Rev. B **86**, 035132 (2012). https://doi.org/10.1103/PhysRevB.86.035132
90. S.L. Cooper et al., Magnetic excitations and phonon anomalies in URu_2Si_2. Phys. Rev. B **36**(10), 5743–5746 (1987). https://doi.org/10.1103/PhysRevB.36.5743
91. J.A. Koningstein, O.S. Mortensen, Experimental observation of an antisymmetric Raman scattering tensor. Nature **217**(5127), 445–446 (1968). https://doi.org/10.1038/217445a0
92. H. Rho, M.V. Klein, P.C. Canfield, Polarized Raman scattering studies of crystal-field excitations in $ErNi_2B_2C$. Phys. Rev. B **69**, 144420 (2004). https://doi.org/10.1103/PhysRevB.69.144420
93. H. Rho, M.V. Klein, P.C. Canfield, Crystal-field excitations and spin-phonon interactions in $DyNi_2B_2C$: Raman scattering study. Phys. Rev. B **82**, 064423 (2010). https://doi.org/10.1103/PhysRevB.82.064423
94. R. Liu et al., Novel Raman-active electronic excitations near the charge transfer gap in insulating cuprates. Phys. Rev. Lett. **71**, 709–3712 (1993). https://doi.org/10.1103/PhysRevLett.71.3709
95. E. Riccardi et al., Gate-dependent electronic Raman scattering in graphene. Phys. Rev. Lett. **116**, 066805 (2016). https://doi.org/10.1103/PhysRevLett.116.066805
96. H.-H. Kung et al., Chiral spin mode on the surface of a topological insulator. Phys. Rev. Lett. **119**, 136802 (2017). https://doi.org/10.1103/PhysRevLett.119.136802
97. K. Haule, G. Kotliar, Complex Landau-Ginzburg theory of the hidden order in URu_2Si_2. Europhys. Lett. **89**(5), 57006 (2010)

98. S. Elgazzar et al., Hidden order in URu_2Si_2 originates from Fermi surface gapping induced by dynamic symmetry breaking. Nat. Mater. **8**(4), 337–341 (2009). https://doi.org/10.1038/nmat2395
99. P.M. Oppeneer et al., Spin and orbital hybridization at specifically nested Fermi surfaces in URu_2Si_2. Phys. Rev. B **84**(24), 241102 (2011). https://doi.org/10.1103/PhysRevB.84.241102
100. P. Thalmeier, T. Takimoto, Signatures of hidden-order symmetry in torque oscillations, elastic constant anomalies, and field-induced moments in URu_2Si_2. Phys. Rev. B **83**(16), 165110 (2011). https://doi.org/10.1103/PhysRevB.83.165110
101. J.G. Rau, H.-Y. Kee, Hidden and antiferromagnetic order as a rank-5 superspin in URu_2Si_2. Phys. Rev. B **85**(24), 245112 (2012). https://doi.org/10.1103/PhysRevB.85.245112
102. P.S. Riseborough, B. Coqblin, S.G. Magalhães, Phase transition arising from the underscreened Anderson lattice model: a candidate concept for explaining hidden order in URu_2Si_2. Phys. Rev. B **85**(16), 165116 (2012). https://doi.org/10.1103/PhysRevB.85.165116
103. H. Ikeda et al., Emergent rank-5 nematic order in URu_2Si_2. Nat. Phys. **8**(7), 528–533 (2012). https://doi.org/10.1038/nphys2330
104. C. Thomas et al., Three-dimensional modulated spin liquid model applied to URu_2Si_2. Phys. Rev. B **87**, 014422 (2013). https://doi.org/10.1103/PhysRevB.87.014422
105. T. Das, Spin-orbit density wave induced hidden topological order in URu_2Si_2. Sci. Rep. **2**, 596 (2012). https://doi.org/10.1038/srep00596
106. R. Flint, P. Chandra, P. Coleman, Basal-plane nonlinear susceptibility: a direct probe of the single-ion physics in URu_2Si_2. Phys. Rev. B **86**, 155155 (2012). https://doi.org/10.1103/PhysRevB.86.155155
107. P. Chandra, P. Coleman, R. Flint, Hastatic order in URu_2Si_2: hybridization with a twist. Phys. Rev. B **91**, 205103 (2015). https://doi.org/10.1103/PhysRevB.91.205103
108. T.D. Matsuda et al., Details of sample dependence and transport properties of URu_2Si_2. J. Phys. Soc. Jpn. **80**(11), 14710 (2011). https://doi.org/10.1143/JPSJ.80.114710
109. R.E. Baumbach et al., High purity specimens of URu2Si2 produced by a molten metal UX technique. Philos. Mag. **94**(32–33), 3663–3671 (2014). https://doi.org/10.1080/14786435.2014.895876
110. Y. Kasahara et al., Exotic superconducting properties in the electron-hole-compensated heavy-fermion "Semimetal" URu_2Si_2. Phys. Rev. Lett **99**, 116402 (2007). https://doi.org/10.1103/PhysRevLett.99.116402
111. A.P. Mackenzie, Y. Maeno, The superconductivity of Sr_2RuO_4 and the physics of spin-triplet pairing. Rev. Mod. Phys. **75**, 657–712 (2003). https://doi.org/10.1103/RevModPhys.75.657
112. R. Okazaki et al., Rotational symmetry breaking in the hidden-order phase of URu_2Si_2. Science **331**(6016), 439–442 (2011). https://doi.org/10.1126/science.1197358
113. S.C. Riggs et al., Evidence for a nematic component to the hidden-order parameter in URu_2Si_2 from differential elastoresistance measurements. Nat. Commun. **6**, 6425 (2015)
114. A. Menovsky, J.J.M. Franse, Crystal growth of some rare earth and uranium intermetallics from the melt. J. Crystal Growth **65**(1), 286–292 (1983). https://doi.org/10.1016/0022-0248(83)90062-3
115. Y. Haga et al., Purification of uranium metal using the solid state electrotransport method under ultrahigh vacuum. Jpn. J. Appl. Phys. **37**(6R), 3604 (1998)
116. T.D. Matsuda et al., Super clean sample of URu_2Si_2. J. Phys. Soc. Jpn. **77**(Suppl. A), 362–364 (2008). https://doi.org/10.1143/JPSJS.77SA.362. eprint
117. M. Pluta, Nomarski's DIC microscopy: a review. *Proc. SPIE* **1846** (1994). https://doi.org/10.1117/12.171873
118. W. Lang, *Nomarski Differential Interference-Contrast Microscopy* (Zeiss Information, Oberkochen, 1968)
119. K. Huang, *Statistical Mechanics* (Wiley, London, 1963)
120. M.W. Zemansky, R.H. Dittman, *Heat and Thermodynamics: An Intermediate Textbook*. International Series in Pure and Applied Mathematics (McGraw-Hill, New York, 1997)
121. R.K. Pathria, P.D. Beale, *Statistical Mechanics* (Elsevier Science, Oxford, 1996)

122. W. Hayes, R. Loudon, *Scattering of Light by Crystals* (Courier Corporation, Chelmsford, 2012)
123. E. Kroumova et al., Bilbao crystallographic server: useful databases and tools for phase-transition studies. Phase Trans. Multinat. J. **76**(1–2), 155–170 (2003). https://doi.org/10.1080/0141159031000076110
124. M.I. Aroyo et al., Crystallography online: Bilbao crystallographic server. Bulgar. Chem. Commun. **43**(2), 183–197 (2011)
125. J. Buhot et al., Lattice dynamics of the heavy-fermion compound URu_2Si_2. Phys. Rev. B **91**, 035129 (2015). https://doi.org/10.1103/PhysRevB.91.035129
126. J. Buhot et al., Anisotropic Kondo pseudogap in URu_2Si_2. Phys. Rev. B **101**, 245103 (2020). https://doi.org/10.1103/PhysRevB.101.245103
127. J. Buhot et al., Raman scattering study of the lattice dynamic of URu_2Si_2 and sample's preparation. English. J. Korean Phys. Soc. **62**(10), 1427–1430 (2013). https://doi.org/10.3938/jkps.62.1427
128. P.G. Niklowitz et al., Parasitic small-moment antiferromagnetism and nonlinear coupling of hidden order and antiferromagnetism in URu_2Si_2 observed by Larmor diffraction. Phys. Rev. Lett. **104**(10), 106406 (2010). https://doi.org/10.1103/PhysRevLett.104.106406
129. S. Tonegawa et al., Direct observation of lattice symmetry breaking at the hidden-order transition in URu_2Si_2. Nat. Commun. **5** (2014). https://doi.org/10.1038/ncomms5188
130. A. Zawadowski, M. Cardona, Theory of Raman scattering on normal metals with impurities. Phys. Rev. B **42**, 10732–10734 (1990). https://doi.org/10.1103/PhysRevB.42.10732
131. C. Pfleiderer, J.A. Mydosh, M. Vojta, Pressure dependence of the magnetization of URu_2Si_2. Phys. Rev. B **74**, 104412 (2006). 10.1103/PhysRevB.74.104412
132. J. Buhot et al., Symmetry of the excitations in the hidden order state of URu_2Si_2. Phys. Rev. Lett. **113**, 266405 (2014). https://doi.org/10.1103/PhysRevLett.113.266405
133. J. Levallois et al., Hybridization gap and anisotropic far-infrared optical conductivity of URu_2Si_2. Phys. Rev. B **84**, 184420, 184420 (2011). https://doi.org/10.1103/PhysRevB.84.184420
134. P.G. Niklowitz et al., Role of commensurate and incommensurate low-energy excitations in the paramagnetic to hidden-order transition of URu_2Si_2. Phys. Rev. B **92**, 115116 (2015). https://doi.org/10.1103/PhysRevB.92.115116
135. L.N. Ovander, The form of the Raman tensor. Opt. Spectrosc. **9**, 302 (1960)
136. M. Cardona, Resonance phenomena, in *Light Scattering in Solids II*, ed. by M. Cardona, G. Güntherodt (Springer, Berlin, 1982), pp. 45–49
137. G.F. Koster, *Properties of the Thirty-Two Point Groups*. Massachusetts Institute of Technology Press Research Monograph (M.I.T. Press, Cambridge, 1963)
138. H.-H. Kung et al., Surface vibrational modes of the topological insulator Bi_2Se_3 observed by Raman spectroscopy. Phys. Rev. B **95**, 245406 (2017). https://doi.org/10.1103/PhysRevB.95.245406
139. M. Sundermann et al., Direct bulk-sensitive probe of 5f symmetry in URu_2Si_2. Proc. Natl. Acad. Sci. U. S. A. **113**(49), 13989–13994 (2016). https://doi.org/0.1073/pnas.1612791113
140. M. Sundermann et al., Impact of Fe substitution on the electronic structure of URu_2Si_2 (2020). arXiv. Preprint arXiv:2005.03946
141. A. Amorese et al., From antiferromagnetic and hidden order to Pauli paramagnetism in UM_2Si_2 compounds with $5f$ electron duality. Proc. Natl. Acad. Sci. **117**(48), 30220–30227 (2020). https://doi.org/10.1073/pnas.2005701117
142. S. Ghosh et al., One-component order parameter in URu_2Si_2 uncovered by resonant ultrasound spectroscopy and machine learning. Sci. Adv. **6**(10), eaaz4074 (2020). https://doi.org/10.1126/sciadv.aaz4074
143. H. Kusunose, H. Harima, On the hidden order in URu_2Si_2 – antiferro hexadecapole order and its consequences. J. Phys. Soc. Jpn. **80**(8), 084702 (2011). https://doi.org/10.1143/JPSJ.80.084702
144. C. Bareille et al., Momentum-resolved hidden-order gap reveals symmetry breaking and origin of entropy loss in URu_2Si_2. Nat. Commun. **5** (2014). https://doi.org/10.1038/ncomms5326

145. P.A. Sharma et al., Phonon thermal transport of URu_2Si_2: broken translational symmetry and strong-coupling of the "Hidden Order" to the lattice. Phys. Rev. Lett. **97**, 156401 (2006). https://doi.org/10.1103/PhysRevLett.97.156401
146. L. Boyer, V.M. Yakovenko, et al., A model for metastable magnetism in the hidden-order phase of URu_2Si_2. Ann. Phys. (New York) **388** (2018). https://doi.org/10.1016/j.aop.2017.11.024
147. K. Matsuda et al., Spatially inhomogeneous development of antiferromagnetism in URu_2Si_2: evidence from ^{29}Si NMR under pressure. Phys. Rev. Lett. **87**, 087203 (2001) https://doi.org/10.1103/PhysRevLett.87.087203
148. M. Yokoyama et al., Competition between hidden order and antiferromagnetism in URu_2Si_2 under uniaxial stress studied by neutron scattering. Phys. Rev. B **72**, 214419 (2005). https://doi.org/10.1103/PhysRevB.72.214419

Chapter 4
Secondary Emission in Bi_2Se_3

Abstract In this chapter I will present our study of the secondary emissions (radiation) in a prototypical 3D topological insulator (TI) material, Bi_2Se_3. I will begin with a short overview on the physics of 3D TIs, with focus on the properties of the spin polarized surface states (Sect. 4.1.2). In Sect. 4.2, I will present our Raman spectroscopic results on the bulk and surface phonon modes in Bi_2Se_3, for which we identified the symmetries and self-energies and discussed the excitation dependence of them. In Sect. 4.3, I will present our electronic Raman scattering results in the much higher energy regime, where we observed a chiral spin mode as collective spin–flip excitations of the surface Dirac fermions. In Sect. 4.4, I will present our photoluminescence study of Bi_2Se_3, where we show a highly circular polarized surface exciton as a result of chiral spin texture in the surface states. Finally, I will make a brief summary of our study and its implications in Sect. 4.5.

The work in this chapter was done in collaboration with S.-W. Cheong (Rutgers), A.F. Kemper (NCSU), L.S. Levitov (MIT), S. Maiti (U. Florida), D.L. Maslov (U. Florida and NHMFL), R. Merlin (U. Michigan), and S. Oh (Rutgers).

The phononic study that focused on the surface phonons of Bi_2Se_3 is published in [1]. The Raman study of the chiral spin mode on the surface of Bi_2Se_3 is published in [2], and the photoluminescence (PL) study of the chiral excitons in Bi_2Se_3 is published in [3].

4.1 Introduction

Since ancient times, humans have been using physical properties, such as electrical conductivity, magnetism, hardness, color, etc., to classify the materials around us. Based on the electrical transport properties, we can broadly divide materials into two classes, i.e., the conducting metals where the conductivity increases as temperature decreases, and the nonconductive insulators with exponentially decreasing conductivity with decreasing temperature. One of the great triumphs of

H.-H. Kung, *Collective Excitations in the Antisymmetric Channel of Raman Spectroscopy*, Springer Theses, https://doi.org/10.1007/978-3-030-89332-3_4

condensed mater physics in the last century is to understand the differences between these two classes of materials based on the electronic "band theory."

If we think about it, the electronic properties in solids are not trivial to describe at all. For example, the electrical transport is not well-defined in a single atom or molecule and only becomes relevant when we consider many atoms closely packed such that their electronic wave functions overlap with each other. To describe such "collective" behaviors, we need to consider the many electrons as a whole instead of ascribing them to each atoms, and this is the central idea behind the band theory. In a solid, the electrons lose their single particle identities due to the very high density and long ranged Coulomb interaction with each other. The collective movements of electrons renormalizes into propagating waves called "Bloch waves," where the eigen-energies are neither discrete local states nor a simple parabola as free electrons. Instead, the atomic energy levels merge to form electronic bands, where dispersions are determined by the crystal symmetry, orbital characters, chemical bondings, etc.

Within this context, if the chemical potential (Fermi energy, E_F) situates inside an energy gap where no states are allowed, then it is a band insulator, otherwise a metal [4]. The most well-known insulator is perhaps the vacuum, with a gap of about 1 MeV below the electron–hole continuum, i.e., the minimum energy required for the "interband transition" that creates an electron–hole pair (Fig. 1.2). Whereas in a metal, it only requires infinitesimally small amount of energy to create an electron–hole pair through the "intraband transition," which is responsible for the charge transport. That is, there exists a connection between the band structures and electrical transport properties, which justifies the classification of metal versus insulator.

The band theory can be verified experimentally by angle resolved photoemission spectroscopy (ARPES), which directly maps the dispersion of quasiparticle states [5]. In the past century, physicists have been able to explain most of the material properties using the band theory. The development of such microscopic theories has allowed us to precisely control materials' properties by fine-tuning the band structure and chemical potential. It is fair to say that the foundation of the digital revolution we enjoy today is built upon the deep understanding of the band theory. Although physicists are still struggling to understand the metal-insulator transition and some other exotic phenomena due to the strong electron–electron interaction in some transitional metal oxides [6], the situation of the band theory is in general satisfactory for most systems.

However, toward the end of the past century, the discovery of quantum Hall effect opened the gateway to an entirely new phase of matter that does not fit into our traditional classification of materials [7]. In this new phase of matter, the bulk can be made insulated while the boundary is metallic by simply tuning carrier density and magnetic field, without breaking any space group symmetry. These materials are termed "topological" where the conductive boundary states are protected by global symmetries, rather than the detailed crystalline surface structures and electron filling as in non-topological band insulators.

In particular, the proposal and realization of 3D topological insulators (TIs) have attracted major interest in the condensed matter physics community into this emerging field. In these materials, the conductive 2D surface states are protected by the time reversal symmetry (TRS) and guaranteed to exist without any external magnetic field. Microscopically, these topological boundary states originate from strong spin–orbit coupling (SOC) in the bulk, which abruptly terminates at the crystalline boundary. Due to the symmetry protection, these states are extremely robust unlike their non-topological counterparts, and the transport properties are mainly governed by the "topological invariant" rather than the fine prints in the bulk band structures [8].

Make no mistake, the topological invariant themselves are also derived from the bulk band structures, but in a much more abstract way such that the details of the bands do not affect the outcome. In other words, we may put the materials through extreme conditions (such as magnetic field, temperature, pressure, chemical doping, etc.), and the topological boundary states and their associated transport properties would often remain unchanged, as long as the "topology" of the bands is unchanged. This is obviously a very nice property for electronic device design and applications. However, the original proposals have been limited to 2D systems, which are more difficult to manufacture and verify experimentally.

In the past decade, major developments have been made in the understanding of 3D materials' boundary properties based on their bulk band topology near the Fermi energy [8], immediately raising great interests in the condensed matter physics community. Materials that possess nontrivial band topologies and surface states are termed "topological materials." Just like metal versus insulators, topological materials can be further classified into different categories based on the topological invariants, and many of them are already experimentally verified [9–11]. Broadly speaking, depending on the bulk band structure at the Fermi level, topological materials can be separated into topological insulators and topological semimetals. The interesting aspects of the topological boundary states in three-dimensional (3D) TIs will be discussed in more detail later; see Sect. 4.1.2.

During the transition from a trivial band insulator to a topological insulator, there must exist a critical point at which the conduction and valence band intersect each other at a few nodal points or 1D nodal lines (either open arcs or closed loops) in the Brillouin zone. Depending on the symmetry and dimension of the system under study, the topology of these band crossing points can be further classified. For example, graphene can be considered as one of the simplest kinds of topological semimetal, i.e., a 2D Dirac semimetal. Around the nodes, the quasiparticles are described by the linearly dispersing bulk bands and topological boundary states. A variety of anomalous transport and optical properties have been predicted and some realized in various kinds of 3D topological semimetals [10, 11]. These exotic properties not only provide test beds for our current understanding of condensed matter theory but also possess promising applications in optoelectronics and spintronics.

The subject of topological band theory studies the relationships between topological boundary states and the bulk using only first principle calculations and lattice symmetries. This is a very active and largely unexplored field, with new results appearing almost every day and many challenges lying ahead. In the following sections, we will focus on a particular kind of topological material, i.e., the 3D topological insulators.

4.1.1 Band Topology in Solids

The topological invariants are ultimately derived from the *bulk* band structures, which originate from the chemistry and symmetries of the lattice. Just as in geometrical topology that studies the properties that are preserved during deformation of objects, band topology is concerned with properties that are conserved as we continuously deform the bulk band structures. The widely used example is that a doughnut and a coffee mug both contain one open hole in it, which corresponds to the "genus" $g = 1$ that remains unchanged as we continuously deform the two objects into each other. Therefore, that hole is a topological property that cannot be easily removed, unless we cut the doughnut open with a knife or break off the coffee mug handle. In topological materials, the "doughnut hole" is reflected on the boundary, where *metallic* states must exist regardless of how we deform the bulk band structure through external perturbations or charge doping (within the range of bulk band gap). Note that not all boundary states in a topological material are topologically protected, there could be trivial states that coexist. Just like cutting the doughnut into halves, removing the topological boundary states also requires something dramatic, which is different for each topological class. In 3D topological insulators, this means applying magnetic field strong enough such that the Zeeman splitting is comparable to the spin–orbit coupling (SOC) or decrease the SOC strength significantly, which are not physically possible in most cases. Therefore, we often refer to the topological surface states as protected by time reversal symmetry.

Note that not all topological materials require strong SOC, but in the case of TIs, SOC plays a central role and therefore we should do a small detour and discuss it briefly now. The SOC is a relativistic effect that is present both in a single atom and also on the extended Bloch waves in solids. That is, SOC can be derived from the relativistic Dirac equations by keeping terms on the order of $1/c^2$. However, we can capture the essence of SOC in the atomic limit by the following simple arguments. In the rest frame of the electron in an atom, the SOC originates from the effective magnetic field acting on the electron by the nucleus.

$$\mathbf{B} = -\frac{\mathbf{v} \times \mathbf{E}}{c^2}$$

Let $\mathbf{E} = -\nabla_r V\hat{r} = \frac{\mathbf{r}}{er}\frac{\partial U}{\partial r}$ and $\mathbf{v} = \hat{p}/m$, and we can further simplify the equation to the more familiar form:

$$\mathbf{B} = \frac{1}{emc^2}(-\frac{1}{r}\frac{\partial U}{\partial r})\mathbf{r} \times \mathbf{p}.$$

Note that the electric potential gradient $\partial U/\partial r$ is proportional to the nucleus charge to the 4th-power, Z^4, and therefore tend to be important only in heavy elements. The last term $\mathbf{r} \times \mathbf{p}$ is nothing other than the orbital angular momentum, L, and therefore SOC is not present in s-orbitals. The SOC interaction Hamiltonian for atomic orbitals is then

$$H_{SOC} = \frac{\mu_B}{\hbar m e c^2}\left(\frac{1}{r}\frac{\partial U}{\partial r}\right)\mathbf{L}\cdot\mathbf{S}. \tag{4.1}$$

While the SOC in solids must ultimately be related to the atomic limit and thus have the same order of magnitude, it is not straightforward to derive a general equation for solids. This is because of the extremely diverse chemical compositions and spatial symmetries in solids. It is however possible to formulate effective models for many general cases, such as the Dresselhaus effect and the Bychkov–Rashba effect [12]. A detailed discussion and derivation of these effects are beyond the scope of this thesis, and we will only use the results throughout the rest of the chapter.

In 3D topological insulators, there are 4 topological invariants, $(\nu_0, \nu_1, \nu_2, \nu_3)$, each taking value 0 or 1, with 1 being topologically nontrivial. Fu, Kane, and Mele came up with a simple method to compute these invariants [13, 14]:

$$(-1)^{\nu_i} = \prod_{j=1}^{n} \delta_j, \tag{4.2}$$

where $n = 8$ for ν_0 and $n = 4$ for the rest, and the product $\delta_{j1} \times \delta_{j2}$ determines the time reversal polarization at time reversal invariant momentum (TRIM) Λ_j. Here, Λ_{j1} and Λ_{j2} are 2 momenta that project onto the same point on a particular surface, on which we are interested in the surface states' spectra. For example, the k-points (1,0,0) and (1,0,1) both project onto the same point on the (0,0,1) surface. We also see that ν_0 is a special invariant, which involves all 8 TRIM in 3D and characterizes a "strong TI." That is, any of the surfaces is guaranteed to host topological surface states, which are also robust against weak time reversal invariant perturbations. If $\nu_0 = 0$, the other 3 invariants can still independently take nonzero values and predict at least one surface with topological states.

Although $\delta_j = \pm 1$ is in general not easy to compute, for lattices with inversion symmetry, δ_j simplifies to the product of the parity eigenvalues of all valence bands evaluated at TRIM Λ_j [13]. Nontrivial surface states arise due to the "band inversion" at TRIM points in the BZ as depicted in Fig. 4.1. The bulk conduction and valence bands are separated by a band gap, as derived from DFT calculations without including SOC. Assuming that the two bands have different parity eigenvalues (with respect to the inversion symmetry operator), denoted by

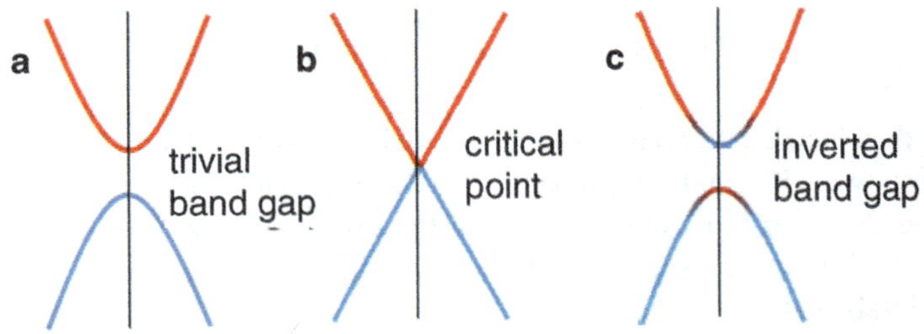

Fig. 4.1 From left to right are illustrations of "band inversion" due to increasing spin orbit coupling strength (figure adapted from [8]). (**a**) Band insulator where the conduction and valence bands with opposite parity eigenvalues are separated by a trivial band gap. With increasing SOC, the gap closes in (**b**) and eventually crosses each other and opens up a nontrivial band gap due to avoided crossing. As a result, the orbital character at the high symmetry point is inverted

red and blue colors, the inclusion of SOC will decrease the band gap at some high symmetry k-points given the proper orbital characters [15]. Once the gap becomes negative, the band crossing points can be hybridized and open up a gap in the bulk due to "avoided crossing." The parity at the high symmetry point is then inverted, giving an additional negative sign to δ_j, which leads to nonzero ν_i. Outside the crystal, the topological invariant is necessarily zero, because vacuum is a trivial "insulator" for electrons. As a consequence of the "bulk–boundary correspondence" [16, 17], the band gap must goes to zero at the bulk–vacuum interface, i.e., gapless surface states are guaranteed by the topology of the bulk bands.

4.1.2 Topological Surface States

It should be emphasized that every bulk material has its surface (boundary) states, which arise simply due to the abrupt change of electronic potential at the bulk–vacuum boundary. This argument applies both to the electronic and lattice degree of freedom, as illustrated in Sect. 4.2, comparing the surface to bulk states. In general, the long lived Bloch states inside the bulk of the crystal are solutions to the periodic Schrödinger's equation. These solutions are no longer valid at the crystal interface due to the breakdown of lattice periodicity [18, 19]. The eigenstates (electronic or phononic) satisfying the surface boundary conditions are localized states at the surface and exponentially decay into the bulk, as illustrated in Fig. 4.2. Namely, the eigen-energies of the surface states are entirely within the bulk band gap. Note that this is very different than a bulk state that extends to the surface. The main difference is that the bulk states will more often than not have finite k_z dependence, while surface states do not. If the surface states are not energetically well-separated from the bulk bands, then the hybridization between surface and bulk bands results in the so-called surface resonance, as illustrated in the third column of Fig. 4.2. Typically

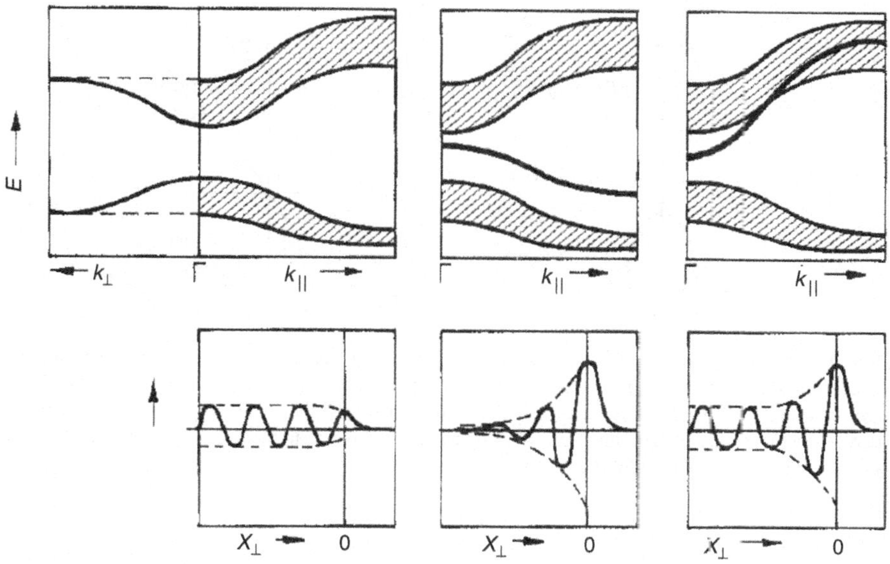

Fig. 4.2 Illustrations of the electron bands/states and wave functions at the crystal–vacuum interface, reproduced from [19]. (Top row) From left to right are the electronic band structures for bulk bands, surface states, and surface resonance, respectively. The shaded areas are the bulk bands of all k_z values projected onto the surface BZ, where k_z is the momenta along the out-of-plane direction. They are reproduced in the center and right columns to illustrate the positions of the bulk bands, where the surface state/resonance is denoted by the thick line. (Bottom row) The electronic wave functions associated with the top band structures

this happens if the surface state and bulk band satisfy certain symmetry constraints (see, for example, [20]). In a surface resonance, the state has finite intensity through the whole bulk but is enhanced at the surface.

The non-topological surface states or resonances are highly sensitive to the details of the crystal's surface structures, such as the termination of the atomic layers, surface reconstruction and dangling bonds. Whereas the topological surface states are a necessity of the material's band topology that does not care about the details of the surface structure. In fact, the topological surface states are extremely robust and not easily destroyed by surface doping, which is useful to distinguish the topological from trivial surface states that could coexist in topological insulators [21, 22].

The 3D TIs and trivial (band) insulators are similar to each other in the sense that both of their Fermi energies (E_F) are within the bulk band gap (Fig. 4.3). But in the case of a 3D TI, there are additional gapless surface states inside the bulk band gap that intersects E_F at the "Dirac point." These additional states support 2D conducting channels at the surface and are protected from nonmagnetic impurity scatterings due to time reversal symmetry [13]. The topological surface states in a 3D TI are described by the Hamiltonian:

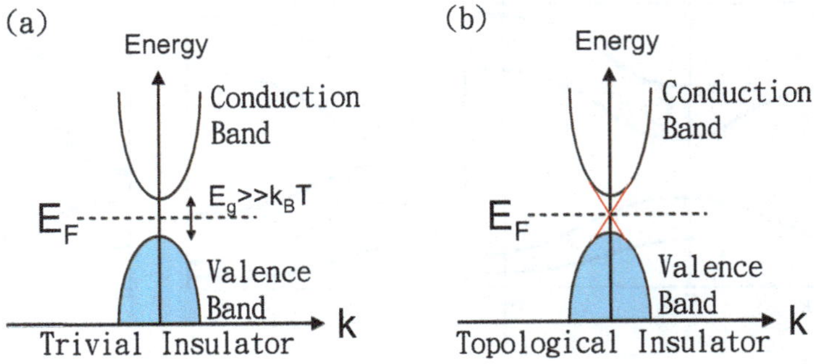

Fig. 4.3 (**a**) Illustration of the typical band structure of a trivial insulator. Colored area denotes bands filled by electrons. Valence band is totally filled, while the conduction band is empty. They are separated by a band gap E_g, which is much larger than the thermal energy $k_B T$. The dash line depicts the position of Fermi level E_F. (**b**) The band structure of an ideal topological insulator. The gapless surface states are denoted by the red lines

$$H(\mathbf{k}) = \frac{k^2}{2m^*}\sigma_0 + v_F \boldsymbol{\sigma} \cdot \mathbf{k}, \tag{4.3}$$

where m^* is the effective mass, v_F is the Fermi velocity, $\boldsymbol{\sigma} = (\sigma_x, \sigma_y, \sigma_z)$ are the Pauli matrices, σ_0 is the 2×2 unit matrix, and $\mathbf{k} = (-k_y, k_x, \frac{\lambda}{2v_F}[k_+^3 + k_-^3])$ with $k_\pm \equiv k_x \pm ik_y$. The z component of $\mathbf{k}$ describes hexagonal warping of the surface states away from the Dirac point [23].

The energy dispersion of the surface states described by Eq. (4.3) is

$$E_\pm(\mathbf{k}) = \frac{k^2}{2m^*} \pm \sqrt{v_F^2 k^2 + \lambda^2 k^6 \cos^2(3\theta)}, \tag{4.4}$$

where ± corresponds to the upper and lower Dirac cones, and θ is the azimuth angle of in-plane momentum $\mathbf{k}$ with respect to the x axis ($\Gamma - K$). To the lowest order expansion in k, the dispersion is linear with the wave vector k, $E = v_F \hbar k$, where v_F is Fermi velocity. Due to its linear dispersion, the *rest mass* of Dirac Fermion near Dirac point is zero. Their wave function must be described by Dirac equation rather than Schrödinger equation. The charge carriers in the surface states are also called "Dirac fermions." One direct consequence of zero effective mass is the well-defined chirality

$$h \equiv \sigma \cdot \mathbf{p} = \pm 1, \tag{4.5}$$

where σ is Pauli matrix and $\mathbf{p}$ is momentum unit vector. For typical particles with nonzero mass, $\mathbf{p}$ depends on the observer's reference frame and may even change sign if the observer is moving faster than the particle. Therefore, chirality is not well-defined for massive particles. But the eigenvalue of chirality operator is irrelevant to the observer's reference frame when acting on wave functions described by Dirac equation. Therefore, the chirality of Dirac fermion is well-defined and the surface states are referred to as "chiral states."

A consequence of well-defined chirality is the suppression of the $2k_F$-scatterings (backscattering). Since chirality is a conserved physical quantity in topological insulators, electrons moving in opposite direction are required to have opposite spins (from Eq. 4.5). Therefore, backscattering is suppressed since it requires higher energy to flip electron spins [24]. This predicts a higher mobility for the conductive topological surface states as compared to usual surface states. However, one should note that the spin-momentum texture does *not* exclude scattering from any other angles, e.g., 175 degrees. To date, the observed electronic transport properties of the topological surface states are in general quite similar to usual 2D electronic states, except for the weak anti-localization [25, 26], and quantum interference effects [27].

Another consequence of the linear dispersion is the π-shift in Berry's phase, γ [28]. We know that electrons in crystals moving in a closed trajectory at constant energy in momentum space gain an additional phase called Berry's phase that is zero in typical conductors and π in massless Dirac materials (without considering spin–orbit coupling). Further consideration of spin–orbit interaction (which is non-negligible in topological insulators) will modify this π shift in Berry's phase [25]. The exact value of this phase shift can be experimentally determined through Shubnikov–de Haas oscillations [26, 29].

4.1.3 The Band Structure in Bi_2Se_3

$Bi_{1-x}Sb_x$, Bi_2Se_3, and Bi_2Te_3 were first theoretically predicted to be 3D topological insulators [13, 15]. Later on, the 2D surface state of $Bi_{1-x}Sb_x$ was experimentally confirmed using Angle Resolved Photoemission Spectroscopy (ARPES) [30]. Soon after, Bi_2Se_3 [31] and Bi_2Te_3 [32] were also experimentally confirmed by ARPES. Calculated band structure of Bi_2Se_3 is shown in Fig. 4.4a, where the warmer color denotes a higher density of states (DOS). The gapless states with linear dispersion can be seen crossing bulk band gap. The center white line denotes the position of Fermi surface. Experimental result of Bi_2Se_3 from ARPES is shown in Fig. 4.4b. The color code denotes ARPES spectra height, which is proportional to electron DOS. The yellow areas are bands with high DOS, indicating the position of bulk bands, and the lower one and upper one are valence band and conduction band, respectively. The white dotted line marks the position of the Fermi level. A red colored, gapless band with almost linear dispersion can be seen inside the bulk band gap. The electron cyclotron mass can be estimated from the energy dispersion in Fig. 4.4b. The cyclotron effective mass is usually defined as [4]

$$m_c = \frac{\hbar^2}{2\pi}\frac{\partial A}{\partial E}, \tag{4.6}$$

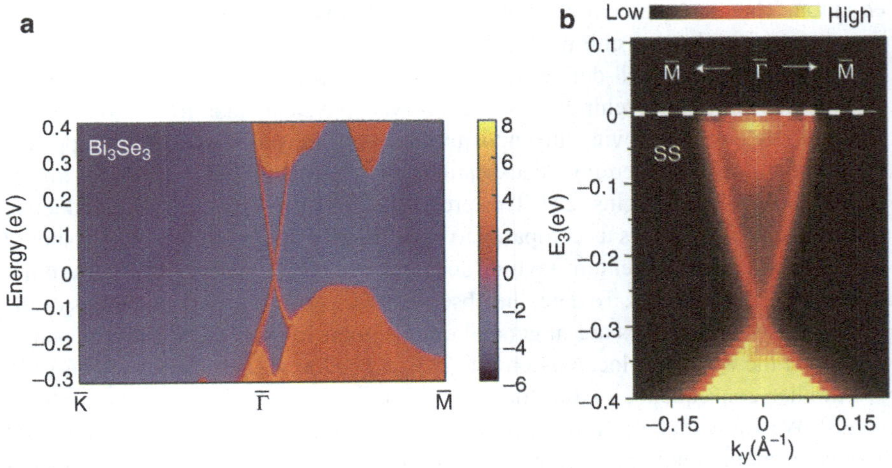

Fig. 4.4 (**a**) Theoretical work based on first principle calculations from [15]. The warmer color indicates higher electron DOS, and the blue regions are bulk band gap. The upper and lower irregular red regions are the bulk conduction and valence bands, respectively. A linear dispersed gapless surface state can be seen inside the bulk band gap. (**b**) The ARPES measurement of surface electronic band structure from [31]. The warmer color denotes higher electron DOS, and the black region is bulk band gap. The upper and lower yellow regions are bulk conduction and valence bands, respectively. The white dash line marks the position of Fermi level. A red colored, gapless band with nearly linear dispersion can be seen inside the bulk band gap

where A is k-space area enclosed by the orbit and E is the energy. Combine the above equation with the linear dispersion relation of Dirac fermions, $E = v_F \hbar k$, it can be shown that $m_c = \hbar k / v_F$. Therefore, the cyclotron mass is zero at Dirac point. The Fermi velocity estimated from Fig. 4.4b is $\sim 4.6 \times 10^5$ m/s, and the cyclotron mass in surface states near E_F is $\sim 0.25\,m_0$ ($k_F \sim 0.1\,\text{Å}^{-1}$ and $E_F \sim 300$ meV), where m_0 is the free electron mass. For the electrons at the bottom of conduction band in Fig. 4.4b (assume parabolic dispersion, $E = \hbar^2 k^2/(2m_c)$), the estimated cyclotron mass $m_c \sim 0.12\,m_0$ for electrons near E_F ($k_F \sim 0.04\,\text{Å}^{-1}$ and $E_F \sim$ 50 meV).

The crystal structure of Bi_2Se_3 is shown in Fig. 4.5 (figure from [15]). The primitive lattice vectors are denoted as t_1, t_2, and t_3. It has layered structure with a triangular lattice within one layer. Each unit cell consists of five-atom layers along the **c**-axis, which is known as a quintuple layer (indicated by the red square). The lattice atoms in a quintuple layer have three different positions, denoted as A, B, and C. The thickness of each quintuple layer is about 1 nm, and the coupling force between quintuple layers is weak (van der Waals type).

One of the most important feature of the surface state is its spin texture. The electron spins of the surface states are predicted to be in-plane and perpendicular to the momentum (see Fig. 4.6a). The red arrows denote spin direction of each point on the intersection of Fermi surface and Dirac cone. Those spin textures have been verified by ARPES (Fig. 4.6b–d). Figure 4.6b is the top view of the Dirac

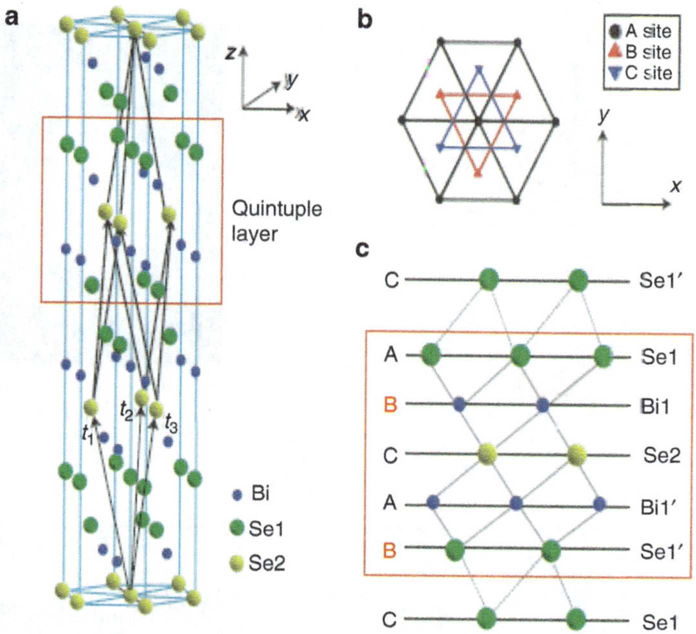

Fig. 4.5 (**a**) Crystal structure of Bi_2Se_3 with three primitive lattice vectors denoted as t_1, t_2, and t_3. A quintuple layer is indicated by the red square. (**b**) Top view along the z-direction. The lattice atoms in a quintuple layer have three different positions, denoted as A, B, and C. (**c**) Side view of the quintuple layer, which is about 1 nm thick. Figure adapted from [15]

cone intersection with Fermi surface, where the yellow arrows are the expected spin orientation. The inner white area is the bottom of bulk conduction band, but a darker ring with radius $\sim 0.1\,\text{Å}^{-1}$ is the intersection of Dirac cone with Fermi level. Figure 4.6c and d shows the measured spin polarizations along the k_x direction (z direction is defined as out of plane). No clear signal can be seen in x and z components. But in the y component, a clear peak of opposite sign and equal value at $k_x \sim \pm 0.1\,\text{Å}^{-1}$ can be seen, which is in accord with the position of Dirac cone edge. This observation agrees well with the theoretical prediction [14]. We can understate this spin texture in a simple way. We know that, due to relativistic effect, moving charge carriers in their rest frame will experience an effective magnetic field $\mathbf{B}_{eff} \sim \mathbf{v}_F \times \mathbf{E}$, where $\mathbf{v}_F$ is Fermi velocity of the charge carriers, and $\mathbf{E}$ is the electric field exerted on the charge carriers in crystal's rest frame. This effect is more pronounced in systems with strong spin–orbit coupling. When considering the carriers confined on the surface of a crystal, those carriers will only experience an effective electric field $\mathbf{E}_s$ along out-of-plane direction from symmetry argument. Therefore, the effective magnetic field will be pointing along in-plane direction and perpendicular to $\mathbf{v}_F$. Electrons moving on the surface will then tend to line up their spins with $\mathbf{B}_{eff}$ in order to minimize Zeeman energy, which is proportional to $\sigma \cdot \mathbf{B}_{eff}$. From this point of view, charge carriers moving in opposite direction

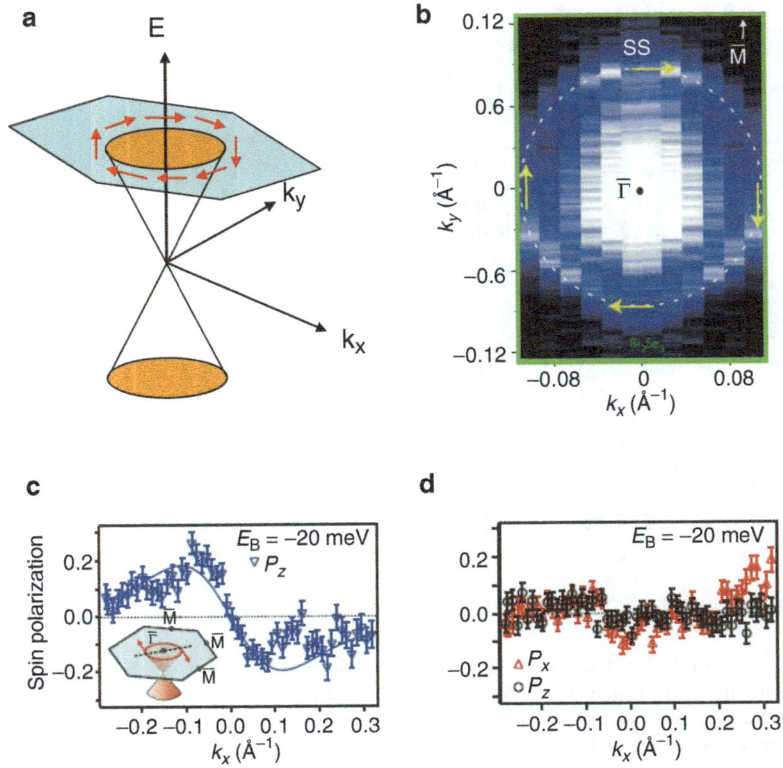

Fig. 4.6 (**a**) Schematic drawing of Bi_2Se_3 surface state spin texture near Fermi energy (denoted by the blue hexagon). The red arrows denote spin direction of each point on the intersection of Fermi surface and Dirac cone. (**b**) ARPES result from [31]. The inner white area is the bottom of bulk conduction band. A darker ring with radius ~0.1 Å^{-1} is the intersection of Dirac cone with Fermi surface. The yellow arrows denote the spin directions. (**c**) and (**d**) are the spin polarizations along k_x, with z defined as the out-of-plane direction. Figure adapted from [21]. No clear signal can be seen in x and z components. But in the y component, a clear peak of opposite sign and equal value at $k_x \sim \pm 0.1$ Å^{-1} can be seen, which is in accord with the position of Dirac cone

will have their spin polarization inverted. Because the effective field is pointing along opposite direction on the opposite surface, spin polarization of charge carriers moving in the same direction but on opposite surface will also be inverted.

Kane and Mele developed an easy way to distinguish nontrivial topological insulators (such as Bi_2Se_3) from trivial ones (band insulator) using a special topological invariant, Z_2 [8, 13]. They found that for a material with Fermi level inside band gap, Z_2 can only be 0 (trivial insulator) or 1 (nontrivial topological insulator). Z_2 can be easily determined by counting the number of Dirac pairs inside the band gap and then take modulo 2. For materials with $Z_2 = 1$, the gapless surface states are protected from weak disorder and nonmagnetic impurities by time reversal symmetry (TRS). For materials with $Z_2 = 0$, the surface states can be easily destroyed by a small perturbation and become trivial insulators.

4.1.4 *Materials and Methods*

Table 4.1 lists 6 Bi_2Se_3 single crystals and films measured in this study. The single crystals were grown by modified Bridgman method. Mixtures of high-purity bismuth (99.999%) and selenium (99.999%) with the mole ratio $Bi : Se = 2 : 3$ were heated up to 870 °C in sealed vacuum quartz tubes for 10 h and then slowly cooled to 200 °C with rate 3 °C/h, followed by furnace cooling to room temperature. The thin film samples were epitaxially grown on Al_2O_3 (0001) substrates in a custom designed molecular beam epitaxy (MBE) chamber [29, 33]. They were immediately transferred into a cryostat after taking out of MBE chamber. Notice that the light penetration depth in Bi_2Se_3 within energy range of current study is about 10 nm [34]. Therefore, the signal is dominated by scattering from the first few QLs of Bi_2Se_3, and the scattering volume in the superlattice samples is practically the same as bulk.

Bi_2Se_3 has a rhombohedral crystal structure with the D_{3d} point group symmetry. The irreducible representations and Raman selection rules are given in Table 4.2. With five atoms in a primitive unit cell, there are a total of three acoustic and 12 optical bulk phonon branches. At the Γ-point, the irreducible representations of the Raman active phonons are $2A_{1g} + 2E_g$, and the infrared active phonons are $2A_{2u} + 2E_u$ [35, 36], where the corresponding atomic displacements are shown in Fig. 4.7. These bulk phonon modes have been measured by Raman and infrared spectroscopies [35–46], and the values reported in [37] and [40] are listed in Table 4.3.

The crystal naturally cleaves along the (111) surface terminated at Se atoms, forming optically flat QLs weakly bonded by van der Waals force [35]. The surface QL has the symmorphic $P6mm$ wallpaper group symmetry (two-dimensional crystallographic point group C_{6v}) [48–50]. Since the surface layer phonon modes in Bi_2Se_3 are not perfectly localized and decay into the bulk, it is more appropriate to analyze our experimental results within the layer group $P3m1$ (crystallographic

Table 4.1 The list of single crystal and films measured in this study

Sample #	Composition	Description	Growth
#2	Bi_2Se_3	50 QL thick film	MBE
#8	$(Bi_2Se_3)_m(In_2Se_3)_n$	50 nm superlattice with (m,n)=(5,5)	MBE
#10	$(Bi_2Se_3)_m(In_2Se_3)_n$	50 nm superlattice with (m,n)=(10,5)	MBE
#13	$Bi_{1.95}In_{0.05}Se_3$	Single crystal with indium doping	Bridgman
#14	Bi_2Se_3	Pristine single crystal	Bridgman
#A	Bi_2Se_3	Pristine single crystal	Bridgman

Table 4.2 The Raman selection rules in the bulk and on the surface of Bi_2Se_3 [51, 52]. Upon the reduction of symmetry from point group D_{3d} to C_{3v}, the A_{1g} and A_{2u} irreducible representations merge into A_1, A_{2g} and A_{1u} merge into A_2, and E_g and E_u merge into E [53]. Note that the surface phonon and electron wave functions are characterized by different symmetry groups [49] and thus follow different selection rules

Scattering geometry	Bulk (D_{3d})	Surface phonon (C_{3v})	Surface electron (C_{6v})
RR	$A_{1g} + A_{2g}$	$A_1 + A_2$	$A_1 + A_2$
RL	$2E_g$	$2E$	$2E_2$
XX	$A_{1g} + E_g$	$A_1 + E$	$A_1 + E_2$
YX	$A_{2g} + E_g$	$A_2 + E$	$A_2 + E_2$

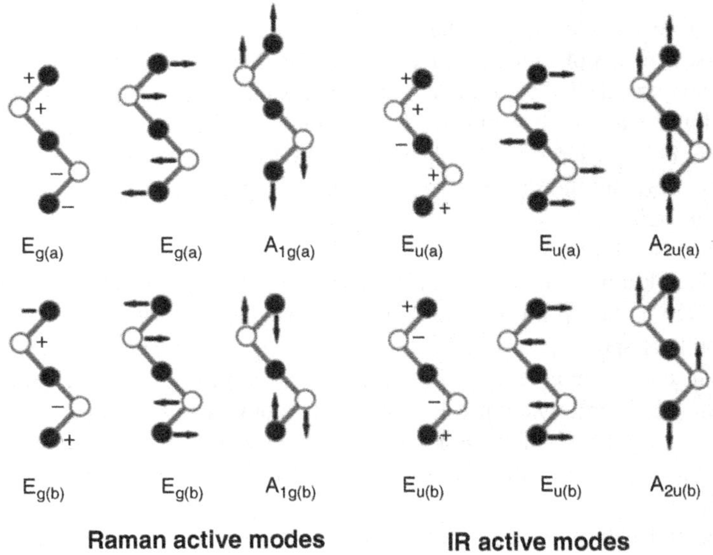

Fig. 4.7 Schematic drawing of the atomic displacements corresponding to the optical phonon modes in Bi_2Se_3 at the BZ center. The black and white spheres correspond to Bi and Se atoms, respectively. The ± signs denotes movements in the direction perpendicular to the paper. Figure adapted from [47]

point group C_{3v}, which is a subgroup containing common symmetry operators of D_{3d} and C_{6v} groups) [49].

All Raman scattering measurements are taken from *ab* surfaces freshly cleaved or grown immediately prior to the measurements. Samples #2–14 are measured in a quasi-backscattering geometry in a continuous He-flow optical cryostat. A glove bag with controlled dry nitrogen gas environment was sealed to the cryostat loading port. After purging the bag to the desired conditions, the single crystals were cleaved in the glove bag immediately before loading into the cryostat for cooling, without exposure to air. We use λ_L =532 nm solid state laser for excitation, where the spot

Table 4.3 The summary of the bulk and surface phonon mode energies. This work's data is compared to the spectroscopic and DFT computation studies reported in the literature. All values are given in units of cm^{-1}

	Experiment		Calculation	
Symmetry	This work	Literature	LDA+SOI [75]	GGA+SOI [47]
$A_{1g}^{(1)}$	75	73 [39, 40, 43, 46]	77	64
$A_{1g}^{(2)}$	180	175 [39, 40, 43, 46]	176	157
$E_{g}^{(1)}$	39	39 [39, 40, 43, 46]	41	39
$E_{g}^{(2)}$	137	133 [39, 40, 43, 46]	139	124
$A_{2u}^{(1)}$	–	N/A	139	137
$A_{2u}^{(2)}$	–	N/A	161	156
$E_{u}^{(1)}$	–	61 [37]	30	65
$E_{u}^{(2)}$	–	133 [37]	131	127
$A_{1}^{(1)}$	60	68 [68]	N/A	N/A
$A_{1}^{(2)}$	136	129 [40]	N/A	N/A
$A_{1}^{(3)}$	158	160 [40, 71]	N/A	N/A
$A_{1}^{(4)}$	173	N/A	N/A	N/A
$E^{(1)}$	67	68 [40]	N/A	N/A
$E^{(2)}$	126	125 [40]	N/A	N/A

size is roughly 50 μm. The scattered light was analyzed and collected by a custom triple-grating spectrometer equipped with a liquid nitrogen cooled CCD detector.

The irreducible representations of the D_{3d} and C_{3v} groups corresponding to these scattering geometries are listed in Table 4.2. Notice that in both the D_{3d} and C_{3v} groups, the phonon intensities do not depend on the orientation of the crystallographic axis. The notations X and Y have no reference to the crystallographic a and b axes.

4.2 Surface Phonons

Theoretical modeling of surface lattice dynamics was first developed by Lifshitz and Rosenzweig [54, 55] and later expanded by various workers [56–59]. The basic idea is to consider the free surface as a perturbation of an infinite lattice and therefore to derive the surface modes from the spectrum of bulk vibrations. As a result, the frequencies of atomic vibration modes at the surface are modified to a smaller value than in the bulk at the Brillouin zone center (Γ point). If there is a gap in the phonon density of state (DOS) and with large enough distortion, the surface phonon DOS can be entirely separated from the bulk [54, 57]. Such modes are long lived and localized at the surface, where the dispersion can be quite different than the bulk [60]. However, it is often experimentally challenging to distinguish surface signal from the overwhelmingly stronger intensity contribution of the bulk. Moreover, if the surface vibration mode is not completely gapped out from the bulk spectrum, then the surface and bulk modes are indistinguishable. Instead, the "bulk phonon" acquires only a slight energy shift near the crystal surface. Notice that the surface modes originate from abrupt termination of the lattice restoring force across bulk–vacuum interface in a semi-infinite crystal and should not be confused with the phonons in quasi-2D ultrathin samples, which are almost decoupled from the underlying substrate of a different material [38, 39, 42, 44, 61].

While the bulk phonon modes have been extensively studied in Bi_2Se_3 single crystals [35–46], only a few papers have reported studies of the surface vibration modes. Zhu and coworkers observed strong Kohn anomaly at about $2k_F$ using helium atom scattering (HAS) [62] and deduced the interaction between surface phonon and the Dirac electrons to be much stronger than the values previously reported by angle-resolved photoemission spectroscopy (ARPES) measurements [63–66], suggesting that the electron–phonon coupling on TI surface may be more complex than anticipated. Time-resolved ARPES study of single crystals reported the observation of one A_{1g} bulk phonon at about 74 cm^{-1}, and an additional mode with slightly lower energy that couples much stronger to the surface states, consistent with what was suggested by transport measurements [67]. This mode was interpreted as a surface phonon associated with the observed A_{1g} bulk phonon [68]. However, alternative results have also been reported [65, 66, 69, 70], suggesting the existence of multiple phononic decaying channels that may depend on the details of sample preparation. Electron energy loss spectroscopy (EELS) study has distinguished a weak mode at about 160 cm^{-1} in Bi_2Se_3, which was assigned to the surface vibration mode associated with an A_{1g} bulk phonon [71]. The Raman scattering work on bulk single crystal [40] and exfoliated nano-crystals reported several additional features, and they were attributed to infrared active phonon modes becoming Raman active due to inversion symmetry breaking at crystal surface [38, 44].

4.2.1 Results

Raman spectroscopy is a conventional tool for studying surface phonon modes [72, 73]. The advantages of Raman spectroscopy is the wider accessible energy range compared with HAS, much higher energy resolution compared to EELS, and the additional information on the symmetries of the phonon modes. We focus our study on the bulk single crystals, which are unexposed to air or any chemicals. In addition to the four Raman active bulk phonons, we observed 6 additional modes with about 20–100 times weaker intensities compared to the bulk phonons (Fig. 4.8). By comparing the data to the results obtained by the complementary spectroscopic techniques and the calculations, we assign the observed additional modes to surface phonons arising from out-of-plane lattice distortion near the crystal–film interface.

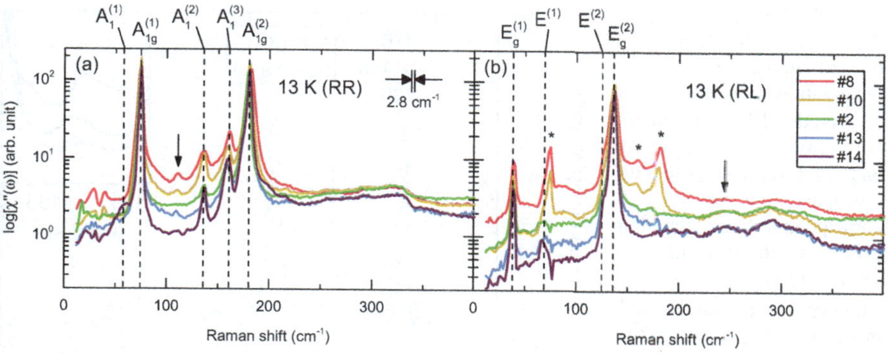

Fig. 4.8 The Raman response function $\chi''(\omega)$ measured in the (**a**) RR and (**b**) RL scattering geometry at 13 K with 532 nm excitation from various Bi_2Se_3 samples as described in Table 4.1, plotted on a semi-log scale. The dashed lines label the observed phonon modes as tabulated in Table 4.3. (**a**) The mode at $110\,cm^{-1}$ indicated by arrow is due to the phonon signal from α-In_2Se_3 layers [74]. The asterisks mark the phonon modes with A_{1g} and A_1 symmetries, which appear in RL geometry due to indium atom diffusion. The instrumental resolution of $2.8\,cm^{-1}$ is shown

Sample Dependences

Figure 4.8 shows the Raman response function $\chi''(\omega)$, taken at 13 K with 532 nm excitation, plotted on a semi-log scale. In order to confirm the tiny features of surface modes, we compared the results from bulk crystals and MBE films. Figure 4.8a and b is measured with the RR and RL scattering geometries, respectively (Table 4.2). The dashed lines label the observed phonons as tabulated in Table 4.3. The strong modes at 72 and $174\,cm^{-1}$ in RR scattering geometry are the bulk A_{1g} phonons of Bi_2Se_3 (Fig. 4.8a), and the strong modes centered at 37 and $132\,cm^{-1}$ in RL are the bulk E_g phonons (Fig. 4.8b), consistent with the previous Raman studies [39, 40] and calculations [75].

The broad feature at about $330\,cm^{-1}$ in RR is possibly due to second-order scattering of the $A_{1g}^{(2)}$ phonon, broadened due to the large downward dispersion of the phonon branch [75]. Similarly, the broad feature observed around $300\,cm^{-1}$ in RL is assigned to two-phonon excitation, $A_{1g}^{(2)} + E_g^{(2)}$. The broad feature at about $245\,cm^{-1}$ (Fig. 4.8b, marked by arrow) was previously assigned to the 2D stretching mode of Se atoms on the surface [76]. However, we do not observe the reported resonance effect of this mode with near-infrared excitation (Fig. 4.9). Notice that this mode's energy is also consistent with the two-phonon excitation of $A_{1g}^{(2)} + E_g^{(1)}$.

In order to distinguish the broad features from electronic origin, such as excitations from the topological surface states, we compared the results with indium doped Bi_2Se_3 in Fig. 4.8. Indium doping was shown to increase the carrier density and suppress the topological surface states in Bi_2Se_3 [29, 77]. Here, we collected data from bulk single crystals and MBE grown In_2Se_3/Bi_2Se_3 superlattices, where indium doping is achieved through diffusion in the superlattices [78]. In all indium

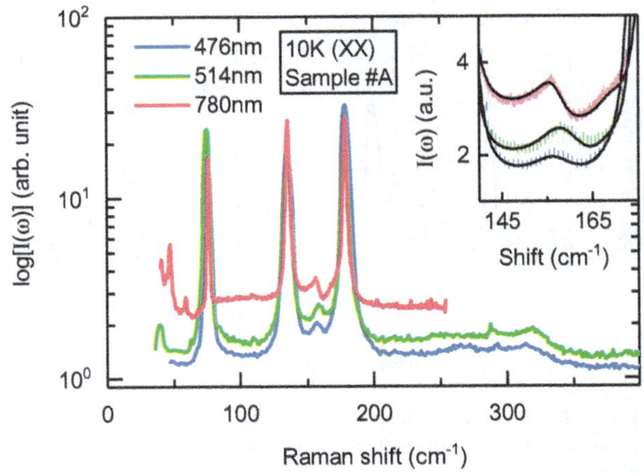

Fig. 4.9 The excitation dependence of scattering intensity in the XX polarization geometry, measured at 10 K from a bulk Bi_2Se_3 single crystal, plotted on a semi-log scale. The blue, green, and pink lines correspond to laser excitation energy of 476, 514, and 780 nm, respectively. Inset: enlarged plot around the $A_1^{(3)}$ mode. The black lines are fit to Fano line shape in Eq. (4.7)

doped samples, the broad features show the same intensity, suggesting their origin unrelated to the topological surface states. This feature is slightly weaker in the superlattice sample #8, despite that the first-order phonon modes are still sharp and strong. However, this is likely mainly due to the indium atom diffusion into the Bi_2Se_3 layer, which breaks the translation symmetry and therefore further broadens the multi-phonon mode. The diffused indium atoms also lower the local crystal symmetry in the Bi_2Se_3 layers, which therefore allows vibration modes with A_{1g} and A_1 symmetries to appear in the RL geometry, which is otherwise forbidden for the crystal symmetry of Bi_2Se_3 (Fig. 4.8b, marked by asterisks). The small feature at 110 cm^{-1} in RR is due to a strong phonon of α-In_2Se_3 layers [74] (indicated by arrow in Fig. 4.8a).

In addition to the strong bulk first-order Raman phonons and the broad features, we see additional sharp modes that are about 20 times weaker than the bulk phonons. In Fig. 4.8a, two such features at 136 and 158 cm^{-1} are seen in all samples in RR scattering geometry, labeled $A_1^{(2)}$ and $A_1^{(3)}$, respectively. In the bulk single crystal sample #14, we observed a mode at about 60 cm^{-1}, which we label as $A_1^{(1)}$. We associate these three features with vibration modes at the crystal surface, to be discussed in the RR polarization for the Sample #14 in the next section. We also noticed several sharp features below 50 cm^{-1} in samples #8 and #10 in RR, which are possibly zone folded phonons. In the RL scattering geometry, we observed two weak features at 67 and 126 cm^{-1}, labeled $E^{(1)}$ and $E^{(2)}$, respectively (Fig. 4.8b). The energies of these modes are close to the strong bulk phonons and therefore require higher resolution to distinguish them.

Excitation Dependences

In Fig. 4.9 are the Raman spectra of the bulk sample at different excitation wavelengths at 10 K. The spectra were obtained in the XX polarization. As in Fig. 4.8, we observe an additional peak at 158 cm^{-1}, which we refer to as $A_1^{(3)}$. However, note that the mode is more asymmetric when 780 nm excitation wavelength is used. While the bulk phonons show little resonance effect, the $A_1^{(3)}$ phonon displays antisymmetric line shape with 780 nm excitation, reminiscent of a Fano line shape [79] (Fig. 4.9, inset). This is an indication that the $A_1^{(3)}$ phonon is interacting with a continuum. This was overlooked in the previous Raman studies and may be related to the 20 meV "kink" in the topological surface state's energy dispersion curve reported by some ARPES measurements [69, 70].

The observation of Fano line shape is a clear evidence for the existence of underlying electronic continuum in the A_1 symmetry channel, which interacts with the $A_1^{(3)}$ phonon [79, 80]. The excitation dependence also suggests resonance enhancement of the electronic continuum with near-infrared wavelength, consistent with the reported surface states at about 1.6 eV above the Fermi energy [81, 82]. Fitting the 780 nm data with Eq. 4.48 in [80]:

$$I(\omega) = \frac{\pi \rho T_e^2 (\omega_0 - \omega - V T_p / T_e)^2}{(\omega_0 - \omega + V^2 R)^2 + (\pi V^2 \rho)^2} \tag{4.7}$$

yields electron–phonon interaction strength $V \approx 2.6\,\text{cm}^{-1}$, and phonon energy $\omega_0 \approx 158\,\text{cm}^{-1}$. Here we assumed that the electron DOS ρ is a constant in the relevant energy window and neglected the real part of the electronic Green's function R. T_p and T_e are the phonon and electronic continuum Raman transition matrix elements, respectively.

Symmetries of the Phonon Modes

To further understand the observed phonon modes, we measure the Raman response in four scattering geometries of the D_{3d} and C_{3v} point group as listed in Table 4.2 (Fig. 4.10a). The intensity contributed by each symmetry channel in different scattering geometries is dictated by the Raman tensors [51, 52], and the results for D_{3d} and C_{3v} groups are listed in Table 4.2. Therefore, by obtaining polarized Raman spectra in four proper scattering geometries, we can separate the measured Raman response from each symmetry channel.

$$\begin{aligned}
\chi''_{A1g}(\omega) + \chi''_{A1}(\omega) &= \chi''_{\text{XX}}(\omega) - \frac{1}{2}\chi''_{\text{RL}}(\omega) \\
\chi''_{A2g}(\omega) + \chi''_{A2}(\omega) &= \chi''_{\text{YX}}(\omega) - \frac{1}{2}\chi''_{\text{RL}}(\omega) \\
\chi''_{Eg}(\omega) + \chi''_{E}(\omega) &= \frac{1}{2}\chi''_{\text{RL}}(\omega).
\end{aligned} \tag{4.8}$$

The results are shown in Fig. 4.10b. We notice that no lattice vibrational mode is observed in the A_{2g} and A_2 symmetry channels. This is because the Raman

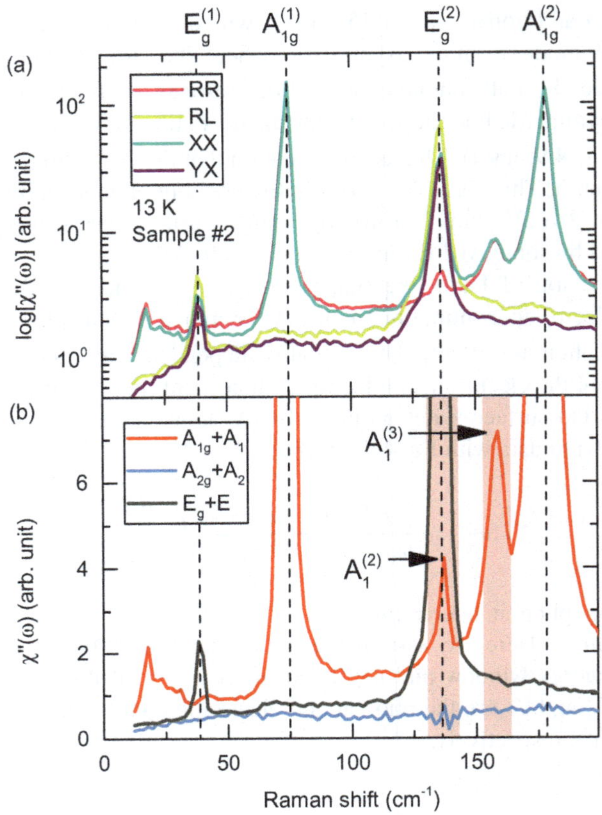

Fig. 4.10 (**a**) The Raman spectra taken in all four scattering geometries at 13 K with 532 nm excitation from a Bi_2Se_3 thick film, plotted on a semi-log scale. (**b**) The Raman response of different symmetry channels, obtained from data in (**a**). The bulk phonons are marked by dashed lines, whereas the surface modes are indicated by arrows and shaded in red

tensors for these two channels are antisymmetric and commonly correspond to pseudovector-like excitations [51, 83, 84], which is forbidden for phononic Raman scattering in Bi_2Se_3. Since the signals in A_{2g} and A_2 channels are expected to be zero, we can claim that all vibration modes appearing in RR have either A_{1g} or A_1 symmetry (Table 4.2).

The $A_1^{(2)}$ mode happens to have energy very close to the $E_g^{(2)}$ phonon, making it particularly difficult for spectroscopic experiments to distinguish. Here, we utilize the symmetry properties to separately detect them with polarized light. The polarization leakage of optical elements is precisely measured and removed (Appendix), and thereby excluding the possibility of $A_1^{(2)}$ being a trivial polarization leakage from the $E_g^{(2)}$ phonon.

To distinguish surface modes that are particularly weak and close in energy to the bulk phonons, we take high resolution spectra from a carefully prepared bulk crystal

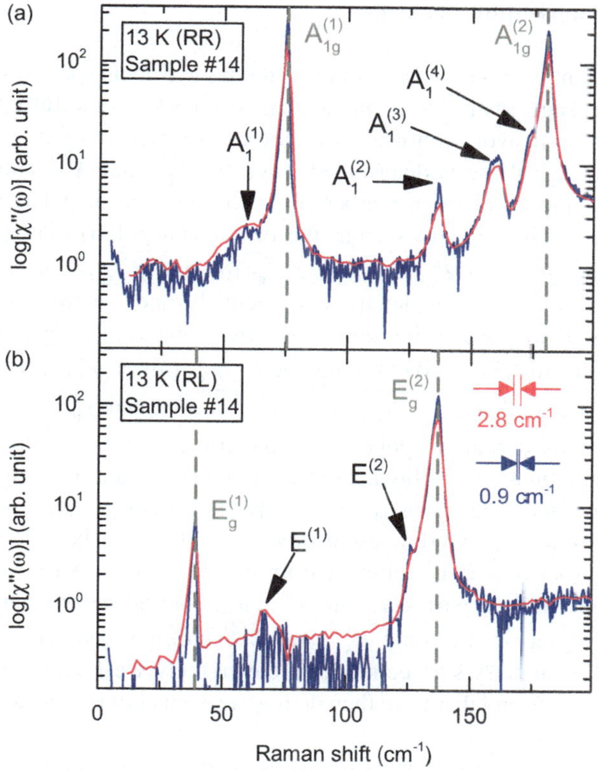

Fig. 4.11 The Raman spectra taken in the (**a**) RR and (**b**) RL scattering geometry at 13 K with 532 nm excitation from a bulk Bi_2Se_3 single crystal are plotted on a semi-log scale. The red and blue curves correspond to instrumental resolution of 2.8 and 0.9 cm^{-1} (as shown in (**b**)), respectively. The bulk phonons are marked by gray dashed lines

#14, cleaved in nitrogen environment. We show in Fig. 4.11 the spectra taken at 13 K in RR and RL scattering geometries, where the smoother low resolution (2.8 cm^{-1}) data is overlapped with the high resolution (0.9 cm^{-1}) spectra. Besides the more pronounced $A_1^{(2)}$ and $A_1^{(3)}$ modes already visible in Fig. 4.10, we see a few additional features in the high resolution data: (1) A mode centered at 173 cm^{-1} appearing as a shoulder to the $A_{1g}^{(2)}$ bulk phonon in RR geometry (Fig. 4.11a), which we designate as $A_1^{(4)}$. (2) Another mode centered at 126 cm^{-1} appearing as a shoulder to the $E_g^{(2)}$ bulk phonon in RL geometry (Fig. 4.11b), which we designate as $E^{(2)}$. (3) The mode $A_1^{(3)}$ shows broadened peak structure. This cannot be due to splitting of an A-symmetry phonon, e.g., lowering of symmetry, since A_1 is a one-dimensional representation. This can be explained as due to Fano interference, which become more pronounced with infrared excitation (Fig. 4.9).

Removal of Polarization Leakage

Due to the weakness of the surface vibration modes in Raman spectroscopy, possible polarization leakage arising from optical elements has to be carefully measured and removed, in order to avoid confusion with the weak surface modes.

The degree of leakage is determined from the $A_{1g}^{(1)}$ and $A_{1g}^{(2)}$ bulk phonons of single crystal samples at room temperature. The removal of polarization leakage is done by subtracting intensity from the orthogonal polarization geometry, i.e., $\chi''_{YX}(\omega) = \overline{\chi''_{YX}(\omega)} - \alpha \cdot \overline{\chi''_{XX}(\omega)}$, where $\overline{\chi''_{YX}(\omega)}$ and $\overline{\chi''_{XX}(\omega)}$ are raw data taken in YX and XX polarization geometries, respectively, and α is the leakage ratio due to the limitations of polarization optics. It is reasonable to suggest that the same ratio also applies to XX polarization geometry: $\chi''_{XX}(\omega) = \overline{\chi''_{XX}(\omega)} - \alpha \cdot \overline{\chi''_{YX}(\omega)}$. Similarly, we have $\chi''_{RL}(\omega) = \overline{\chi''_{RL}(\omega)} - \beta \cdot \overline{\chi''_{RR}(\omega)}$ and $\chi''_{RR}(\omega) = \overline{\chi''_{RR}(\omega)} - \beta \cdot \overline{\chi''_{RL}(\omega)}$ for the circularly polarized geometries, where β is the leakage ratio due to the limitations of the broadband quarter wave plate and alignment of the Berek compensator. The ratios α and β are in general a weak function of ω, but in a narrow energy window as in this study, they can be safely assumed as constants. In order to avoid confusion from contributions of surface phonons, we chose YX and RL geometries as our reference for the determination of α and β. In these two geometries, only $E_g^{(1)}$ and $E_g^{(2)}$ bulk phonons are expected to be present, the E symmetry surface modes are extremely weak and close to the bulk phonons (Fig. 4.11), and therefore they do not raise concern for the determination of α and β.

In Fig. 4.12, we show spectra of unprocessed raw data and polarization leakage removed results taken at 300 K from the *ab* surface of a Bi_2Se_3 thick film in black and red lines, respectively. The leakage intensity of $A_{1g}^{(1)}$ and $A_{1g}^{(2)}$ bulk phonons in raw data taken with YX and RL geometries can be fully removed with leakage ratios $\alpha = 0.004$ and $\beta = 0.015$, respectively. These values are within the specification of used broadband polarization optics.

The value of α depends only on the wavelength of light, and therefore the same value $\alpha = 0.004$ is used for all samples and temperatures measured with 532 nm excitation. The value of β depends critically on the alignment of the Berek compensator, which may vary between experiments and has to be determined using the method described above in each experiment. In this study, the value of β is always within the range 0.015 ± 0.005.

4.2.2 *Discussion*

At the crystal surface of Bi_2Se_3, the lattice structure is distorted along *c*-axis due to the abrupt reduction of the interlayer van der Waals force that binds the crystal together and is calculated by density functional theory (DFT) to be about 10% along *c*-axis [68]. Additionally, the observation of two-dimensional electron gas

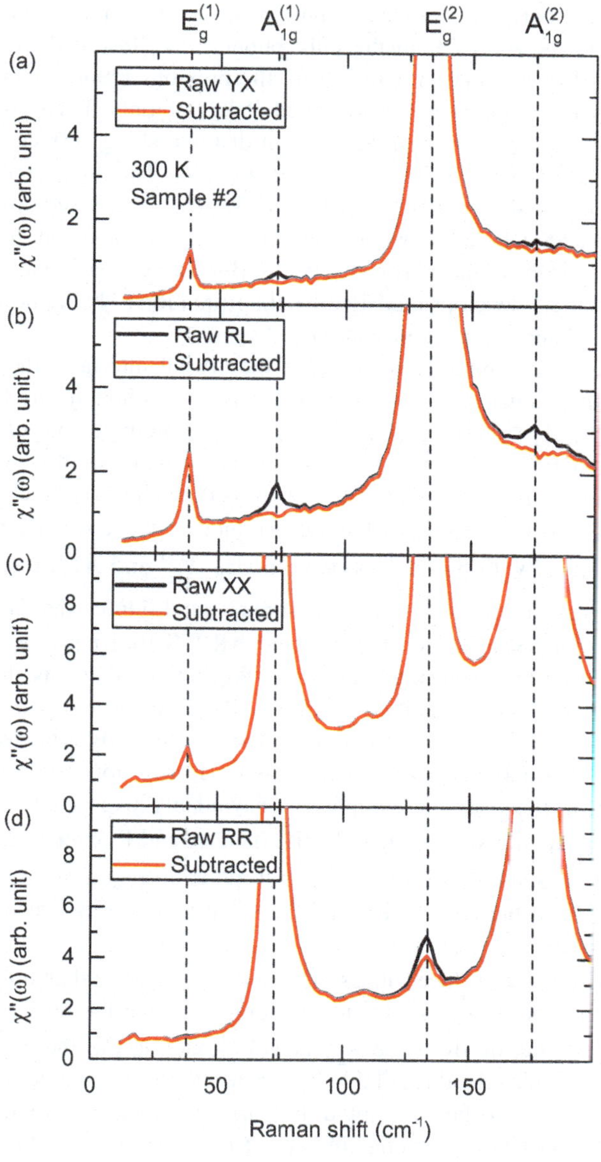

Fig. 4.12 Comparison of raw data and polarization leakage removed spectra, taken in (**a**) YX, (**b**) RL, (**c**) XX, and (**d**) RR polarization geometry from the *ab* surface of sample #2 at 300 K, with 532 nm excitation

formed on Bi_2Se_3 surface also supports the picture of subsurface van der Waals gap expansion [85–87]. However, finite phonon DOS exist across the entire energy range in Bi_2Se_3 [75], allowing the surface modes to decay into bulk phonon modes. Therefore, the surface mode is not entirely "peeled off" from the bulk bands. Then the hybridization between surface and bulk bands results in a "surface resonance" with slightly lower energy than the bulk phonon, as illustrated in Fig. 4.2. This is markedly a different interpretation from the previous reports, where additional phonon peaks were observed in exfoliated flakes of only a few QLs thickness. In these thin samples, there is little sense in distinguishing surface from bulk, as signal from each layer contributes to distinct peaks. This is well-studied in few layers of graphite and graphene samples [88]. We can clearly see this effect in our thinnest sample #8, where the phonons are broadened to the extent that one cannot distinguish bulk from surface resonance peaks. But with increasing thickness, a clear intensity "shoulder" develops that signifies the formation of bulk phonon bands and the separation of surface resonance at the BZ center.

Due to inversion symmetry breaking at the crystal interface, the surface resonances from the Raman active A_{1g} and IR active A_{2u} phonons are both expected to appear in the A_1 symmetry (C_{3v} group), corresponding to out-of-plane atomic motion. The energies of such surface modes are usually slightly lower than the corresponding bulk phonons. This is consistent with the four A_1 modes we observed (Fig. 4.11a). From the energies of these A_1 modes, we conclude that $A_1^{(1)}$ and $A_1^{(4)}$ are associated with the bulk phonon modes $A_{1g}^{(1)}$ and $A_{1g}^{(2)}$, respectively. The measured energy of the $A_1^{(1)}$ mode is somewhat different than the previously reported value of 68 cm^{-1} by time-resolved ARPES [68] but close to what was suggested by transport measurements [67]. We believe that this difference may be partly due to surface quality variation. ARPES measured sample is usually cleaved in ultrahigh vacuum, whereas the surface in this study is cleaved in nitrogen environment. This may also explain why this mode was not observed in the MBE samples (Fig. 4.8), where the sample is unavoidably exposed to air for a few minutes during the transfer between MBE chamber and Raman cryostat. The $A_1^{(4)}$ mode appears as a shoulder to the $A_{1g}^{(2)}$ bulk phonon, requiring higher resolution to distinguish from the bulk mode, and therefore was overlooked in the previous Raman study [40].

In comparison, the surface modes $A_1^{(2)}$ and $A_1^{(3)}$ have higher intensity and are better resolved. One possibility for this difference is that the bulk counterparts of these resonances are the IR active $A_{2u}^{(1)}$ and $A_{2u}^{(2)}$ phonons, as the measured energy is close to the calculated values (Table 4.3). Since these bulk modes are not Raman active, we were able to better resolve the surface resonance. Another possibility is that the phonon DOS is practically zero at these energies in the A_1 symmetry channel, and the surface vibration modes are truly localized. Distinguishing these two scenarios is in fact experimentally nontrivial, especially since the experimental values of the $A_{2u}^{(1)}$ and $A_{2u}^{(2)}$ bulk phonon energies are yet unknown.

Since the in-plane symmetries are mainly preserved as the DFT calculated atomic surface distortion is purely out of plane [68], one would not expect surface phonon with E symmetry (C_{3v} group) for Bi_2Se_3. However, the in-plane bonding potential is also modified by having distortion along c-axis, and therefore the phonon frequency at surface is still slightly different than the bulk value. If the modification is tiny, the E modes are expected to be weak and close to the bulk phonons. In Fig. 4.8b and 4.11b, we can see hints of two additional modes, labeled by $E^{(1)}$ and $E^{(2)}$. The energies of these modes are in fact close to the measured values of $E_u^{(1)}$ and $E_u^{(2)}$ bulk phonons [36, 37] and are consistent with the previous Raman study [40] (Table 4.3). However, the frequency of E_1 is slightly higher than $E_u^{(1)}$, which is against the expectation from a surface resonance. This may reflect the fact that this is an in-plane mode, orthogonal to the lattice distortion direction. Or, this may be indicative of nontrivial electron–phonon interaction with the surface states, and worth further studying.

4.3 Chiral Spin Modes

Ferromagnetic systems and partially spin-polarized Fermi liquids support collective spin excitations (spin waves), in which all electron spins respond coherently to external fields, and the "glue" that correlates the phases of individual spins is provided by the exchange interaction. In nonmagnetic materials where inversion symmetry is broken but time-reversal invariance remains intact, strong spin-orbit coupling (SOC) may play the role of an effective magnetic field, which locks electron spins and momenta into textures. This phenomenon is encountered in three-dimensional (3D) topological insulators (TIs), which harbor topologically protected surface states [13, 15, 30, 32, 89]. The surface band structures of TIs have been a focus of intense studies, both from the fundamental point-of-view [8, 90–95] and for potential spintronics device applications [96–103]. However, the many-body interactions leading to collective effects still remain largely unexplored. An essential aspect of this physics is an interplay between the Coulomb interaction and SOC, which is expected to give rise to a new type of collective spin excitations—chiral spin waves [104–110].

The spectrum of topological surface states (Eq. 4.4) consists of hexagonally warped electron- and hole-like Dirac cones of opposite chirality. A light-induced excitation from the occupied state in the hole cone to an empty state in the electron cone is accompanied by a spin-flip of the quasiparticle (Fig. 4.13b). Such direct transitions form a continuum that starts at the threshold energy ω_- (Fig. 4.13c) [111, 112].

Due to the Pauli exclusion principle, two electrons in the triplet state avoid each other, thus reducing the energy of the Coulomb repulsion. Therefore, the repulsive Coulomb interaction between electrons translates into an attractive exchange interaction between their spins, leading to bound states below the continuum of spin-flip excitation, i.e., chiral spin modes. In general, there are three such modes (red curves in Fig. 4.13c), which correspond to linearly polarized oscillations of the magnetic moment in the absence of the external magnetic field [106, 107, 109]. At $q = 0$, there is a doubly degenerate mode with an in-plane magnetic moment and a transverse mode with an out-of-plane moment, with energies $\omega_{s,||}$ and $\omega_{s,\perp}$,

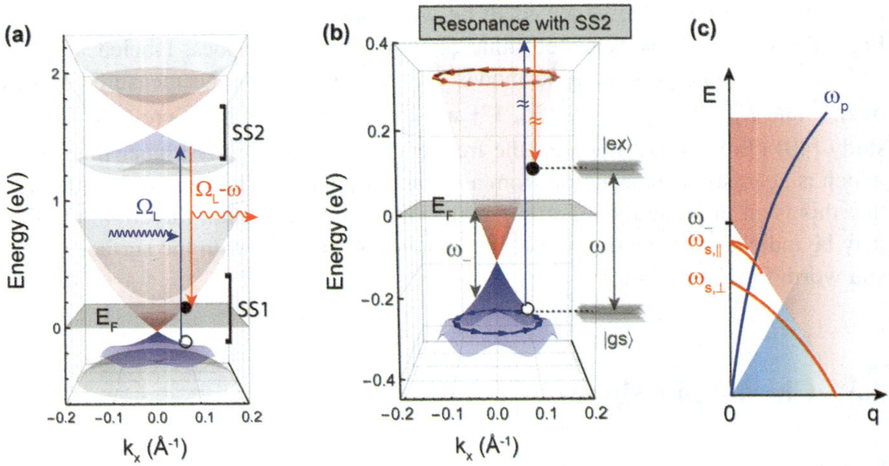

Fig. 4.13 (**a**) The band structure of Bi_2Se_3 around the Brillouin zone center, reconstructed from ARPES results in [81, 113]. The low-energy Dirac cones of surface states (SS1) are described by Eq. (4.4) with parameters $m^* \approx 0.066\,\text{eV}^{-1}\text{Å}^{-2}$, $v_1 \approx 2.4\,\text{eVÅ}$, and $v_w \approx 25\,\text{eVÅ}^3$. The upper and lower Dirac cones of opposite chirality are shown in red and blue, whereas the bulk bands are shown in gray. A pair of unoccupied surface Dirac cones (SS2) resides about 1.8 eV above SS1 [81]. The arrows illustrate a resonant Raman process stimulated by an incoming photon with energy Ω_L, in which an electron is promoted from the lower to the upper Dirac cone of SS1, through resonant scattering via the intermediate states, SS2. The energy difference between the excited and ground state defines the Raman shift, ω. (**b**) Enlarged view of SS1 around E_F with arrows showing the spin textures, where ω_- is the threshold energy for direct transitions between occupied states in the lower Dirac cone and available states of the upper cone. (**c**) Schematic dispersions of the particle–hole continua in the spin (red) and charge (blue) channels, surface plasmon [114] (blue line), and collective spin modes [106] (red lines)

respectively. Because the chiral spin modes are below the continuum, they are expected to remain long-lived even at elevated temperatures [115]. These modes are in essence (zero-field) spin waves that can be measured by resonant Raman scattering because they couple to the electromagnetic field through antisymmetric Raman tensors [109].

4.3.1 Results

As-grown Bi_2Se_3 is usually electron doped due to naturally formed Se vacancies [116]. In this study, we use characterized samples with low carrier concentration. The Fermi energy (E_F) is determined by scanning tunneling spectroscopy to be about 150 meV above the Dirac point of SS1 (Fig. 4.13b) [117].

The polarized Raman spectra were acquired in a quasi-backscattering geometry from the *ab* surface of Bi_2Se_3 single crystals grown by modified Bridgman method.

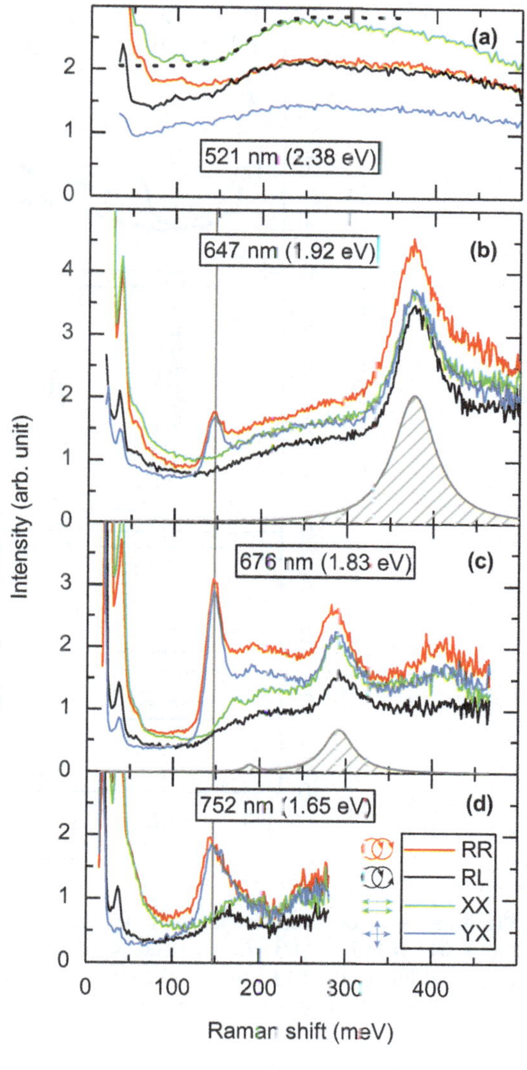

Fig. 4.14 The intensities of secondary emission are measured in circular and linear (with respect to crystallographic axes) scattering polarization geometries, as shown pictorially next to the figure legend in panel (**d**), at 24 K with (**a**) 521, (**b**) 647, (**c**) 676, and (**d**) 752 nm excitations. The vertical solid gray line indicates the sharp 150 meV peak, most pronounced for the incident photon energy of 1.83 eV, which is the closest to the energy difference between SS1 and SS2 (Fig. 4.13a). The dotted black line in (**a**) is a guide to the eye showing the energy threshold for the surface-to-bulk excitations. The hatched areas in (**b**) and (**c**) are Lorentzian fits to the excitonic photoluminescence peaks

We use 521, 647, 676, and 752 nm lines of a Kr^+ laser for excitation, where the spot size is roughly $50 \times 50\,\mu m^2$ and the power is about 10 mW. The scattered light is analyzed by a custom triple-grating spectrometer.

Excitation Profile

In Fig. 4.14, we show spectra of secondary emission for four scattering geometries employing both linear and circular polarizations. Of the four excitations, the 521 nm one is the farthest from near-resonant transition between SS1 and SS2, while the

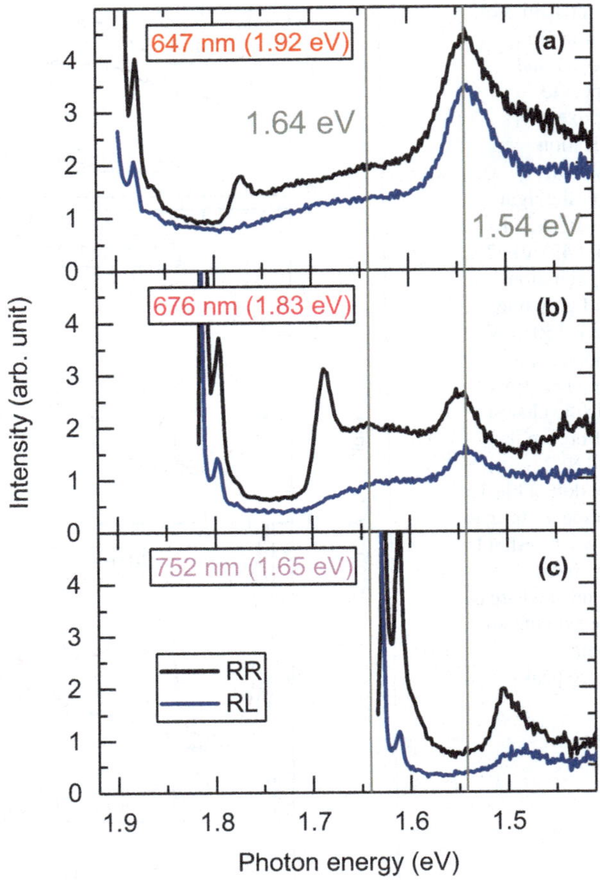

Fig. 4.15 Secondary emission intensity measured in RR and RL polarization geometries at 24 K for (**a**) 647, (**b**) 676, and (**c**) 752 excitation energies, plotted against scattered photon energy. The gray solid line marks 1.54 and 1.64 eV, where the exciton peaks centers' coincide for 647 and 676 nm excitations, indicating that the peaks are due to photoluminescence of excitonic states

676 nm one is the closest. The spectra contain contributions from electronic Raman scattering and exciton photoluminescence. The latter is present for all polarizations and can be subtracted from the spectra. The signal below 50 meV is dominated by phonon modes that are discussed elsewhere [1].

Luminescent Background Subtraction

Figure 4.15 shows the intensity of secondary emission measured with RR and RL polarizations at 24 K for 647, 676, and 752 excitation energies, plotted as function of emission photon energy. The exciton emission centers at 1.54 eV for 647 and

676 nm excitations and has about the same intensity for both RR and RL scattering geometries. Another small peak is at 1.64 eV for both excitation and is likely another state related to the 1.54 eV exciton.

In typical materials, exciton states have energies lower than the interband transition edge by the amount of the binding energy. The excited electron–hole pairs relax to the exciton states, which eventually decay back to the ground state through photon emission if symmetry allows. In some cases, if the photon emission is not allowed and the exciton is said to be "dark." The exciton peaks are absent for 752 nm excitation spectra, suggesting that the corresponding interband transition has a threshold of about 1.8 eV, which puts the binding energy to about 0.3 eV.

Bi_2Se_3 is commonly understood as a small gap semiconductor with the Fermi surface close to the bulk conduction band edge, therefore the observation of a well-defined exciton peak in the near-infrared range is quite surprising, suggesting an optical band gap of about 1.8 eV, much larger than the semiconductor gap. Since direct gap is required for exciton radiative emission, the most probable interband transition is between the M-shaped bulk valence band and a W-shaped conduction band about 1.5 eV above the Fermi energy [81] (Fig. 4.13a). The exciton most likely forms below the band minima at $|k| \approx 0.1\,\text{Å}^{-1}$.

To remove photoluminescence background from the measured spectra, we fit the 1.54 and 1.64 eV exciton peaks with a Lorentzian function, as shown by the hatched peaks in Fig. 4.14. We also subtract a small constant background from all spectra to account for other photoluminescence contribution.

Temperature Dependence

The most interesting feature of the spectra is a sharp peak around 150 meV observed in the XY and RR geometries. The peak is the strongest for the 676 nm excitation, which is in near resonance with the transition between SS1 and SS2, thus confirming the surface origin of the observed signal.

In order to better understand the origin of the 150 meV peak, we subtract the photoluminescence contributions and then utilize the symmetry properties of the Raman tensors to separate the measured spectra into the E_2, A_1, and A_2 symmetry channels of C_{6v} point group (Table 4.2). In Fig. 4.16, we plot the temperature dependence of Raman response $\mathcal{R}(\omega, T)$ in three symmetry channels. It is clearly seen that the 150 meV peak is associated with the A_2 symmetry channel. The continuum broadens and becomes invisible above 150 K, but the peak is still well-defined even at $T = 300$ K. The temperature dependences of the chiral spin mode intensity (area under the peak), FWHM, and energy (peak center) are shown in Fig. 4.17. The parameters are obtained from fitting the data to the Lorentzian line shape. The width of the spin mode saturates below about 150 K, indicating that the main damping mechanism of the spin mode is likely due to disorder. In general, the energy and width of the chiral spin mode is well-described by the theoretical simulation that incorporates finite temperature effects only via thermal smearing of the Fermi functions and neglects inelastic damping mechanisms. The decreasing

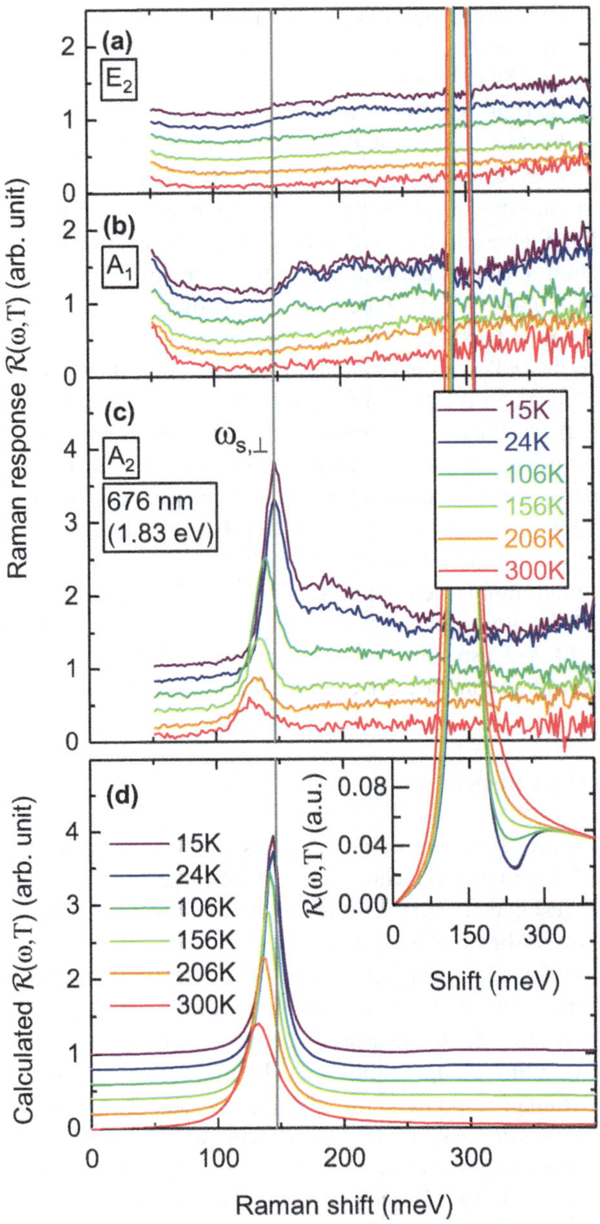

Fig. 4.16 Temperature dependence of the symmetry-separated Raman response $\mathcal{R}(\omega, T)$ in the (**a**) E_2, (**b**) A_1, and (**c**) A_2 symmetry channels, measured with the near-resonant 676 nm excitation. The photoluminescence background was subtracted from the raw data. The gray solid line locates the position of the sharp peak that is present only in the A_2 channel. (**d**) Calculated $\mathcal{R}(\omega, T)$ in the A_2 channel with the dimensionless interaction coupling constant $u \approx 0.6$ and impurity broadening of $\Gamma = 8$ meV was extracted from the data and used in the calculation. The lines are shifted vertically for clarity. Inset: Zoom-in of the calculated $\mathcal{R}(\omega, T)$ without vertical shift, showing the spin-flip continuum with a threshold energy of 260 meV

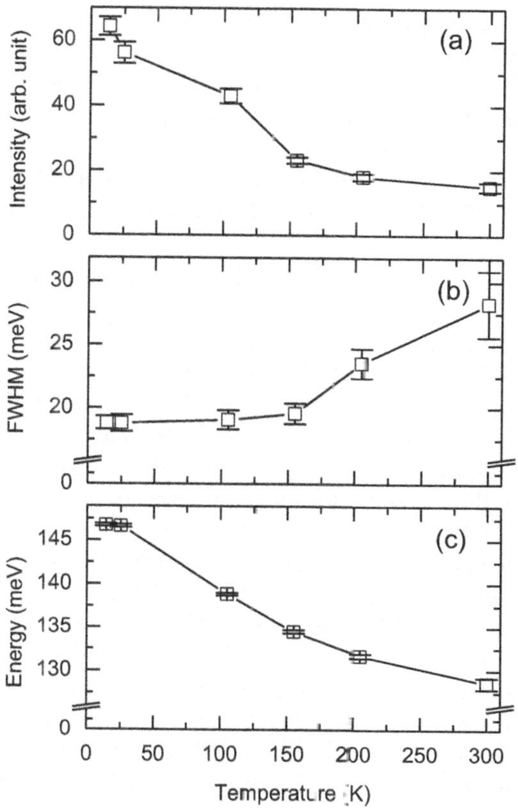

Fig. 4.17 Temperature dependence of (**a**) the intensity, (**b**) full width at half maximum (FWHM), and (**c**) peak center of the chiral spin mode in Fig. 4.16 by fitting to a Lorentzian line shape. The error bars reflect one standard deviation of the fit

intensity at elevated temperatures is not captured in the above temperature effects and is likely an indication of unaccounted spin decay channels, such as through the surface phonons [1].

4.3.2 Discussion

The basis functions of the A_2 representation of C_{6v} transform as the z component of the angular momentum, which is a pseudovector [53]. This suggests that the observed peak in the A_2 channel corresponds to a spin mode (marked by $\omega_{s,\perp}$ in Fig. 4.13c) with an out-of-plane magnetic moment (also a pseudovector).

To quantify the assignment of the 150 meV peak to the out-of-plane chiral spin mode, we calculate the Raman response of surface chiral states. We are interested in spin-flip resonant Raman processes between states near the Fermi level in SS1 and the states in SS2. Two resonance transitions are possible: an electron from the lower cone of SS1 can be transferred into either the lower cone or upper cone of SS2 and can come back to the upper cone of SS1, producing a particle–hole pair.

A characteristic feature of Bi_2Se_3 is that the lower cones of SS1 and SS2 are almost perfectly nested: a fit to the ARPES data in [81, 113] gives $v_1 = 2.4\,\text{eVÅ}$ and $v_2 = 2.0\,\text{eVÅ}$. Therefore, the "hole-to-hole" transition in the Raman vertex is essentially dispersionless, and the corresponding transition probability is enhanced by a factor of $\approx 1/(E_g - \Omega_L)^2$, whereas the "hole-to-electron" term is small. This explains why only one resonance is observed in the experiment.

The Raman response function in the A_2 symmetry channel can now be written as

$$\mathcal{R}(\omega, T) \propto \chi''_{zz}(\omega, T)/(E_g - \Omega_L)^2, \tag{4.9}$$

where χ''_{zz} is the imaginary part of the zz component of the spin susceptibility tensor and has a continuum of spin-flip excitations and a pole corresponding to a transverse collective mode, $\omega_{s,\perp}$ (Fig.4.13). A simple result for the frequency of this mode can be obtained if one neglects hexagonal warping and considers the weak-coupling limit. In this case, $\omega_{s,\perp} = 2E_F\left[1 - 2\exp(-4/u)\right]$, where $u \equiv UE_F/2\pi\hbar v_1^2 \ll 1$ is the dimensionless coupling constant [2].

For a more general case, which includes the realistic band structure and finite temperature, the Raman response has to be evaluated numerically. The results of this calculation are shown in Fig. 4.16d. With the band structure parameters obtained from ARPES measurements [113], the only two fitting parameters are the exchange coupling constant, fixed at $u \approx 0.6$ to reproduce the mode frequency at 15 K, and the impurity scattering rate chosen as $\Gamma = 8$ meV. Comparison of the measured and computed spectra (Fig. 4.16c and d, correspondingly) shows that the model well describes not only the overall shape of the signal but also its evolution with the temperature. In the sample measured, the threshold for the spin-flip continuum is supposed to be at $\omega_- \approx 260\,\text{meV}$. However, the onset of this continuum is difficult to observe because its spectral weight is transferred into the collective mode; in the inset of Fig. 4.16d, we show a zoom into the computed crossover region between the collective mode and the continuum.

A quantitative agreement between the theory and experiment gives us confidence in that the observed 150 meV sharp peak in the A_2 symmetry channel is indeed a transverse chiral spin mode. The fact that the measured intensity decreases with increasing temperature faster than the calculated spectra is an indication of unaccounted spin decay channels at elevated temperatures, e.g., through interaction with surface phonons [1].

Transitions Between Surface States and Bulk Bands

In Fig. 4.14, the spectrum for the non-resonant 521 nm excitation shows no sharp peaks but a broad feature with an onset at about 200 meV. This energy scale corresponds to the onset of transitions between the surface Dirac cone and the bulk states. Such transitions do not need to flip spins and thus appear also in the XX scattering geometry, which contains the fully symmetric channel.

Figure 4.18 is a realistic illustration of the Bi_2Se_3 band structure reconstructed from ARPES measurements [113]. The surface state dispersion is [23]

$$E_{SS}(\mathbf{k}) = \Delta + \frac{k^2}{2m^*} \pm \sqrt{v_F^2 k^2 + \left(\frac{2v_3}{v_F}\right)^2 k^6 \cos^2 3\theta}$$
$$\approx \Delta + \frac{k^2}{2m^*} \pm v_F k + 2\left(\frac{v_3^2}{v_F}\right) k^5 \cos^2(3\theta) \qquad (4.10)$$

where $\pm$ denote the upper and lower Dirac cones and θ is the azimuth angle of momentum $\mathbf{k}$ with respect to the x axis $(\Gamma - K)$. Fitting data in [113] to Eq. (4.10) gives $m^* \approx 0.066\,\text{eV}^{-1}\text{Å}^{-2}$, $v_F \approx 2.4\,\text{eVÅ}$, and $v_3 \approx 25\,\text{eVÅ}^3$. One can readily see that the energy of a direct transition from the lower to upper Dirac cone is $2\sqrt{v_F^2 k^2 + (\frac{2v_3}{v_F})^2 k^6 \cos^2(3\theta)}$. In samples measured, Δ is determined by tunneling spectroscopy [117] to be about $-150\,\text{meV}$, and therefore the Fermi momentum $k_F \approx 0.054\,\text{Å}^{-1}$ along k_x, thus resulting in a threshold energy $\omega_- \approx 260\,\text{meV}$.

The direct transition energy between SS1 and the bulk conduction band is given by $\epsilon(\mathbf{k}) = E_{CB}(\mathbf{k}) - E_{SS}(\mathbf{k})$, where $E_{SS}(\mathbf{k})$ is given by Eq. (4.10), and the bulk conduction band dispersion follows a quasi-2D parabolic model [118]:

$$E_{CB}(\mathbf{k}) = E_0 + \frac{k_{||}^2}{2m_{||}^*} + \frac{k_\perp^2}{2m_\perp^*}, \qquad (4.11)$$

where $E_0 \approx 130\,\text{meV}$ is determined by E_F and the relative position between SS1 and bulk conduction band minimum [113, 117], $m_{||}^* \approx 0.03\,\text{eV}^{-1}\text{Å}^{-2}$ is the in-plane effective mass, determined from fitting the ARPES data in [113] to quadratic

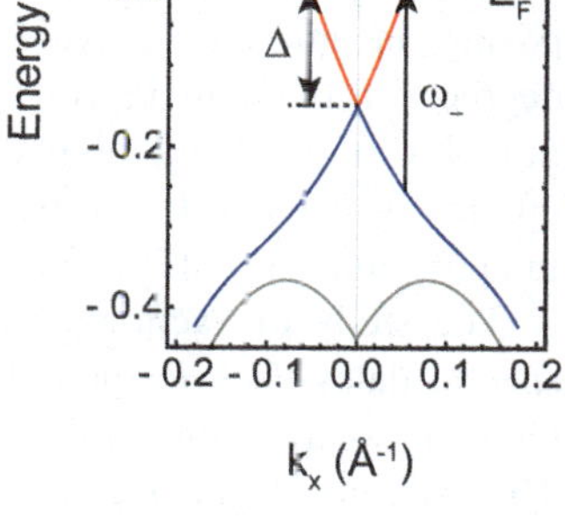

Fig. 4.18 Transitions between surface states and bulk conduction band. Band structure near the Γ point and Fermi surface, reconstructed from ARPES measurements [113]. The blue and red lines denote the lower and upper Dirac cones, respectively, whereas the bulk bands are shown in gray. In the measured sample, the Dirac point is about 150 meV below the Fermi energy E_F

dispersion. In the measured sample where $k_F \approx 0.054\,\text{Å}^{-1}$ along k_x, the threshold energy $E_{\min} \approx 180\,\text{meV}$, similar to what was observed in Fig. 4.14a.

4.4 Photoluminescence

As we have demonstrated in the previous section, spin–orbit coupling (SOC) can play a central role in emerging phenomena in condensed matter physics, e.g., 3D topological insulators (TIs) [13, 15, 21, 32], valleytronics [119–121] and Ising superconductivity [122, 123]. In this section, we show surprising results where two photoluminescence (PL) peaks are observed in the visible range in Bi_2Se_3. Figure 4.20a shows the right and left circularly polarized (CP) secondary emission spectra, indicated by $I_{RR}(\omega, T)$ and $I_{RL}(\omega, T)$, respectively. At low temperature, two PL peaks were detected in the visible range, where the peak center energies remain fixed with changing excitation. While broad PL features are not uncommon even in metals, a well-defined PL peak usually requires a minimum in the *e*–*h* band, which is unexpected to appear in the visible range for a small gap semiconductor, such as Bi_2Se_3.

The peak structure in the PL spectra is usually an evidence of fixed energy *e*–*h* bound states, known as excitons, that couple to the electromagnetic field through radiative decays. Due to phase relaxation during the thermalization of *e*–*h* pairs and the formation of excitons, the PL is usually unpolarized.

4.4.1 Circular Polarized Photoluminescence

Polarized PL is an indication of excitons favoring certain "optical orientation" than the other, which is usually due to underlying symmetry breaking in the system, leading to removal of band degeneracies. For example, spin polarization of electron bands can be induced by SOC through Rashba [124] and Dresselhaus [125] effects in solids without inversion symmetry, or in crystals with specific atomic site asymmetries [126]. The spin polarized energy bands lead to various novel collective phenomena, e.g., chiral spin modes [115] and chiral excitons [127], posing an emerging playground for studying interplay between spin and other degrees of freedom in solids [120]. In particular, the chiral excitons in TMD monolayers display a high degree of "optical orientation," allowing them to be efficiently and selectively excited by CP photons [120, 128, 129], with applications in optospintronics [129, 130].

The study of exciton's "optical orientation" through measuring polarized PL spectra dates back to the early 1970s [131–133], as an efficient way of probing the spin polarizations of electron bands in III–V and II–VI semiconductors [134]. The excitons originating from spin polarized electrons and holes are characterized by its azimuth angular momentum, $S_z = \pm 1$. The oppositely oriented excitons are thus independently excited by circularly polarized photon of opposite helicity [135]. Since the electronic bands are spin polarized, the exciton orientation is preserved during thermalization, eventually leading to CP PL at electron–hole (e–h) recombination. Therefore, studying the polarized PL spectra is particularly interesting in a TI material, where information of the surface state spin dynamics is particularly important due to its potential application in spintronics [96, 103].

4.4.2 Results

In Bi_2Se_3, the large Rashba SOC on the surface results in three spin polarized bands at the Brillouin zone center (Γ-point), shown as red lines in Fig. 4.19a: the topologically protected surface Dirac cone at the Fermi level (SS1) [21, 89], the

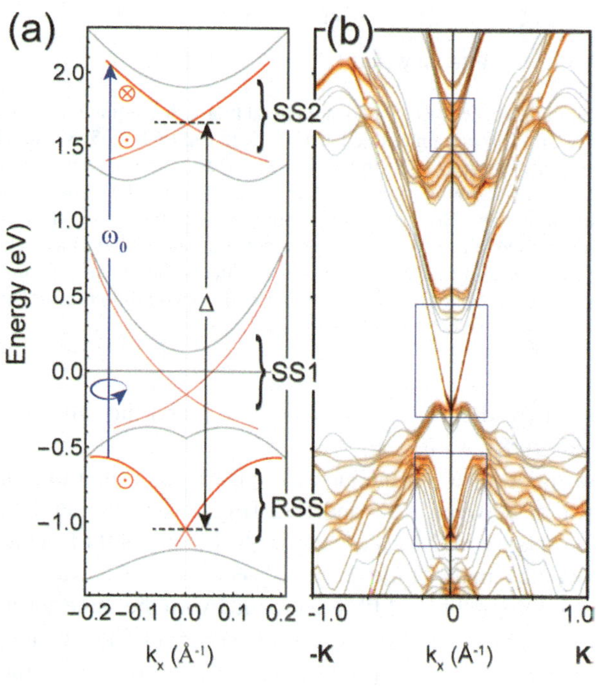

Fig. 4.19 The band structure around BZ center in Bi_2Se_3. (**a**) The electronic band structure near the Γ point was inferred from ARPES measurements and DFT calculations [81, 87]. The Rashba surface states (RSS) and the unoccupied topological surface states (SS2) are in red, with the in-plane spin orientations denoted by ⊙ and ⊗. The topological surface states (SS1) near Fermi energy (E_F) are shown in light red. The bulk bands do not contribute to the CP PL and are shown in gray. (**b**) DFT calculated band structure in a slab geometry with 32 Åvacuum layer, shown along the Γ-K cut of the Brillouin zone, with spectral weight projected onto the top quintuple layer for $J_z = 1/2$. The blue squares highlight the three surface bands labeled in (**a**)

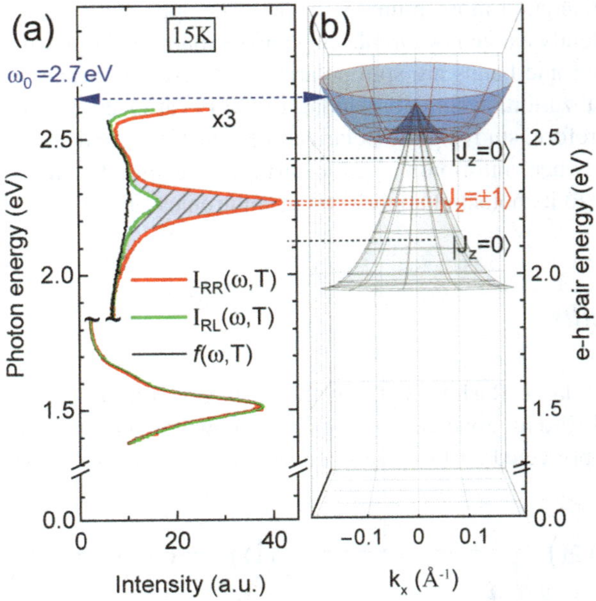

Fig. 4.20 Photoluminescence spectra and noninteracting e–h continuum of relevant interband transitions. (**a**) The CP PL spectra measured at 14 K, with right-CP 2.65 eV excitation. Right- and left-CP PL are designated by $I_{\mathrm{RR}}(\omega, T)$ in red and $I_{\mathrm{RL}}(\omega, T)$ in green, respectively. The black line shows unpolarized PL background, $f(\omega, T)$. The intensity in the 1.8–2.6 eV range is multiplied by factor of 3 for clarity. (**b**) The dispersion relation of non-interacting electron–hole pairs for possible transitions from RSS to SS2, with zero momentum transfer. The only transition consistent with excitation energy used in this study is shown in blue, while others are shown in gray. With finite interaction, the excitonic bound states form below the band minimum and are denoted by the total angular momenta of the electron–hole pairs, J_z

high energy unoccupied surface states (SS2) [81], and the fully occupied Rashba surface states (RSS) [87, 136].

In the following, we use continuous wave visible laser light to promote resonant interband transition between RSS and SS2 (Fig. 4.19a) and study polarized secondary emission spectra in the backscattering geometry. All of the results reported below are reproducible in the "time reversed" geometry, i.e., switching the excitation polarization leads to switching of PL polarization with the same spectral line shape, and so we will henceforth focus on the results with right-CP excitation.

In Fig. 4.20a, we can further separate $I_{\mathrm{RR}}(\omega, T)$ and $I_{\mathrm{RL}}(\omega, T)$ into two spectral contributions: (1) a broad unpolarized emission band, $f(\omega, T)$, and (2) a narrower peak that is almost fully polarized, with intensity defined as $\mathcal{L}_R(\omega, T) = I_{\mathrm{RR}}(\omega, T) - f(\omega, T)$. We note that $f(\omega, T)$ and $\mathcal{L}_R(\omega, T)$ inherit distinct line shapes and therefore likely originate from different photoemission processes. We will henceforth focus on the polarized PL, $\mathcal{L}_R(\omega, T)$.

The broad unpolarized background $f(\omega, T)$ can be easily removed from the spectra as detailed in the following. With right circularly polarized (CP) excitation, the measured photoluminescence (PL) intensities can be decomposed into two parts:

$$I_{\mathrm{RR}}(\omega, T) = \mathcal{L}_R(\omega, T) + f(\omega, T),$$
$$I_{\mathrm{RL}}(\omega, T) = \mathcal{L}_L(\omega, T) + f(\omega, T), \tag{4.12}$$

where $\mathcal{L}_R(\omega, T)$ and $\mathcal{L}_L(\omega, T)$ denote right- and left-CP PL, respectively, and $f(\omega, T)$ denotes the featureless unpolarized broad background. We assume an energy independent depolarization ratio $r(T) = \frac{\mathcal{L}_L(\omega,T)}{\mathcal{L}_R(\omega,T)}$. Inserting $r(T)$ into the above expression of $I_{\mathrm{RL}}(\omega, T)$, we can write the unpolarized emission as

$$f(\omega, T) = \frac{I_{\mathrm{RL}}(\omega, T) - r(T) \cdot I_{\mathrm{RR}}(\omega, T)}{1 - r(T)}. \tag{4.13}$$

Then, $r(T)$ was estimated by imposing that $f(\omega, T)$ is a smooth function, and we arrive at

$$\mathcal{L}_R(\omega, T) = \frac{I_{\mathrm{RR}}(\omega, T) - I_{\mathrm{RL}}(\omega, T)}{1 - r(T)}. \tag{4.14}$$

Excitation Profile

In Fig. 4.20a, we see that a small fraction of $\mathcal{L}_R(\omega, T)$ is also present in the orthogonal polarization, $\mathcal{L}_L(\omega, T) = I_{\mathrm{RL}}(\omega, T) - f(\omega, T) \equiv r(T)\mathcal{L}_R(\omega, T)$, where $r(T)$ is a "depolarization ratio." We find that the depolarization ratio $r(T) \approx 0.1$ for single crystals measured with 2.6 eV excitation and increase to about 0.3 with 2.8 eV excitation (Fig. 4.21c). Such high degree of CP PL can only be explained by spin polarized bands resulting from SOC, which is only possible on the surface states due to the inversion symmetric bulk crystal structure of Bi_2Se_3.

Figure 4.21b shows the integrated PL intensity, $\int_{1.95}^{2.46\,eV} [I_{\mathrm{RR}}(\omega, T) - I_{\mathrm{RL}}(\omega, T)]d\omega$, versus excitation measured at about 15 K. Within the energy range where we see the CP PL, the only possible surface transition is from RSS to SS2. Figure 4.21c plots the depolarization ratio $r(T)$ against excitation energy. Assuming that the depolarization process is largely due to energy relaxation in the *e–h* continuum, then a linear extrapolation of $r(T)$ to zero suggests the *e–h* band minimum at 2.48 eV, which is very close to the noninteracting *e–h* energy estimated from the band structure (Fig. 4.20b). Below this energy, 2D excitons can only be resonantly excited, and the "hot luminescence" from the excitons is in principle 100% polarized. In this study, we will only discuss the PL from excitons created through *e–h* pairs.

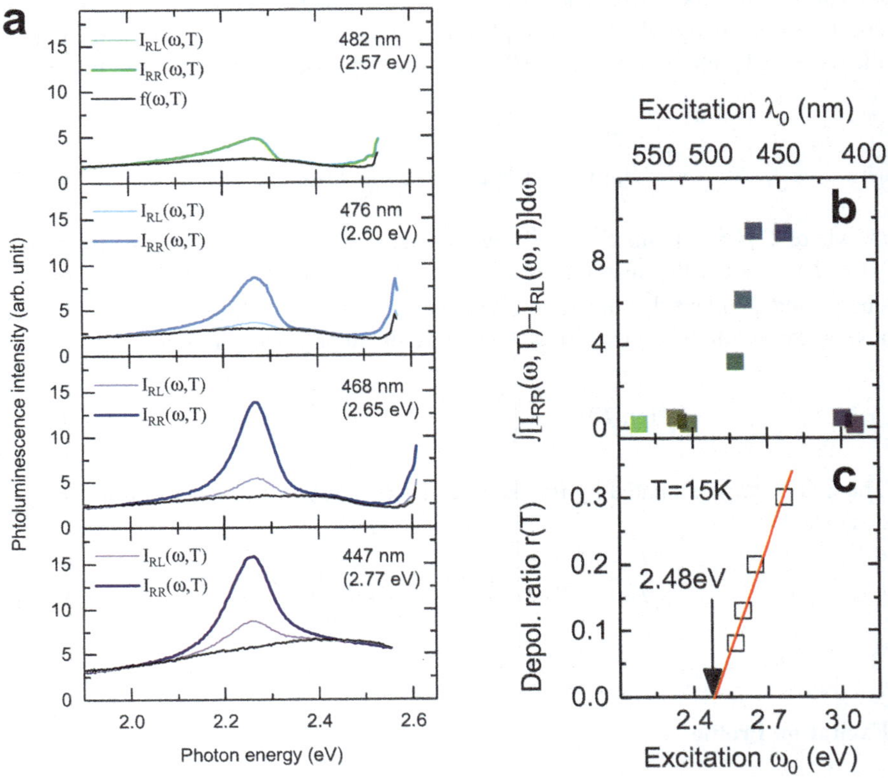

Fig. 4.21 (**a**) PL intensity measured at about 15 K is plotted against photon energy for 4 different excitations: ω_0 =2.57, 2.60, 2.65, and 2.77 eV. The light and dark colored lines denote $I_{RL}(\omega, T)$ and $I_{RR}(\omega, T)$, respectively. The smooth background $f(\omega, T)$ is plotted as black lines. (**b**) The integrated polarized PL intensity, $\int_{1.95}^{2.46\,eV}[I_{RR}(\omega, T)-I_{RL}(\omega, T)]d\omega$ (shaded in blue in Fig. 4.20a) versus excitation energy measured at about 15 K. (**c**) The depolarization ratio $r(T \approx 15\,K)$ is plotted against excitation energy. The red line is a linear extrapolation to $r(T) = 0$, suggesting a minimum excitation energy of 2.48 eV

The band minimum suggests the binding energy for the exciton states, denoted by $|E_1^{R/L}\rangle$ in Fig. 4.20b, to be about 0.2 eV. However, this is an underestimation of the electron–hole continuum edge. Because part of the depolarization is likely due to magnetic impurity scattering or local strain on the sample [137]. Thus the binding energy could be slightly higher than 0.2 eV.

To illustrate the unusual behaviors of the polarized exciton PL, it is perhaps suitable to contrast it with the PL data of a more common exciton. Figure 4.22 shows excitation dependence of PL of Bi_2Se_3 in a more extended energy regime, measured at about 15 K with 7 different excitations. We see that besides the highly polarized exciton PL emission at about 2.3 eV, there is another (much stronger) PL peak centered at about 1.5 eV, emitting with the same intensity of right- and left-CP

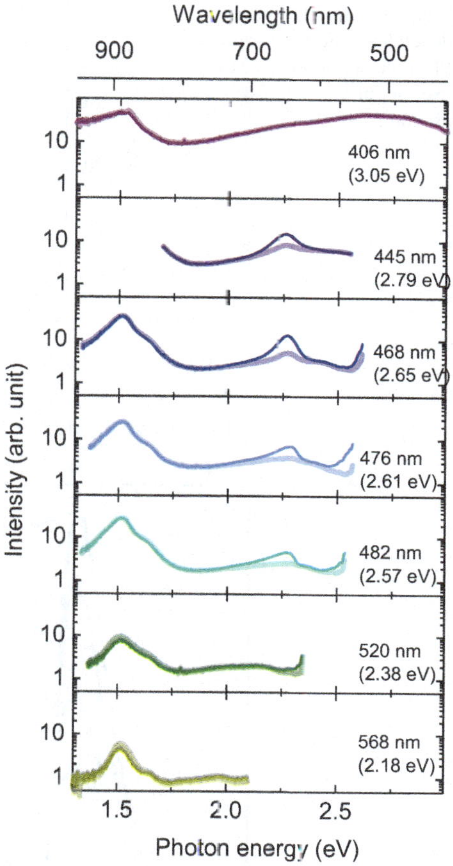

Fig. 4.22 PL intensity measured at about 15 K is plotted against photon energy for 7 different excitations. The intensities are shown in log scale. The light and dark colored lines denote $I_{RL}(\omega, T)$ and $I_{RR}(\omega, T)$, respectively

light, and it appears in all excitations. The line shape of this exciton is Lorentzian (symmetric), which is in contrast to the highly asymmetric line shape of the CP PL near 2.6 eV excitations. Figure 4.23 shows the excitation dependence of the integrated PL intensity of the 1.52 eV exciton, which shows monotonic increase of intensity with higher energy excitations. This is quite from the excitation profile of the 2.26 eV CP PL in Fig. 4.21b, which shows clear resonance around 2.7 eV excitation. The above comparisons show that the CP PL is unusual even within Bi_2Se_3, and new theory is required to understand this phenomenon.

Temperature Dependence

Figure 4.24a shows temperature dependence of $\mathcal{L}_R(\omega, T)$ for Bi_2Se_3 single crystal. While the PL is much stronger at low temperature, the polarized emission is present even at room temperature. Interestingly, we find that $r(T)$ remains about 0.1 up to room temperature. This demonstrates that while temperature has effects on the

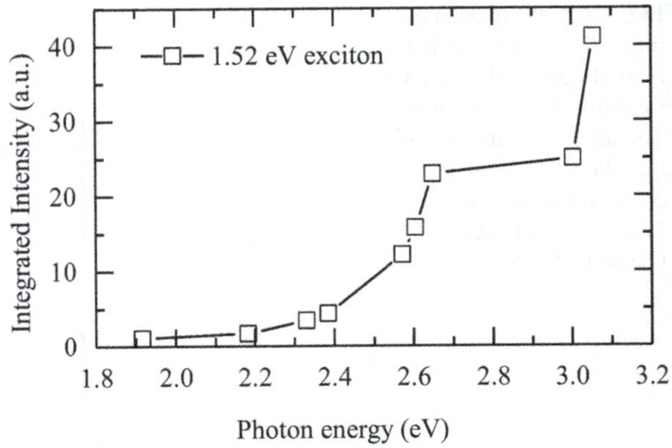

Fig. 4.23 Excitation dependence of the integrated PL intensity of the 1.52 eV exciton, plotted against photon energy

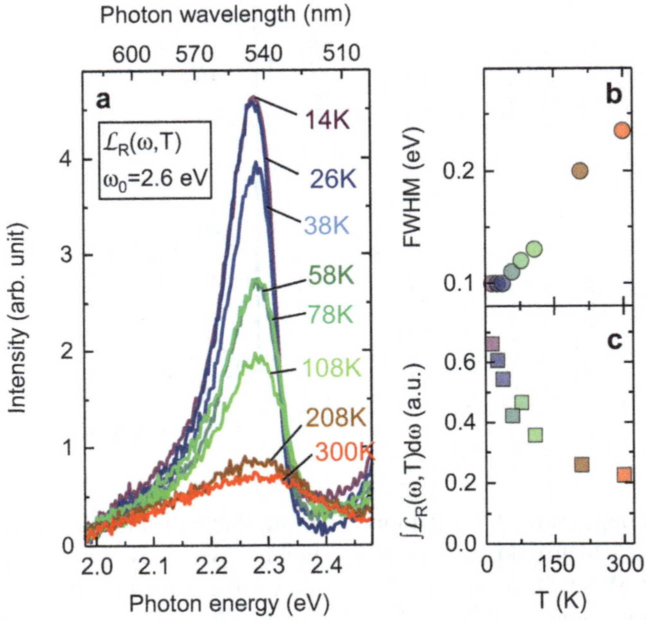

Fig. 4.24 Temperature dependence of CP PL measured with 2.6 eV right-CP excitation. (**a**) The intensity of the polarized PL, $\mathcal{L}_R(\omega, T)$, is plotted against photon energy for various temperatures. (**b**) The temperature dependence of the FWHM of the polarized PL. (**c**) The temperature dependence of the integrated intensity of the polarized PL, $\int \mathcal{L}_R(\omega, T)\, d\omega$

exciton lifetime and population, it has little impact on the polarization of the exciton emission.

Another interesting anomaly of the observed exciton emission is its line shape. Conventional phonon assisted exciton emission results in asymmetric broadening toward the high energy tail, and the extent of the broadening is larger at elevated temperatures [138], where both of them are in contrast to our observations in Bi_2Se_3.

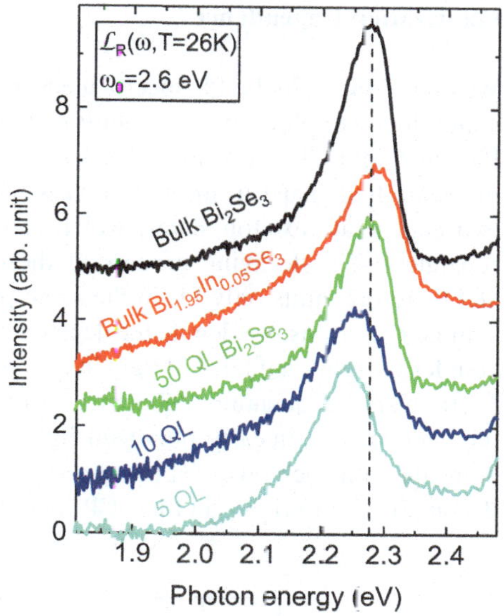

Fig. 4.25 $\mathcal{L}_R(\omega, T)$ is plotted against photon energy for different Bi_2Se_3 samples, measured with 2.6 eV right-CP excitation at 26 K. Spectra are shifted vertically for clarity. Dashed line is a guide to the eye indicating the PL peak center of the bulk Bi_2Se_3 single crystal

Sample Dependence

Further experiments done on MBE grown Bi_2Se_3 thin films of different thickness show qualitatively the same PL line shape, intensity and the depolarization ratio $r(T)$ as in single crystals (Fig. 4.25). This shows that the observed polarization preserving PL is indeed an intrinsic property of Bi_2Se_3.

The PL peak in bulk $Bi_{1.95}In_{0.05}Se_3$ crystal shows slight broadening compared to the pristine Bi_2Se_3 crystals, which could be due to increased inhomogeneity level from indium substitution. The PL peaks in the superlattice samples show lower energy compared to the bulk Bi_2Se_3, which could be due to changes of the band structures in the thin samples. The superlattice samples also present slight broadening, which could be due to increased inhomogeneity from indium diffusion into the Bi_2Se_3 layers.

The fact that the polarized PL intensity is unaffected by Bi_2Se_3 thickness seems to suggest a 3-dimensional (3D) nature of the exciton. However, the penetration depth of Bi_2Se_3 around 2.6 eV is about 10 nm [34]. In the backscattering geometry, the PL or scattered light intensity is again attenuated inside the crystal. The effective penetration depth is then about 5 nm (5 QL), reaching the 2D limit of Bi_2Se_3 [15]. Therefore, the optical experiment in the visible range is effectively a surface probe for Bi_2Se_3, and we cannot distinguish the surface from bulk contributions by studying sample thickness dependence.

Polarization Dependence

We plot in Fig.4.26 the PL spectra measured with different polarization geometries, which helps to elucidate the nature of the observed PL in Bi_2Se_3. The polarized PL with right-CP excitation (Fig. 4.26a) and left-CP excitation (Fig. 4.26b) show the same line shape in the time reversed channels. This suggests that there exist two degenerate exciton states, which can be excited by right- and left-CP light, separately. The decaying process of these exciton preserves angular momentum and therefore emits only PL in the same polarization as the excitation photon. For symmetry reasons, we denote the degenerate exciton states as $|E_1^R\rangle$ and $|E_1^L\rangle$, which couples to right- and left-CP light, respectively.

To test the quantum coherence between states $|E_1^R\rangle$ and $|E_1^L\rangle$, we excite the sample with linearly polarized light. In Fig. 4.26c, $I_{XR}(\omega, T)$ and $I_{XL}(\omega, T)$ coincide with each other. This suggests that the linearly polarized excitation being decomposed into right- and left-CP light can independently excite $|E_1^R\rangle$ and $|E_1^L\rangle$.

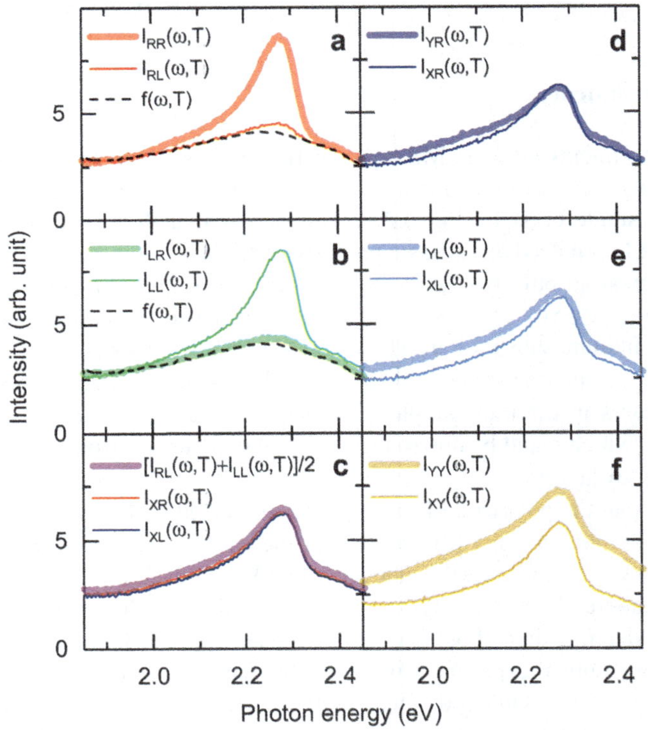

Fig. 4.26 Polarization dependence of PL measured at 26 K, with 2.6 eV right-CP excitation. (**a**) The thick and thin curves show the right- and left-handed PL spectra under right-CP and (**b**) left-CP excitation. The black curve shows a featureless unpolarized broad background. Panels (**c**)–(**e**) compare CP PL spectra excited with linearly polarized light. (**f**) Comparison of the spectra with excitation polarization parallel and cross to the PL polarization

The two types of excitons have independent but similar decay channels, such that the emitted PL acquires the same line shape and intensity. Such process does not allow any dependence on the orientation of the linearly polarized excitation, as linear light can be decomposed into circular light up to a total phase difference. The above argument is confirmed in Fig. 4.26d and e, where orthogonal linear light is used for excitation. The resulted CP PL show roughly the same intensity and line shape for both circular polarizations.

Moreover, the fact that $I_{\mathrm{XL}}(\omega, T)$ coincides with $[I_{\mathrm{RL}}(\omega, T) + I_{\mathrm{LL}}(\omega, T)]/2$ suggests that the quantum coherence is not preserved in the decay process of the exciton (Fig. 4.26c). With linearly polarized light, both $|E_1^R\rangle$ and $|E_1^L\rangle$ are simultaneously excited. The left-CP emission will therefore mainly come from $|E_1^L\rangle$, with small proportion from depolarized emission of $|E_1^R\rangle$. If $|E_1^L\rangle$ and $|E_1^R\rangle$ are phase coherent, then due to interference, the combined PL intensity will not be the same as the intensity obtained by individually exciting them, i.e., $I_{\mathrm{XL}}(\omega, T) \neq [I_{\mathrm{RL}}(\omega, T) + I_{\mathrm{LL}}(\omega, T)]/2$. In our experiment, we have never observed such interference phenomena.

This effect is particularly evident by exciting with the linearly polarized light while also probing the linearly polarized PL. Figure 4.26f shows comparison of spectra with the excitation polarization being parallel and cross to the PL polarization. Apart from the overall intensity change, the PL peak is present in both spectra with roughly the same strength. This is in contrast to the PL of valley excitons in some TMD monolayers, where the PL polarization is always parallel to the polarization of excitation photon [139], which is explained as a demonstration of quantum phase coherence in the decaying channels of both types of valley excitons. Here, the polarization dependence of PL in Bi_2Se_3 implies incoherent decay processes. As a result, $|E_1^L\rangle$ and $|E_1^R\rangle$ act as independent but polarization preserving emitters.

4.4.3 Chiral 2D Excitons

The observation of polarization preserving PL resonance can be readily understood in terms of a 2D chiral exciton, i.e., an electron–hole bound state originating from spin polarized surface states, as illustrated in Fig. 4.27. To explain how gapless Dirac particles can result in a formation of a discrete two-particle bound state, we present a model that involves an optically excited electron in SS2 and a hole state in RSS, as shown in Fig. 4.19a. Neglecting the high order warping terms, we model the Hamiltonians as

$$
\begin{aligned}
H_{SS2}(\mathbf{p}) &= \Delta + v\mathbf{p}\cdot\boldsymbol{\sigma}_e,\\
H_{RSS}(\mathbf{p}) &= -\frac{\mathbf{p}^2}{2m_h} - \alpha\mathbf{p}\cdot\boldsymbol{\sigma}_h,
\end{aligned}
\tag{4.15}
$$

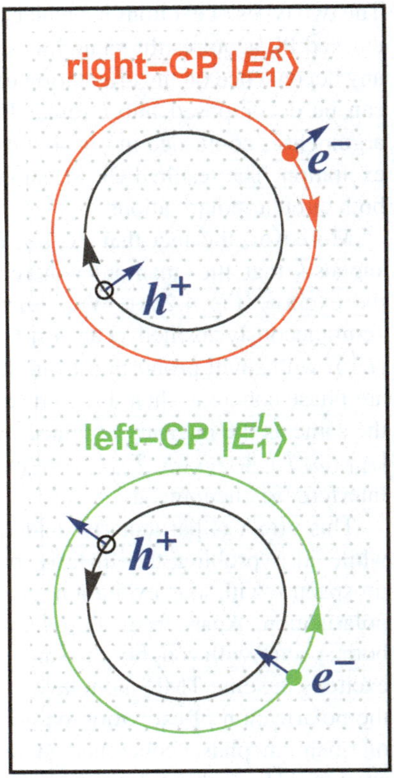

Fig. 4.27 Illustration of the chiral 2D excitons. The doubly degenerate excitonic states ($|E_1^R\rangle$ and $|E_1^L\rangle$) are shown, where a Dirac fermion from SS2, denoted in red and green, orbits a quasiparticle hole from RSS (shown in black open circle), with the spin polarization denoted by blue arrows. The exciton emissions from $|E_1^R\rangle$ and $|E_1^L\rangle$ are right- and left-CP, respectively. Interchange between these two states requires backscattering and spin flip and therefore is not allowed

where Δ is the energy difference between RSS and SS2, v is the Dirac velocity, m_h is the effective mass of the hole, α is Rashba coefficient, and 2D momentum $\mathbf{p}$ is measured from the Γ point. We note that Rashba spin-orbit interaction is of the form $\alpha(\boldsymbol{\sigma}_h \times \mathbf{p}) \cdot \hat{z}$ (similarly for the Dirac states with the substitution $\alpha \to v$ and $\boldsymbol{\sigma}_h \to \boldsymbol{\sigma}_e$). However this structure is equivalent to that in Eq. (4.15) after relabeling spin components as $\sigma_x \to \sigma_y$, $\sigma_y \to -\sigma_x$.

We model the non-interacting electron–hole pair by the two-body Hamiltonian $H_{eh}(\mathbf{p}) = H_{SS2}(\mathbf{p}_+) - H_{RSS}(\mathbf{p}_-)$, where $\mathbf{p}_\pm = \mathbf{p} \pm \frac{1}{2}\mathbf{k}$ and $\mathbf{k}$ is the exciton momentum. For zero momentum transfer ($\mathbf{k} = 0$), the eigenvalues of $H_{eh}(\mathbf{p})$ have W-shaped dispersion resembling a multi-layer Mexican hat (Fig. 4.20b). Crucially, the two-body bands originating from $H_{eh}(\mathbf{p})$ are bounded from below for any parameter values v and α due to the presence of the $\mathbf{p}^2$ term with $m_h > 0$. We therefore expect the two-particle bound state originating due to Coulomb attraction between electron and hole to be a discrete state positioned below the Mexican hat bottom layer band edge. This somewhat a counter-intuitive behavior, which arises despite the Dirac conduction band being unbounded, is illustrated in Fig. 4.20b. As we will see below, the characteristic momenta of this bound state are defined by the Mexican hat rim radius.

Excitonic Bound States

The formation of a discrete bound state may seem to be at odds with the fact that the energy spectrum of massless Dirac fermions spans the entire continuum $-\infty < \epsilon < \infty$, leaving no room for discrete states at any energy. The only known exception to this picture are the discrete complex-energy states arising in the so-called atomic collapse regime [140]. Here, in contrast, we predict bound states with discrete real-valued energies. Such states arise because an electron in the conduction band can bind to a hole in the valence band such that the electron and the hole are moving with roughly the same velocity, $\mathbf{v}_e(\mathbf{p}) = \mathbf{v}_h(\mathbf{p})$. In this case the (possibly large) kinetic energies of the electron and the hole have no impact on the formation of the bound state. The condition $\mathbf{v}_e(\mathbf{p}) = \mathbf{v}_h(\mathbf{p})$ defines the Mexican hat bottom, in agreement with the discussion above. As we will see, if the condition $\mathbf{v}_e(\mathbf{p}) = \mathbf{v}_h(\mathbf{p})$ is fulfilled on a circle in momentum space, the bound states can arise even for an arbitrarily weak attraction potential.

The quantum state and the binding energy of this chiral 2D exciton can be obtained with the help of the two-particle Schrödinger equation:

$$\epsilon\psi_{eh}(\mathbf{r}) = (H_{eh}(\mathbf{p}) + V(\mathbf{r}))\,\psi_{eh}(\mathbf{r}), \tag{4.16}$$

where $V(\mathbf{r})$ is the electron–hole interaction potential. Here, $\psi_{eh}(\mathbf{r})$ describes the exciton wave function, and for the sake of simplicity, we focus on the case of zero net exciton momentum ($\mathbf{k} = 0$), which is sufficient for the direct optical transitions discussed here.

To understand the general properties of the resulting discrete states, in particular their spin structure, it is instructive to replace the interaction potential by a model short-range attraction $V(\mathbf{r}) = -\lambda\delta(\mathbf{r})$, which in fact provides a good approximation for the Coulomb interaction screened by conduction electrons in Bi_2Se_3. The bound states for this problem can be readily obtained using the T-matrix approach, giving an equation for eigen-energies of the form:

$$1 = \lambda Z(\epsilon, \mathbf{k}), \tag{4.17}$$

$$Z(\epsilon, \mathbf{k}) = -\sum_{\mathbf{p}} \frac{1}{\epsilon - \Delta - \frac{\mathbf{p}_+^2}{2m_h} - v\mathbf{p}_+ \cdot \boldsymbol{\sigma}_e - \alpha\mathbf{p}_- \cdot \boldsymbol{\sigma}_h + i0},$$

where $\sum_{\mathbf{p}}$ denotes integration $\int \frac{d^2p}{(2\pi\hbar)^2}$. Equation (4.17) can be evaluated explicitly, giving a closed-form algebraic equation for exciton spectrum. In the Supplementary Materials of [3], we show that the discrete states exist for any attraction strength $\lambda > 0$, no matter how weak.

At $\mathbf{k} = 0$, $H_{eh}(\mathbf{p})$ possesses azimuthal symmetry, allowing us to understand the spin structure of Eq. (4.17) on general symmetry grounds. The eigenstates must have definite projection of the angular momenta z component, $\sigma_e^z = \pm 1$ and $\sigma_h^z = \pm 1$. Strictly speaking, exciton states have to be classified within the surface symmetry group, which is C_{6v} for actual Bi_2Se_3 or $C_{\infty v}$ for a rotationally invariant Hamiltonian. At Γ point, both RSS and SS2 transform as the Γ_7 double group representation of the C_{6v} point group. Extending the symmetry argument to states near the Γ point, the 2D exciton states arising from Eq. (4.17) must have the symmetries [53]: $\Gamma_7 \times \Gamma_7 = \Gamma_1 + \Gamma_2 + \Gamma_5$, or equivalently in Mulliken symbols: $A_1 + A_2 + E_1$.

Inspecting the exciton wave functions, we find that the two $|J_z = 0\rangle$ states are fully symmetric with respect to all symmetry operators of both C_{6v} and $C_{\infty v}$ groups and therefore belong to the A_1 irreducible representation. On the contrary, the doubly degenerate $|J_z = \pm 1\rangle$ states belong to the E_1 representation, which transform as an in-plane electric dipole. Here, neither the orbital angular momentum nor the spin is a good quantum number, and the states should be labeled by the total angular momentum. The lowest energy state with $J_z = 0$ takes the form: $|A_1^1\rangle = \frac{1}{\sqrt{2}}(|\uparrow_e\downarrow_h\rangle - |\downarrow_e\uparrow_h\rangle)$. The excited state with the $J_z = 0$ component has the form: $|A_1^2\rangle = \frac{1}{\sqrt{2}}(|\uparrow_e\downarrow_h\rangle + |\downarrow_e\uparrow_h\rangle)$. The A_1 representation transforms as an electric dipole along c-axis, and therefore $|A_1\rangle$ is optically active only when excited from the ac surface.

In contrast, the states with $J_z = \pm 1$ have the E_1 symmetry, which transforms as in-plane electric dipoles and therefore couples to the circularly polarized light in backscattering geometry from the ab surface. Hence, we label the $J_z = \pm 1$ states as $|E_1^R\rangle = |\uparrow_e\uparrow_h\rangle$ and $|E_1^L\rangle = |\downarrow_e\downarrow_h\rangle$, populated by right- and left-CP light, respectively. It can be shown that these two states are degenerate and intermediate to the energies of the $|J_z = 0\rangle$ states, as we illustrated in Fig. 4.20b. Crucially, only the $|E_1^R\rangle$ and $|E_1^L\rangle$ are optically active, whereas both $|A_1^1\rangle$ and $|A_1^2\rangle$ are "invisible" in our experiment. This possibly explains the unusual line shape of exciton resonance with a pronounce tail on the long wavelength side, which can be due to emission from a "dark" state through high order terms beyond dipole approximation.

Optical Orientation of Chiral 2D Excitons

The polarization preserving PL arises as a direct result of the symmetry properties of the constructed 2D excitons. Within the backscattering geometry of our experiment, circularly polarized light can only promote excitations with $\Delta J_z = \pm 1$. Assuming no cross-relaxation between $J_z = 0$ and $J_z = \pm 1$ states, we expect a single PL peak arising from recombination of the $J_z = \pm 1$ exciton, which is consistent with our data. However, one would expect that inelastic scattering with phonons would couple the $|J_z = +1\rangle$ and $|J_z = -1\rangle$ states, which could cause an increase of $r(T)$ with increasing incident photon energy. In Fig. 4.22, we see only a moderate increase of $r(T)$ even if the incident photon energy is about 300 meV above the Mexican hat minimum (Fig. 4.20). This is because the interchange between the two states involves spin–flip as illustrated in Fig. 4.27 and therefore is unfavored for interaction with nonmagnetic impurities or full symmetric phonons.

We knew from the previous Sect. 4.2 that nonsymmetric surface phonons in Bi_2Se_3 are weak, and hence J_z remains approximately conserved during the energy relaxation. Hence this explains the exceptionally high degree of polarization observed in all samples and up to room temperature. It would be interesting to study in the future the interaction between the chiral exciton and other asymmetric collective modes, such as the Dirac plasmons and chiral spin modes, which might be the main relaxation channels for the chiral excitons.

4.5 Summary

In summary, we have studied the secondary emissions in the 3D TI Bi_2Se_3 single crystal and thin films, in the energy range of 0.5 meV to 1.5 eV. Within this energy regime, we have observed three surface originated phenomena, which is a particularly interesting topic in this material (Sect. 4.1.2).

Below 50 meV, the spectra is dominated by Raman scattering from lattice vibration normal modes (phonons, second-order phonons, etc.). Compared to the DFT calculated bulk phonon bands, we observed in total four additional out-of-plane and possibly 2 in-plane surface vibrational modes, where we tabulate the energies and symmetries in Table 4.3. These modes all have energies very close to the bulk values where the bulk phonon DOS is not zero, and therefore they were assigned to surface resonances. Thanks to the high energy resolution and sensitivity, we were able to resolve an additional $A_1^{(4)}$ mode that has not been previously observed. This completes all of the possible surface resonances that can be observed in Bi_2Se_3, and a proper symmetry analysis has been given.

Above 100 meV, the Raman spectra is basically flat with little signal, except for a new collective mode at 150 meV in the chiral A_2 symmetry channel of C_{6v} group. This mode is only observed under surface resonance energy of 1.8 eV and is identified as transverse chiral spin mode: a collective spin-flip excitation of the surface Dirac fermions. The high energy secondary emission spectra are overlapped with two PL peaks at about 1.5 and 2.3 eV. The 2.3 eV peak is almost fully circularly polarized parallel to the polarization of the excitation photon. We propose a 2D excitonic model where two energetically degenerate electron–hole bound states form on the surface of Bi_2Se_3.

The results we presented here demonstrate a way of studying the dynamical response of Dirac fermions and their many-body interactions through optical measurement. In the future, experiments should be done on the excitation dependence of the surface resonances, which might further reveal the relation between phononic and electronic degrees of freedom on the surface of TIs. In addition, the magnetic field dependence of the chiral spin mode and exciton would help understand the spin–spin correlations in TIs. Due to the particular interests this material has in spintronics, magnonics, optoelectronics, and quantum computing [129, 130, 141–147], our results may have important applications beyond scientific curiosities.

References

1. H.-H. Kung et al., Surface vibrational modes of the topological insulator Bi_2Se_3 observed by Raman spectroscopy. Phys. Rev. B **95**, 245406 (2017). https://doi.org/10.1103/PhysRevB.95.245406
2. H.-H. Kung et al., Chiral spin mode on the surface of a topological insulator. Phys. Rev. Lett. **119**, 136802 (2017). https://doi.org/10.1103/PhysRevLett.119.136802
3. H.-H. Kung et al., Observation of chiral surface excitons in a topological insulator Bi_2Se_3. Proc. Natl. Acad. Sci. **116**(10), 4006–4011 (2019). https://doi.org/10.1073/pnas.1813514116
4. N.W. Ashcroft, N.D. Mermin, *Solid State Physics*. HRW international editions. Holt, Rinehart and Winston (1976)
5. A. Damascelli, Probing the electronic structure of complex systems by ARPES. Physica Scripta **2004**(T109), 61 (2004). https://doi.org/10.1238/Physica.Topical.109a00061
6. N.F. Mott, The basis of the electron theory of metals, with special reference to the transition metals. Proc. Phys. Soc. A **62**(7), 416 (1949)
7. K.v. Klitzing, G. Dorda, M. Pepper, New method for high-accuracy determination of the fine-structure constant based on quantized hall resistance. Phys. Rev. Lett. **45**, 494–497 (1980). https://doi.org/10.1103/PhysRevLett.45.494
8. A. Bansil, H. Lin, T. Das, *Colloquium*: Topological band theory. Rev. Mod. Phys. **88**, 021004 (2016). https://doi.org/10.1103/RevModPhys.88.021004
9. M.Z. Hasan, S.-Y. Xu, G. Bian, Topological insulators, topological superconductors and Weyl fermion semimetals: discoveries, perspectives and outlooks. Physica Scripta **2015**(T164), 014001 (2015)
10. N.P. Armitage, E.J. Mele, A. Vishwanath, Weyl and Dirac semimetals in three-dimensional solids. Rev. Modern Phys. **90**(1), 015001 (2018). https://doi.org/10.1103/RevModPhys.90.015001
11. B. Yan, C. Felser, Topological materials: Weyl semimetals. Annu. Rev. Condens. Matter Phys. **8**, 337–354 (2017). https://doi.org/10.1146/annurev-conmatphys-031016-025458
12. M.I. Dyakonov, A.V. Khaetskii, *Spin Physics in Semiconductors*, vol. 1 (Springer, 2008). https://doi.org/10.1007/978-3-540-78820-1
13. L. Fu, C.L. Kane, E.J. Mele, Topological insulators in three dimensions. Phys. Rev. Lett. **98**, 106803 (2007). https://doi.org/10.1103/PhysRevLett.98.106803
14. L. Fu, C.L. Kane, Topological insulators with inversion symmetry. Phys. Rev. B **76**, 045302 (2007). https://doi.org/10.1103/PhysRevB.76.045302
15. H. Zhang et al., Topological insulators in Bi_2Se_3, Bi_2Te_3 and Sb_2Te_3 with a single Dirac cone on the surface. Nature Phys. **5**, 438 (2009). https://doi.org/10.1038/nphys1270
16. B.I. Halperin, Quantized Hall conductance, current-carrying edge states, and the existence of extended states in a two-dimensional disordered potential. Phys. Rev. B **25**, 2185–2190 (1982). https://doi.org/10.1103/PhysRevB.25.2185
17. A.M. Essin, V. Gurarie, Bulk-boundary correspondence of topological insulators from their respective Green's functions. Phys. Rev. B **84**, 125132 (2011). https://doi.org/10.1103/PhysRevB.84.125132
18. A. Zangwill. *Physics at Surfaces* (Cambridge University Press, 1988)
19. B. Feuerbacher, R.F. Willis, Photoemission and electron states at clean surfaces. J. Phys. C Solid State Phys. **9**(2), 169 (1976). https://doi.org/10.1088/0022-3719/9/2/007
20. K. Fukutani et al., Symmetry-protected surface state on Mo(112). J. Phys. Condens. Matter **26**(15), 155501 (2014). https://doi.org/10.1088/0953-8984/26/15/155501
21. D. Hsieh et al., A tunable topological insulator in the spin helical Dirac transport regime. Nature **460**(7259), 1101–1105 (2009). https://doi.org/10.1038/nature08234
22. P.D.C. King et al., Large tunable Rashba spin splitting of a two-dimensional electron gas in Bi_2Se_3. Phys. Rev. Lett. **107**, 096802 (2011). https://doi.org/10.1103/PhysRevLett.107.096802
23. L. Fu, Hexagonal warping effects in the surface states of the topological insulator Bi_2Te_3. Phys. Rev. Lett. **103**, 266801 (2009). https://doi.org/10.1103/PhysRevLett.103.266801

24. P. Roushan et al., Topological surface states protected from backscattering by chiral spin texture. Nature **460**(7259), 1106–1109 (2009). https://doi.org/10.1038/nature08308
25. A.A. Taskin, Y. ichi Ando, Berry phase of nonideal Dirac fermions in topological insulators. Phys. Rev. B **84**, 035301 (2011). https://doi.org/10.1103/PhysRevB.84.035301
26. M.J. Brahlek, Atomic scale engineering of topological materials'. PhD thesis. Rutgers University, 2014. https://doi.org/10.7282/T3TM78KB
27. J.G. Checkelsky et al., Quantum interference in macroscopic crystals of nonmetallic Bi_2Se_3. Phys. Rev. Lett. **103**, 246601 (2009). https://doi.org/10.1103/PhysRevLett.103.246601
28. G.P. Mikitik, Yu.V. Sharlai, Manifestation of Berry's phase in metal physics. Phys. Rev. Lett. **82**, 2147–2150 (1999). https://doi.org/10.1103/PhysRevLett.82.2147
29. M. Brahlek et al., Topological-metal to band-insulator transition in $(Bi_{1-x}In_x)_2Se_3$ thin films. Phys. Rev. Lett. **109**, 186403 (2012). https://doi.org/10.1103/PhysRevLett.109.186403
30. D. Hsieh et al., A topological Dirac insulator in a quantum spin Hall phase. Nature **452**(7190), 970–974 (2008). https://doi.org/10.1038/nature06843
31. Y. Xia et al., Observation of a large-gap topological-insulator class with a single Dirac cone on the surface. Nature Phys. **5**(6), 398–402 (2009). https://doi.org/10.1038/nphys1274
32. Y.L. Chen et al., Experimental realization of a three-dimensional topological insulator, Bi_2Te_3. Science **325**(5937), 178–181 (2009). https://doi.org/10.1126/science.1173034
33. N. Bansal et al., Thickness-independent transport channels in topological insulator Bi_2Se_3 thin films. Phys. Rev. Lett. **109**, 116804 (2012). https://doi.org/10.1103/PhysRevLett.109.116804
34. J.W. McIver et al., Control over topological insulator photocurrents with light polarization. Nat. Nanotechnol. **7**(2), 96–100 (2012). https://doi.org/10.1038/nnano.2011.214
35. H. Köhler, C.R. Becker, Optically active lattice vibrations in Bi_2Se_3. Physica Status Solidi (b) **61**(2), 533–537 (1974). https://doi.org/10.1002/pssb.2220610218
36. W. Richter, C.R. Becker, A Raman and far-infrared investigation of phonons in the rhombohedral V_2–VI_3 compounds Bi_2Te_3, Bi_2Se_3, Sb_2Te_3 and $Bi_2(Te_{1-x}Se_x)_3$ $(0 < x < 1)$, $(Bi_{1-y}Sb_y)_2Te_3$ $(0 < y < 1)$. Physica Status Solidi (b) **84**(2), 619–628 (1977). https://doi.org/10.1002/pssb.2220840226
37. A.D. LaForge et al., Optical characterization of Bi_2Se_3 in a magnetic field: Infrared evidence for magnetoelectric coupling in a topological insulator material. Phys. Rev. B **81**, 125120 (2010). https://doi.org/10.1103/PhysRevB.81.125120
38. S.Y.F. Zhao et al., Fabrication and characterization of topological insulator Bi_2Se_3 nanocrystals. Appl. Phys. Lett. **98**(14), 141911 (2011). https://doi.org/10.1063/1.3573868
39. J. Zhang et al., Raman spectroscopy of few-quintuple layer topological insulator Bi_2Se_3 nanoplatelets. Nano Lett. **11**(6), 2407–2414 (2011). https://doi.org/10.1021/nl200773n
40. V. Gnezdilov et al., Helical fluctuations in the Raman response of the topological insulator Bi_2Se_3. Phys. Rev. B **84**, 195118 (2011). https://doi.org/10.1103/PhysRevB.84.195118
41. Y. Kim et al., Temperature dependence of Raman-active optical phonons in Bi_2Se_3 and Sb_2Te_3. Appl. Phys. Lett. **100**(7), 071907 (2012). https://doi.org/10.1063/1.3685465
42. J. Humlíček et al., Raman and interband optical spectra of epitaxial layers of the topological insulators Bi_2Te_3 and Bi_2Se_3 on BaF_2 substrates. Physica Scripta **2014**(T162), 014007 (2014). https://doi.org/10.1088/0031-8949/2014/T162/014007
43. B. Irfan et al., Temperature dependent Raman scattering studies of three dimensional topological insulators Bi_2Se_3. J. Appl. Phys. **115**(17), 173506 (2014). https://doi.org/10.1063/1.4871860
44. M. Eddrief et al., Low-temperature Raman fingerprints for few-quintuple layer topological insulator Bi_2Se_3 films epitaxied on GaAs. Nanotechnology **25**(24), 245701 (2014). https://doi.org/10.1088/0957-4484/25/24/245701
45. Y. Yan et al., Surface-facet-dependent phonon deformation potential in individual strained topological insulator Bi_2Se_3 nanoribbons. ACS Nano **9**(10), 10244–10251 (2015). https://doi.org/10.1021/acsnano.5b04057
46. X. Zhang et al., Review on the Raman spectroscopy of different types of layered materials. Nanoscale **8**, 6435–6450 (2016). https://doi.org/10.1039/C5NR07205K

47. W. Cheng, S.-F. Ren, Phonons of single quintuple Bi_2Te_3 and Bi_2Se_3 films and bulk materials. Phys. Rev. B **83**, 094301 (2011). https://doi.org/10.1103/PhysRevB.83.094301
48. T. Terzibaschian, B. Enderlein, The irreducible representations of the two-dimensional space groups of crystal surfaces. Theory and applications. Physica Status Solidi (b) **133**(2), 443–461 (1986). https://doi.org/10.1002/pssb.2221330202
49. J. Li, J.J. Tu, J.L. Birman, Symmetry predicted transitions in 3D topological insulators. Solid State Commun. **163**, 11–14 (2013). https://doi.org/10.1016/j.ssc.2013.03.010
50. R.-J. Slager et al., The space group classification of topological band insulators. Nature Phys. **9**(2), 98–102 (2013). https://doi.org/10.1038/nphys2513
51. L.N. Ovander, The form of the Raman tensor. Opt. Spectrosc. **9**, 302 (1960)
52. M. Cardona, Resonance phenomena, in *Light Scattering in Solids II*, ed. by M. Cardona, G. Güntherodt (Springer, Berlin, 1982), pp. 45–49
53. G.F. Koster, *Properties of the Thirty-Two Point Groups*. Massachusetts Institute of Technology Press Research Monograph (MIT Press, 1963)
54. I.M. Lifshitz, L.N. Rosenzweig, Dynamics of lattice filling half-space (Russian). Zh. Eksp. Teor. Fiz. **18**, 1012 (1948)
55. I.M. Lifshitz, Some problems of the dynamic theory of non-ideal crystal lattices. Il Nuovo Cimento **3**(4), 716–734 (1956). https://doi.org/10.1007/BF02746071
56. R.F. Wallis, Effect of free ends on the vibration frequencies of one-dimensional lattices. Phys. Rev. **105**, 540–545 (1957). https://doi.org/10.1103/PhysRev.105.540
57. R.F. Wallis, Theory of surface modes of vibration in two- and three-dimensional crystal lattices. Phys. Rev. **116**, 302–308 (1959). https://doi.org/10.1103/PhysRev.116.302
58. G. Benedek, L. Miglio, The Green's function method in the surface lattice dynamics of ionic crystals, in *Surface Phonons*, ed. by W. Kress, F.W. de Wette. (Springer, Berlin, Heidelberg, 1991), pp. 37–66. https://doi.org/10.1007/978-3-642-75785-3_3
59. R.F Wallis, Surface phonons: theoretical developments. Surface Science **299**, 612–627 (1994)
60. M.J. Lagos et al., Mapping vibrational surface and bulk modes in a single nanocube. Nature **543**(7646), 529–532 (2017)
61. V. Chis et al., Vibrations in binary and ternary topological insulators: First principles calculations and Raman spectroscopy measurements. Phys. Rev. B **86**, 174304 (2012). https://doi.org/10.1103/PhysRevB.86.174304
62. X. Zhu et al., Interaction of phonons and Dirac fermions on the surface of Bi_2Se_3: A strong Kohn anomaly. Phys. Rev. Lett. **107**, 186102 (2011). https://doi.org/10.1103/PhysRevLett.107.186102
63. X. Zhu et al., Electron-phonon coupling on the surface of the topological insulator Bi_2Se_3 determined from surface-phonon dispersion measurements. Phys. Rev. Lett. **108**, 185501 (2012). https://doi.org/10.1103/PhysRevLett.108.185501
64. C. Howard et al., Anomalous behavior in the phonon dispersion of the (001) surface of Bi_2Te_3 determined from helium atom-surface scattering measurements. Phys. Rev. B **88**, 035402 (2013). https://doi.org/10.1103/PhysRevB.88.035402
65. R.C. Hatch et al., Stability of the $Bi_2Se_3(111)$ topological state: Electron-phonon and electron-defect scattering. Phys. Rev. B **83**, 241303 (2011). https://doi.org/10.1103/PhysRevB.83.241303
66. Z.-H. Pan et al., Measurement of an exceptionally weak electron-phonon coupling on the surface of the topological insulator Bi_2Se_3 using angle-resolved photoemission spectroscopy. Phys. Rev. Lett. **108**, 187001 (2012). https://doi.org/10.1103/PhysRevLett.108.187001
67. M.V. Costache et al., Fingerprints of inelastic transport at the surface of the topological insulator Bi_2Se_3: Role of electron-phonon coupling. Phys. Rev. Lett. **112**, 086601 (2014). https://doi.org/10.1103/PhysRevLett.112.086601
68. J.A. Sobota et al., Distinguishing bulk and surface electron-phonon coupling in the topological insulator Bi_2Se_3 using time-resolved photoemission spectroscopy. Phys. Rev. Lett. **113**, 157401 (2014). https://doi.org/10.1103/PhysRevLett.113.157401
69. C. Chen et al., Tunable Dirac fermion dynamics in topological insulators. Scientific Reports **3**, 2411 (2013)

70. T. Kondo et al., Anomalous dressing of Dirac fermions in the topological surface state of Bi_2Se_3, Bi_2Te_3, and cu-doped Bi_2Se_3. Phys. Rev. Lett. **110**, 217601 (2013). https://doi.org/10.1103/PhysRevLett.110.217601
71. A. Kogar et al., Surface collective modes in the topological insulators Bi_2Se_3 and $Bi_{0.5}Sb_{1.5}Te_{3-x}Se_x$. Phys. Rev. Lett. **115**, 257402 (2015). https://doi.org/10.1103/PhysRevLett.115.257402
72. N. Esser, W. Richter, Raman scattering from surface phonons, in *Light Scattering in Solids VIII*, ed. by M. Cardona, G. Güntherodt (Springer, Berlin, 1999), pp. 96–168
73. M. Liebhaber et al., Surface phonons of the Si(111)-(7 × 7) reconstruction observed by Raman spectroscopy. Phys. Rev. B **89**, 045313 (2014). https://doi.org/10.1103/PhysRevB.89.045313
74. R. Lewandowska et al., Raman scattering in α-In_2Se_3 crystals. Mater. Res. Bull. **36**(15), 2577–2583 (2001). https://doi.org/10.1016/S0025-5408(01)00746-2
75. B.-T. Wang, P. Zhang, Phonon spectrum and bonding properties of Bi_2Se_3: Role of strong spin-orbit interaction. Appl. Phys. Lett. **100**(8), 082109 (2012)
76. Y.D. Glinka et al., Thickness tunable quantum interference between surface phonon and Dirac plasmon states in thin films of the topological insulator Bi_2Se_3. J. Phys. Condens. Matter **27**(5), 052203 (2015)
77. L. Wu et al., A sudden collapse in the transport lifetime across the topological phase transition in $(Bi_{1-x}In_x)_2Se_3$. Nature Phys. **9**(7), 410–414 (2013)
78. H. Dong Lee et al., Indium and bismuth interdiffusion and its influence on the mobility in In_2Se_3/Bi_2Se_3. Thin Solid Films **556**, 322–324 (2014). https://doi.org/10.1016/j.tsf.2014.01.082
79. U. Fano, Effects of configuration interaction on intensities and phase shifts. Phys. Rev. **124**, 1866–1878 (1961). https://doi.org/10.1103/PhysRev.124.1866
80. M.V. Klein, Electronic Raman scattering, *Light Scattering in Solids I*, ed. by M. Cardona, G. Güntherodt (Springer, Berlin, 1983), pp. 169–172
81. J.A. Sobota et al., Direct optical coupling to an unoccupied Dirac surface state in the topological insulator Bi_2Se_3. Phys. Rev. Lett. **111**, 136802 (2013). https://doi.org/10.1103/PhysRevLett.111.136802
82. D. Niesner et al., Unoccupied topological states on bismuth chalcogenides. Phys. Rev. B **86**, 205403 (2012). https://doi.org/10.1103/PhysRevB.86.205403
83. B.S. Shastry, B.I. Shraiman, Raman scattering in Mott-Hubbard systems. Int. J. Mod. Phys. B **5**, 365–388 (1991). https://doi.org/10.1142/S0217979291000237
84. D.V. Khveshchenko, P.B. Wiegmann, Raman scattering and anomalous current algebra in Mott insulators. Phys. Rev. Lett. **73**, 500–503 (1994). https://doi.org/10.1103/PhysRevLett.73.500
85. M. Bianchi et al., Coexistence of the topological state and a two-dimensional electron gas on the surface of Bi_2Se_3. Nat. Commun. **1**, 128 (2010)
86. M. Bianchi et al., The electronic structure of clean and adsorbate-covered Bi_2Se_3: an angle-resolved photoemission study. Semicond. Sci. Technol. **27**(12), 124001 (2012)
87. T.V. Menshchikova, S.V. Eremeev, E.V. Chulkov, On the origin of two-dimensional electron gas states at the surface of topological insulators. JETP Lett. **94**(2), 106 (2011)
88. A.C. Ferrari, Raman spectroscopy of graphene and graphite: Disorder, electron–phonon coupling, doping and nonadiabatic effects. Solid State Commun. **143**(1-2), 47–57 (2007). https://doi.org/10.1016/j.ssc.2007.03.052
89. M.Z. Hasan, J.E. Moore, Three-dimensional topological insulators. Annu. Rev. Condens. Matter Phys. **2**(1), 55–78 (2011). https://doi.org/10.1146/annurev-conmatphys-062910-140432
90. Y.H. Wang et al., Observation of Floquet-Bloch states on the surface of a topological insulator. Science **342**(6157), 453–457 (2013). https://doi.org/10.1126/science.1239834
91. Z.-H. Zhu et al., Photoelectron spin-polarization control in the topological insulator Bi_2Se_3. Phys. Rev. Lett. **112**, 076802 (2014). https://doi.org/10.1103/PhysRevLett.112.076802

92. L. Wu et al., Quantized Faraday and Kerr rotation and axion electrodynamics of a 3D topological insulator. Science **354**(6316), 1124 (2016). https://doi.org/10.1126/science.aaf5541
93. C. Jozwiak et al., Spin-polarized surface resonances accompanying topological surface state formation. Nat. Commun. **7**, 13143 (2016)
94. W. Li et al., Origin of the low critical observing temperature of the quantum anomalous Hall effect in V-doped $(Bi, Sb)_2Te_3$ film. Scientific Reports **6**, 32732 (2016)
95. Y. Shao et al., Faraday rotation due to surface states in the topological insulator $(Bi_{1-x}Sb_x)_2Te_3$. Nano Letters **17**(2), 980–984 (2017). https://doi.org/10.1021/acs.nanolett.6b04313
96. D. Pesin, A.H. MacDonald, Spintronics and pseudospintronics in graphene and topological insulators. Nat. Mater. **11**(5), 409–416 (2012). https://doi.org/10.1038/nmat3305
97. Y. Xu et al., Observation of topological surface state quantum Hall effect in an intrinsic three-dimensional topological insulator. Nature Phys. **10**(12), 956 (2014)
98. Y. Fan et al., Electric-field control of spin–orbit torque in a magnetically doped topological insulator. Nat. Nanotechnol. **11**, 352 (2016). https://doi.org/10.1038/nnano.2015.294
99. Y. Wang et al., Topological surface states originated spin-orbit torques in Bi_2Se_3. Phys. Rev. Lett. **114**, 257202 (2015). https://doi.org/10.1103/PhysRevLett.114.257202
100. B. Scharf et al., Tunneling planar hall effect in topological insulators: Spin valves and amplifiers. Phys. Rev. Lett. **117**, 166806 (2016). https://doi.org/10.1103/PhysRevLett.117.166806
101. K. Kondou et al., Fermi-level-dependent charge-to-spin current conversion by Dirac surface states of topological insulators. Nature Phys. **12**(11), 1027–1031 (2016)
102. J. Tian et al., Observation of current-induced, long-lived persistent spin polarization in a topological insulator: A rechargeable spin battery. Science Advances **3**(4), (2017). https://doi.org/10.1126/sciadv.1602531
103. Y. Ando, M. Shiraishi, Spin to charge interconversion phenomena in the interface and surface states. J. Phys. Soc. Jpn. **86**(1), 011001 (2017). https://doi.org/10.7566/JPSJ.86.011001
104. A. Shekhter, M. Khodas, A.M. Finkel'stein, Chiral spin resonance and spin-Hall conductivity in the presence of the electron-electron interactions. Phys. Rev. B **71**, 165329 (2005). https://doi.org/10.1103/PhysRevB.71.165329
105. A. Ashrafi, D.L. Maslov, Chiral spin waves in Fermi liquids with spin-orbit coupling. Phys. Rev. Lett. **109**, 227201 (2012). https://doi.org/10.1103/PhysRevLett.109.227201
106. S. Maiti, V. Zyuzin, D.L. Maslov, Collective modes in two- and three-dimensional electron systems with Rashba spin-orbit coupling. Phys. Rev. B **91**, 035106 (2015). https://doi.org/10.1103/PhysRevB.91.035106
107. S. Maiti, M. Imran, D.L. Maslov, Electron spin resonance in a two-dimensional Fermi liquid with spin-orbit coupling. Phys. Rev. B **93**, 045134 (2016). https://doi.org/10.1103/PhysRevB.93.045134
108. F. Perez et al., Spin-orbit twisted spin waves: Group velocity control. Phys. Rev. Lett. **117**, 137204 (2016). https://doi.org/10.1103/PhysRevLett.117.137204
109. S. Maiti, D.L. Maslov, Raman scattering in a two-dimensional Fermi liquid with spin-orbit coupling. Phys. Rev. B **95**, 134425 (2017). https://doi.org/10.1103/PhysRevB.95.134425
110. A. Kumar, D.L. Maslov, Effective lattice model for the collective modes in a Fermi liquid with spin-orbit coupling. Phys. Rev. B **95**, 165140 (2017). https://doi.org/10.1103/PhysRevB.95.165140
111. E. Riccardi et al., Gate-dependent electronic Raman scattering in graphene. Phys. Rev. Lett. **116**, 066805 (2016). https://doi.org/10.1103/PhysRevLett.116.066805
112. O. Kashuba, V.I. Fal'ko, Signature of electronic excitations in the Raman spectrum of graphene. Phys. Rev. B **80**, 241404 (2009). https://doi.org/10.1103/PhysRevB.80.241404
113. M. Nomura et al., Relationship between Fermi surface warping and out-of-plane spin polarization in topological insulators: A view from spin- and angle-resolved photoemission. Phys. Rev. B **89**, 045134 (2014). https://doi.org/10.1103/PhysRevB.89.045134
114. S. Raghu et al., Collective modes of a Helical liquid. Phys. Rev. Lett. **104**, 116401 (2010). https://doi.org/10.1103/PhysRevLett.104.116401

115. S. Maiti, D.L. Maslov, Intrinsic damping of collective spin modes in a two-dimensional Fermi liquid with spin-orbit coupling. Phys. Rev. Lett. **114**, 156803 (2015). https://doi.org/10.1103/PhysRevLett.114.156803
116. F.-T. Huang et al., Nonstoichiometric doping and Bi antisite defect in single crystal Bi_2Se_3. Phys. Rev. B **86**, 081104 (2012). https://doi.org/10.1103/PhysRevB.85.081104
117. J. Dai et al., Toward the intrinsic limit of the topological insulator Bi_2Se_3. Phys. Rev. Lett. **117**, 106401 (2016). https://doi.org/10.1103/PhysRevLett.117.106401
118. E. Lahoud et al., Evolution of the Fermi surface of a doped topological insulator with carrier concentration. Phys. Rev. B **88**, 195107 (2013). https://doi.org/10.1103/PhysRevB.88.195107
119. K.F. Mak et al., Atomically thin MoS_2: A new direct-gap semiconductor. Phys. Rev. Lett. **105**, 136805 (2010). https://doi.org/10.1103/PhysRevLett.105.136805
120. X. Xu et al., Spin and pseudospins in layered transition metal dichalcogenides. Nature Phys. **10**(5), 343–350 (2014)
121. J.R. Schaibley et al., Valleytronics in 2D materials. Nat. Rev. Mater. **1**, 16055 (2016)
122. J.M. Lu et al., Evidence for two-dimensional Ising superconductivity in gated MoS_2. Science **350**(6266), 1353–1357 (2015). https://doi.org/10.1126/science.aab2277
123. X. Xi et al., Ising pairing in superconducting $NbSe_2$ atomic layers. Nature Phys. **12**(2), 139–143 (2016)
124. E.I. Rashba, Properties of semiconductors with an extremum loop 1 cyclotron and combinational resonance in a magnetic field perpendicular to the plane of the loop. Sov. Phys.-Solid State **2**, 1109–1122 (1960)
125. G. Dresselhaus, Spin-orbit coupling effects in zinc blende structures. Phys. Rev. **100**, 580–586 (1955). https://doi.org/10.1103/PhysRev.100.580
126. X. Zhang et al., Hidden spin polarization in inversion-symmetric bulk crystals. Nature Phys. **10**(5), 387–393 (2014)
127. D. Xiao et al., Coupled spin and valley physics in monolayers of MoS_2 and other Group-VI dichalcogenides. Phys. Rev. Lett. **108**, 196802 (2012). https://doi.org/10.1103/PhysRevLett.108.196802
128. Q. Liu, X. Zhang, A. Zunger, Intrinsic circular polarization in centrosymmetric stacks of transition-metal dichalcogenide compounds. Phys. Rev. Lett. **114**, 087402 (2015). https://doi.org/10.1103/PhysRevLett.114.087402
129. K.F. Mak, J. Shan, Photonics and optoelectronics of 2D semiconductor transition metal dichalcogenides. Nat. Photon. **10**(4), 216–226 (2016)
130. M. Gmitra, J. Fabian, Graphene on transition-metal dichalcogenides: A platform for proximity spin-orbit physics and optospintronics. Phys. Rev. B **92**, 155403 (2015). https://doi.org/10.1103/PhysRevB.92.155403
131. R.R. Parsons, Band-to-band optical pumping in solids and polarized photoluminescence. Phys. Rev. Lett. **23**, 1152–1154 (1969). https://doi.org/10.1103/PhysRevLett.23.1152
132. A.I. Ekimov, V.I. Safarov, Optical orientation of carriers in interband transitions in semiconductors. Sov. J. Exp. Theor. Phys. Lett. **12**, 198 (1970)
133. A. Bonnot, R. Planel, C. Benoit à la Guillaume, Optical orientation of excitons in CdS. Phys. Rev. B **9**, 690–702 (1974). https://doi.org/10.1103/PhysRevB.9.690
134. I. Žuti ć, J. Fabian, S. Das Sarma, Spintronics: Fundamentals and applications. Rev. Mod. Phys. **76**, 323–410 (2004). https://doi.org/10.1103/RevModPhys.76.323
135. R. Planel, C.B. Guillaume, Optical orientation of excitons, in *Optical Orientation*, Chap. 8, ed. by F. Meier, B.P. Zakharchenya (Elsevier Science Ltd., 1984), p. 353
136. K. Miyamoto et al., The gigantic Rashba effect of surface states energetically buried in the topological insulator Bi_2Te_2Se. New J. Phys. **16**(6), 065016 (2014)
137. Y. Liu et al., Tuning Dirac states by strain in the topological insulator Bi_2Se_3. Nature Phys. **10**(4), 294–299 (2014)
138. D. Kovalev et al., Free exciton emission in GaN. Phys. Rev. B **54**, 2518–2522 (1996). https://doi.org/10.1103/PhysRevB.54.2518
139. A.M. Jones et al., Optical generation of excitonic valley coherence in monolayer WSe_2. Nat. Nanotechnol. **8**(9), 634–638 (2013)

140. A.V. Shytov, M.I. Katsnelson, L.S. Levitov, Atomic collapse and quasi-Rydberg states in graphene. Phys. Rev. Lett. **99**, 246802 (2007). https://doi.org/10.1103/PhysRevLett.99.246802
141. A. Najmaie, E.Ya. Sherman, J.E. Sipe, Generation of spin currents via Raman scattering. Phys. Rev. Lett. **95**, 056601 (2005). https://doi.org/10.1103/PhysRevLett.95.056601
142. L. Fu, C.L. Kane, Superconducting proximity effect and Majorana fermions at the surface of a topological insulator. Phys. Rev. Lett. **100**, 096407 (2008). https://doi.org/10.1103/PhysRevLett.100.096407
143. V.V. Kruglyak, S.O. Demokritov, D. Grundler, Magnonics. J. Phys. D Appl. Phys. **43**(26), 264001 (2010)
144. J. Alicea, New directions in the pursuit of Majorana fermions in solid state systems. Rep. Progress Phys. **75**(7), 076501 (2012)
145. A.V. Chumak et al., Magnon spintronics. Nature Phys. **11**(6), 453–461 (2015)
146. R.K. Dumas, J. Akerman, Spintronics: Channelling spin waves. Nat. Nanotechnol. **9**(7), 503–504 (2014)
147. S. Urazhdin et al., Nanomagnonic devices based on the spin-transfer torque. Nat. Nanotechnol. **9**(7), 509–513 (2014)

Chapter 5
Conclusion

Abstract In this monograph, we used Raman scattering to identify collective excitations in the "chiral" symmetry channel in two completely different nonmagnetic systems: the HO phase of heavy fermion metal URu_2Si_2 and the surface states of 3D topological insulator Bi_2Se_3.

In URu_2Si_2, the collective mode arises from the uranium *5f* electrons orbital degrees of freedom, crystal field states. However, due to the partly localized and itinerant characters of uranium electrons, the crystal field excitations in URu_2Si_2 are overdamped and difficult to observe experimentally, as in many other uranium compounds. Indeed, due to Kondo hybridization with the conduction band electrons, the local crystal field picture often fails to explain the low temperature phases in heavy electron materials. However, the "arrested Kondo effect" in URu_2Si_2 is a special case where the local degrees of freedom reemerge and become important at low temperature [1]. The spontaneous symmetry breaking results the system in a chiral ground state, where the *5f* electron orbital states reorganize and order in real space to form a "chirality density wave" [1, 2]. One of the contributions in this monograph is to identify a collective mode out of this ground state and how it evolves across the temperature-doping phase diagram [3]. Both temperature and doping evolution of this mode match the low temperature phase diagram of URu_2Si_2, and the mode energy is fully consistent with the Ginzburg–Landau model by Haule and Kotliar [3–6]. The mode observed in Raman scattering of URu_2Si_2 appears only below the phase transition, which is very different from usual crystal field transitions that are already observable above the phase transition, and the linewidth gradually sharpen at low temperature due to reduced relaxation channels. The particular symmetry dependence of this collective mode allows us to identify the symmetry breaking in the HO phase, which strongly constraints the allowed order parameters for the HO phase. Thus, we demonstrate the unique property of Raman spectroscopy in determining the symmetry of collective modes, which is very useful when studying symmetry breaking and phase transitions in solids.

In Bi_2Se_3, the "stage setting" is very different than in URu_2Si_2. Here we do not have any f-orbitals or localized states at the Fermi level to worry about. But instead,

H.-H. Kung, *Collective Excitations in the Antisymmetric Channel of Raman Spectroscopy*, Springer Theses, https://doi.org/10.1007/978-3-030-89332-3_5

the strong SOC on the crystal surface collaborates with time reversal symmetry to produce a pair of 2D surface Dirac cones at Fermi energy around Γ point. The Dirac electrons possess a chiral spin texture, which has intrigued many studies in the recent years [7–10]. While the spin polarization and surface band structure have been well established, the spin correlation between the Dirac fermions is actually very little known. The main contribution in this monograph is to use Raman scattering to identify a mode in the chiral symmetry channel, which corresponds to the collective "spin-flip" excitation out of the surface's chiral ground state [11]. This mode is known as a "chiral spin mode" and is in fact generic to many systems with strong SOC [12, 13]. However, it has never been so cleanly observed in zero field at such high temperature and also never been observed from the surface states of a 3D TI. A chiral spin mode is similar to a magnetic bound state, which peels off from the "spin-flip" transition continuum due to finite interaction, and thus is a fantastic tool to measure the spin correlation between Dirac fermions. In order to selectively enhance the signal from the surface, we tune the excitation energy to resonantly promote interband transitions into unoccupied surface states. This innovative surface selective resonant Raman scattering paves new ways of studying the surface states in Bi_2Se_3 using optical methods [11].

We hope that the work presented in this thesis would inspire future studies in the condensed matter physics community, introducing polarization resolved Raman spectroscopy as a powerful tool to characterize the collective modes in solids. In particular, the sensitivity in the fully antisymmetric channel presents a unique tool to study chiral excitations in a wide range of systems. Besides materials with time reversal symmetry breaking, we argue that antisymmetric Raman excitations could be expected whenever reflection symmetries are broken or if the ground state has a chiral spin texture. These properties encompass a large group of materials that are interesting to the current research, and we believe the studies we present in this thesis would be a useful starting point for future researchers.

References

1. K. Haule, G. Kotliar, Arrested Kondo effect and hidden order in URu_2Si_2. Nature Phys. **5**(11), 796–799 (2009). https://doi.org/10.1038/nphys1392
2. H.-H. Kung et al., Chirality density wave of the "hidden order" phase in URu_2Si_2. Science **347**(6228), 1339–1342 (2015). https://doi.org/10.1126/science.1259729
3. H.-H. Kung et al., Analogy between the "Hidden Order" and the orbital antiferromagnetism in $URu_{2-x}Fe_xSi_2$. Phys. Rev. Lett. **117**, 227601 (2016). https://doi.org/10.1103/PhysRevLett.117.227601
4. K. Haule, G. Kotliar, Complex Landau-Ginzburg theory of the hidden order in URu_2Si_2. Europhys. Lett. **89**(5), 57006 (2010)
5. H. Kusunose, H. Harima, On the hidden order in URu_2Si_2 – antiferro hexadecapole order and its consequences. J. Phys. Soc. Jpn. **80**(8), 084702 (2011). https://doi.org/10.1143/JPSJ.80.084702
6. L. Boyer, V.M. Yakovenko, et al., A model for metastable magnetism in the hidden-order phase of URu_2Si_2. Ann. Phys. (New York) **388** (2018). https://doi.org/10.1016/j.aop.2017.11.024

7. M.Z. Hasan, C.L. Kane, *Colloquium* : Topological insulators. Rev. Mod. Phys. **82**, 3045–3067 (2010). https://doi.org/10.1103/RevModPhys.82.3045
8. M.Z. Hasan, S.-Y. Xu, G. Bian, Topological insulators, topological superconductors and Weyl fermion semimetals: discoveries, perspectives and outlooks. Physica Scripta **2015**(T164), 014001 (2015)
9. A. Bansil, H. Lin, T. Das, *Colloquium*: Topological band theory. Rev. Mod. Phys. **88**, 021004 (2016). https://doi.org/10.1103/RevModPhys.88.021004
10. Y. Ando, M. Shiraishi, Spin to charge interconversion phenomena in the interface and surface states. J. Phys. Soc. Jpn. **86**(1), 011001 (2017). https://doi.org/10.7566/JPSJ.86.011001
11. H.-H. Kung, et al. Chiral spin mode on the surface of a topological insulator. Phys. Rev. Lett. **119**, 136802 (2017). https://doi.org/10.1103/PhysRevLett.119.136802
12. S. Maiti, D.L. Maslov, Intrinsic damping of collective spin modes in a two-dimensional Fermi liquid with spin-orbit coupling. Phys. Rev. Lett. **114**, 156803 (2015). https://doi.org/10.1103/PhysRevLett.114.156803
13. S. Maiti, D.L. Maslov, Raman scattering in a two-dimensional Fermi liquid with spin-orbit coupling. Phys. Rev. B **95**, 134425 (2017). https://doi.org/10.1103/PhysRevB.95.134425

The manufacturer's authorised representative in the EU is Springer Nature Customer Service Centre GmbH, Europaplatz 3, 69115 Heidelberg, Germany. If you have any concerns regarding our products, please contact ProductSafety@springernature.com

Printed and bound by CPI Group (UK) Ltd, Croydon, CR0 4YY
30/09/2026
02225307-0012

Theory and History in the Human and Social Sciences

Series Editor
Jaan Valsiner, Department of Communication and Psychology
Aalborg University
Aalborg, Denmark

Theory and History in the Human and Social Sciences fills the gap in the existing coverage of links between new theoretical advancements in the social and human sciences and their historical roots. Making that linkage is crucial for interdisciplinary synthesis across the disciplines of psychology, anthropology, sociology, history, semiotics, and the political sciences. In contemporary human sciences of the 21st century, there exists increasing differentiation between neurosciences and all other sciences that are aimed at making sense of complex social, psychological, and political processes. This series serves the purpose of (1) coordinating such efforts across the borders of existing human and social sciences, (2) providing an arena for possible inter-disciplinary theoretical synthesis, (3) bringing to the attention of our contemporary scientific community innovative ideas that have been lost in the dustbin of history for no good reason, and (4) providing an arena for international communication between social and human scientists across the world.

Enno Freiherr von Fircks

Culture of Action

Contributions to Ernest Edouard Boesch's Psychology

Springer

Enno Freiherr von Fircks
Department of Psychology
Sigmund Freud University Vienna
Vienna, Austria

ISSN 2523-8663 ISSN 2523-8671 (electronic)
Theory and History in the Human and Social Sciences
ISBN 978-3-031-88807-6 ISBN 978-3-031-88808-3 (eBook)
https://doi.org/10.1007/978-3-031-88808-3

Cover illustration: Julius von Klever (1850–1924)

This Springer imprint is published by the registered company Springer Nature Switzerland AG
The registered company address is: Gewerbestrasse 11, 6330 Cham, Switzerland

Dying and Rising Again *(by Matthias Claudius)*[1]

O child of man, look all around …
Do you know a teaching here,
Greater and more comforting
For us who dwell down here?
There, where victory's palms are waved,
Exists pure being, no becoming,
No dying and no rising again,
As here, on earth, among us.
There, eternal joy prevails,
Eternal light and peace abide,
Life gently springs forth from itself,
And never wearies, never sides.
But to this underworld below,
No such glorious fate is given;
Here, without darkness, there's no glow,
Without death, no life is risen.
The lion decays, swells, and dies—
And food emerges from the beast;
The seed falls into Earth and lies,
Then silently it sprouts, released.
And nature is a mirror clear;
Its truth we are called to see:
What your spirit owes itself,
It must bring forth, to life, to be.
If truly happy you would be,
And build upon a steadfast ground,
You must not fear the thorny path,
Nor trust the rosy road unbound.
Once, a great man sought to teach
And guide us through this mystery;
He led a people through the night,
To find their May Day jubilee.
So save yourself much grief and strife,
O child of man, and heed:
Reflect on self-denial's call,
And on this lofty creed.

[1] The German original is available in the appendix.

Within us dwell two natures bound,
Yet both by one law steered:
Through death and suffering alone,
The way to true joy is cleared.

From Fantasms to Action: The Collective Action Potential (Editor's Preface)

This book is a celebration in two ways. First of all, it is a careful intellectual analysis of the lifetime contribution of Ernest Edouard Boesch to the science of psychology. Boesch was a leading twentieth-century European thinker whose contributions to psychology in the form of the Symbolic Action Theory constitute a synthesis of social and developmental psychology with psychoanalysis. He was the originator of cultural psychology in the decades after the Second World War, and a beloved mentor for a generation of German psychologists who passed through his tutelage at the University of Saarbrücken.

Secondly, this book is a celebration of the transfer of the role of the creator of ideas of an established theory to a dedicated theorist of a younger generation. The author of this book—Enno Freiherr von Fircks—has in a short time emerged in contemporary psychology as the analyst and developer of different classical theories of psychology—those of William Stern (von Fircks, 2024), and now—that of Ernst Boesch. This continuity is important—theories of the past should not be let to die away in the dusty bookshelves of library storage rooms or on the pages of history books where the authors are presented as icons of innovation. The more famous a theorist of the past is presented as part of the history of the field, the more easily is the substance of her or his theories forgotten, or re-told in numerous university courses as finished products. This is not the case—theories in psychology remain unfinished by their authors, and it takes scholars of younger generation to continue where the recognized famous authors finished. In the case of Boesch's theory it was his student Lutz Eckensberger who made the first move to extend it (Eckensberger, 2003). His contributions to the field of developmental psychology constitute a masterful elaboration of Boesch's theory to the understanding of action development in children and adolescents. Now, with the present book, we can observe a fresh new continuation of the effort to build upon Boesch's theory—toward elaborate understanding of the ways in which the internal personal world becomes transformed through potential assistance from others.

The particular innovation that extends Boesch's theory is the concept of *Collective Action Potential.* Von Fircks extends Boesch's Action Potential concept—originally phrased in terms of the individual *psyche*—to the domain of social relations

(Chap. 6). The author emphasizes the role of others in the social network of a person who are providing skills, techniques, and frameworks for developing of abilities by the individual. Both types of action potential—individual and collective—are viewed to be in continuous dialogue. This dialogue entails all forms of regulatory relations that any dialogue does—ranging from full support for the person by the other, to neutralized tolerance, and—in the other extreme—systematic blocking of the individual action potential. This dialogue is also a theoretical parallel to Lev Vygotsky's notion of the Zone of Proximal Development (Valsiner & van der Veer, 2014), which has similar characteristics. The field of opportunities set up for the developing persons by others—the whole cultural history of a given society—is the resource for the Action Potential of a person to expand. Yet, it is the action of the person oneself that makes the dialogue with the Collective Action Potential possible.

The process of co-construction of the Action Potential in dialogue is inherently hostage to the goal orientations of the person and the Other. The person's "amorphous inner Self" is the arena for such co-construction. It is here where the inclusion on the focus on poetry in this book is crucial. Psychology in its efforts to live up to the desired fantasm of science has usually distanced itself from the psychological creativity that over centuries and across societies has been the domain of the dialogical isle of the Collective Action Potential. Poetics involved in ordinary living (Lehmann et al., 2017) is the resource for developing of the Action Potential as well as re-adjusting it in therapeutic domains when needed.

Ernst Boesch left the science of psychology a legacy that is worth careful study and further development of the creative ideas that fill the masterful pages of his writings. He was a poet, a writer, and a psychologist in one person. The present book is a new synthesis of his ideas with our contemporary cultural psychology. As such, it has many new alleys for further elaborations. For science building on the ideas of the past is crucial. History of sciences is the resource of ideas once developed but not implemented for one or another reason by their authors. Eventually it is all based on some personal ideas of the reader that can expand to new breakthroughs in psychology. Reading the present book can be a spring of such experiences.

University of Aalborg, Denmark Jaan Valsiner

References

Eckensberger, L. H. (2003). Wanted: A contextualized psychology based on action theory. In T. S. Saraswathi (Ed.), *Cross-cultural perspectives in human development*. Sage.

Fircks, E. von (2024). *Learning with William Stern: Personology for future*. Information Age Publishers.

Lehmann, O., Chaudhary, N., Bastos, A. C., & Abbey, E. (Eds.). (2017). *Poetry and imagined worlds: Creativity in everyday experience.* Palgrave.

Valsiner, J., & van der Veer, R. (2014). Encountering the border: Vygotsky's zona blizaishego razvitya and its implications for theory of development. In A. Yasnitsky, R. van der Veer & M. Ferrari (Eds.), *The Cambridge handbook of cultural-historical psychology* (pp. 148-173). Cambridge University Press.

E.E. Boesch's Cultural Psychology: A Framework for Theory and Practice (Author's Preface: Enno Freiherr von Fircks)

My Personal Journey into Symbolic Action Theory

At the time, I was a young undergraduate student when I first contacted the prominent cultural psychologist, Jaan Valsiner. My intention was to delve deeper into the topic of myths and society. I had, among other things, studied myths from an existential perspective, drawing on the work of the psychoanalyst Rollo May (May, 1991). However, the existential perspective—often overly individualistic—did not fully satisfy my intellectual curiosity. Jaan Valsiner suggested that I explore the writings of E.E. Boesch and his symbolic action theory. I still recall the profound passion I felt while reading Boesch's works and the preliminary answers his perspective provided to my questions. I shared my enthusiasm for Boesch's writings with Valsiner via email, describing Boesch as an outstanding phenomenologist of everyday life. His autobiographical passages, seamlessly interwoven with his theoretical concepts, left a lasting impression on me. For the first time, I encountered a psychologist who did not turn a blind eye to the critical phenomena of everyday life but instead analyzed them with a holistic perspective.

At the time, I was still deeply influenced by the statistical worldview prevalent in psychology, which sought to express everything in numbers and discouraged engagement with everyday psychological questions (Valsiner, 2017) In my studies, we were told that analyzing such phenomena was unscientific because they could not be controlled. Boesch, however, took a different path, and I am convinced that the everyday psychological perspective enriches the two fundamental disciplinary preconditions for conducting psychology. In other words, psychological questions cannot only be addressed through philosophical principles (Wendt & Funke, 2022) or physical-biological considerations (Funke, 2023) but must also incorporate the culturally conditioned everyday world, yielding critical insights into an individual's psychological system. Boesch was a pioneer of this culturally grounded everyday psychology, and it was precisely this approach that so profoundly impressed my younger self and continues to shape my scientific thinking and practice.

The Power of Myths

Jaan Valsiner opened the door to Boesch's cultural psychology for me, but it was up to me to step through it independently. I began reading Boesch's foundational works, initially focusing on his more recent monographs on symbolic action theory (Boesch, 1991, 1998, 2000, 2005). Through these texts, I gained significant insights into my interest in myths. At that time, my fascination with myths stemmed from the post-World War II "Trümmerfrauen" (rubble women) myth in Germany. Even today, there is considerable debate among historians about whether the phenomenon of the "Trümmerfrauen" was a historical reality or more of a collective idealization (Hagerman, 2015). The historical consensus remains ambivalent (Kramer, 2016). What fascinated me was how both sides fought for the validity of their positions, revealing that myths involve psychological processes that extend beyond the mere pursuit of truth or claims to truth. Whether we believe in a myth depends on our needs, both individual and collective. The "Trümmerfrauen" narrative may have been exaggerated or amplified in historical accounts—a position one could agree upon—but the need for this narrative reveals deeper truths (Brockmeier, 2015; Münkler, 2004). The exaggeration or amplification of the phenomenon, as Boesch might argue, stems from a collective need to emphasize the active participation of all members of the German people in rebuilding the nation, moving beyond passivity in the face of the Nazi past (Reiter, 2016). This played a critical role in fostering a mental reconstruction of the nation, rooted in achievements of which people could ultimately be proud.

Philosophically, Boesch's cultural psychology embodies a pragmatic philosophy. It does not dwell on the metaphysical question of whether myth A or B is true but rather examines the consequences of belief in a myth, how it shapes everyday actions, and how it impacts the relationships between individuals affected by those actions.

Understanding Boesch's Background and the Boesch Reception in the Academic World

Despite the preliminary answers I found in Boesch's symbolic action theory, my fascination with his work remained strong. Independently, I acquired Boesch's earlier works (1971, 1975, 1976) and ventured into previously unfamiliar territory. Through these readings, I discovered that Boesch also introduced his readers to theories of affect and stress, as well as ideas on free association, which he differentiated from Freud's conceptualizations. These early works once again demonstrated Boesch's commitment to a holistic approach and revealed the broad applicability of his theoretical framework to psychology as a discipline. I was astonished to find that, aside from a few notable exceptions, Boesch's theoretical framework has received relatively little attention in contemporary research.

Of course, the Bochum group of cultural and action psychology (Straub et al., 2020), which also founded the *Society for Cultural Psychology* and biennially awards the E.E. Boesch Prize, must be acknowledged here. This group has produced promising approaches for extending Boesch's cultural psychology, though these developments remain largely confined to the German-speaking world (Straub, 2020, 2022). In this regard, I regret the lack of international engagement with the life's work of the Swiss psychologist E.E. Boesch. Even the Society for Cultural Psychology primarily attracts German-speaking psychologists and social scientists, yet science thrives on plurality, which must extend beyond a solely German-language reception. Boesch himself exemplified this international outlook through his work with UNESCO in Thailand, his early scientific endeavors in Geneva (in the Francophone region of Switzerland), his role as a school psychologist in St. Gallen, and his academic career in Saarbrücken, a Franco-German-Luxembourgish border region that reflected his international life trajectory.

I must also highlight Livia M. Simao, an active participant in Jaan Valsiner's cultural psychology network, who has dedicated herself to the reception of Boesch's work[1]. Her numerous publications demonstrate not only a profound understanding of Boesch but also an original body of work that advances his ideas and enriches the international cultural psychology community (Simao, 2005, 2017a, 2017b, 2023). I am particularly impressed by Simao's work on connections between Boesch's cultural psychology and ontology or existential philosophy (Simao, 2020). As we will see in this book, Boesch was consistently concerned with developing a balance between self and world. That is, we must transcend the state of being "thrown into" a pre-existing, culturally shaped world to create a meaningful life. This is one of the core tenets of existential philosophy (Pfeiffer, 1952) and aligns with Boesch's cultural psychology (Boesch, 2021).

Finally, I would like to comment Lonner and Hayes (2007), whose work internationalized Boesch's cultural psychology and offered intriguing insights into both his personality and his oeuvre. Their edited volume includes contributions from Boesch's contemporaries, introducing his ideas to a broader audience and showcasing his legacy. However, it is striking that, for nearly two decades, there has been no significant international research (such as a monograph) advancing Boesch's theories and sharing these developments with a wider academic community.

[1]One needs to note also the Special Issue in the journal *Culture & Psychology* dedicated to Boesch's Symbolic Action Theory.

Stimulating International Boesch Research After 20 Years: Addressing Research Gaps

This monograph aims to address this gap. Through a focused interpretation of Boesch's symbolic action and cultural psychology, I propose revisions to his theory that are not only scientifically valuable but also practical for applied researchers. Boesch's phenomenology of everyday life, in particular, demands that we derive applied implications from his theories. For example, in recent years, my research network has developed a leadership theory based on symbolic action theory (Ghazi & von Fircks, 2024). Boesch's action and cultural psychology also holds transformative potential in clinical settings, which we aim to harness. Accordingly, this monograph offers a diverse collection of chapters: the first three chapters develop the theoretical foundations of Boesch's work, while subsequent chapters explore its clinical applications (Chaps. 4, 5, 6, and 7). Finally, we venture to propose a philosophy of life based on Boesch's action psychology, offering a framework for personal self-reflection (Chap. 8). The concluding chapters address Boesch's conception of poetics (Chap. 9) and literature (Chap. 10), integrating additional thinkers such as Cassirer and Pfeiffer to create interpretative guides that can be applied in schools and the study of poetry. These efforts aim to enhance the dialogical engagement between texts, authors, and readers.

The central message of this monograph is to strengthen our engagement with the cultural and social world and, in doing so, deepen our understanding of the self. In conclusion, the Saarbrücken school of cultural psychology (Valsiner, 1997) remains vibrant, and it is our responsibility to breathe life into it through continuous and dialogical scholarship.

I would like to express my sincere gratitude to Hedwige Boesch (-Achermann), whose support made it possible for me to precede each chapter with poems by Boesch. This collaboration also seems to foster a fruitful exchange of ideas. Furthermore, I extend my heartfelt thanks to my friend, Jaan Valsiner, who tirelessly assisted me in shaping and refining the individual chapters, offering his guidance and inspiration at all times. Finally, I am also grateful to Anna Ghazi, who engaged in valuable discussions with me regarding the content of the chapters and provided insightful suggestions for optimizing my scope of action, both academically and personally.

References

Boesch, E. E. (1971). *Zwischen zwei Wirklichkeiten: Prolegomena zu einer ökologischen Psychologie*. Huber.

Boesch, E. E. (1975). *Zwischen Angst und Triumph: über das Ich und seine Bestätigungen*. Huber.

Boesch, E. E. (1976). *Psychopathologie des Alltags: zur Ökopsychologie des Handelns und seiner Störungen*. Huber.

Boesch, E. E. (1991). *Symbolic action theory and cultural psychology*. Springer.

Boesch, E. E. (1998). *Sehnsucht: Von der Suche nach Glück und Sinn* [Longing: On the search of joy and meaning] (1st ed.). Huber.

Boesch, E. E. (2000). Das lauernde Chaos. *Mythen und Fiktionen im Alltag. Huber*

Boesch, E. E. (2005). *Von Kunst bis Terror: Über den Zwiespalt in der Kultur* [From art to terror: On the cleavage of culture]. Vandenhoeck & Ruprecht.

Brockmeier, J. (2015). *Beyond the archive: Memory, narrative, and the autobiographical process.* Oxford University Press.

Funke, J. (2023). Wundts Irrtum oder: Was ist die bessere Vorbild-Disziplin, Physik oder Biologie?. In G. Jütteman (Ed.) *Wie der Mensch sich selbst entdeckte* (pp. 215–226). Psychosozial-Verlag.

Ghazi, A., & von Fircks, E. (2024). From an emotional autoethnography towards a theory of Symbolic Action Leading (SAL) [Manuscript submitted for publication]. In E. von Fircks (Ed.), *Culture and leadership: From approximation towards symbiosis*. IAP

Hagemann, K. (2015). *Geschichtswissenschaft, Medien und kollektives Gedächtnis. Zum "Mythos Trümmerfrauen"* (Vol. 2015). Peter Lang Academic Publishing. https://www.ingentaconnect.com/content/plg/npl/2015/00002015/00000002/art00002; https://doi.org/10.3726/91505_203

Kramer, N. (2016). Von der ›Volksgenossin‹ zur ›Trümmerfrau‹? – Deutungen von Frauen nach 1945. In *'Volksgemeinschaft': Mythos, wirkungsmächtige soziale Verheißung oder soziale Realität im 'Dritten Reich'?* (pp. 301–317). Brill | Schöningh. https://doi.org/10.30965/9783657771653_017

Lonner, W. J., & Hayes, S. A. (2007). *Discovering Cultural Psychology: A Profile and Selected Readings of Ernest E. Boesch*. IAP.

May, R. (1991). *The cry for myth*. WW Norton.

Münkler, H. (2004). Der Antifaschismus als Gründungsmythos der DDR. In *Mythos und Mythologie* (pp. 221–236). AKADEMIE VERLAG. https://doi.org/10.1524/9783050081533.221

Pfeiffer, J. (1952). *Existenzphilosophie. Eine Einführung in Heidegger und Jaspers* [Exis tential philosophy: An introduction to Heidegger and Jaspers]. Meiner.

Reiter, M. (2016). *Die Generation danach: Der Nationalsozialismus im Familiengedächtnis*. Studienverlag.

Simão, L. M. (2020). Disquieting experiences and conversation. *Theory & Psychology, 30*(6), 864–877. https://doi.org/10.1177/0959354320957402.

Simão, L. M. (2005). Bildung, culture and self. *Theory & Psychology*, *15*(4), 549–574. https://doi.org/10.1177/0959354305054751

Simão. (2017a). Ernst E. Boesch's holistic cultural psychology. In *Striving for the whole* (pp. 131–150). Routledge. https://doi.org/10.4324/9781315130354-8

Simão, L. M. (2017b). Once more about Boesch's ontological predication in focus. *Integrative Psychological and Behavioral Science*, *51*(4), 670–679. https://doi.org/10.1007/s12124-017-9396-4

Straub, J. (2020). *Handlung, interpretation, kritik: grundzüge einer textwissenschaftlichen Handlungs-und Kulturpsychologie* (Vol. 18). Walter de Gruyter GmbH & Co KG.

Straub, J. (2022). *Verstehendes Erklären*. Psychosozial-Verlag.

Straub, J., Chakkarath, P., & Salzmann, S. (Eds.). (2020). *Psychologie der Polyvalenz: Ernst Boeschs Kulturpsychologie in der Diskussion*. Bochum: Westdeutscher Universitätsverlag.

Valsiner, J. (2017). *From methodology to methods in human psychology*. Springer.

Valsiner, J. (1997). The Saarbrucken tradition in cultural psychology, and its legacy. *Culture & Psychology*, *3*(3), 243–245. https://doi.org/10.1177/1354067X9733001

Wendt, A. N., & Funke, J. (2022). *Wohin steuert die Psychologie?: ein Ausrichtungsversuch*. Vandenhoeck & Ruprecht.

Conflict of Interest The author declares no conflict of interests nor competing interests.

Funding The author declares not having received funding.

Ethical Approval The article does not contain any studies with human participants performed by the author.

Informed Consent The article is a theoretical article and does not require consent.

Data Availability Statement Not applicable

Data Deposition Not applicable

Contribution Statement The article is a single work by Enno von Fircks.

Acknowledgments Für meine zukünftigen Studenten und Freunde.

Contents

About the Author

Enno von Fircks is an applied psychologist. His work is concentrated on applying Boeschian cultural psychology to the work setting, as well as showing the philosophical and literary roots of cultural psychology. He also works as a Gestalt practitioner, as a lecturer, and as a tennis instructor in Germany providing him with the opportunity to apply his theoretical insights and develop them in vivo. He has published more than 50 articles and chapters in the fields of clinical, educational, organizational, and general psychology. Further, he has published a monograph in the *SpringerBrief Series* edited by Jaan Valsiner and Pina Marsico called *Conservativism: A cultural-psychological synthesis.* A second monograph was published in 2024 about William Stern's Critical Personalism and its pathways for revolutionizing today's psychology holistically published by IAP (*Learning with William Stern: Personology for Future*). An edited volume—equally published by IAP—called *Culture and Leadership: From Approximation towards Symbiosis* is currently in press. His current projects include two edited volumes about Goethian science and its implications for psychology as a whole as well as a volume in French about cultural-psycholcgical theories and methodologies for the francophone area.

Chapter 1
From What-Is to What-Should-Be: A Systematic Approach to Understanding Action in Cultural Psychology

Evening and Morning *(E.E. Boesch, Enno von Fircks translation)*
In the evening, when so quietly
The heavens close themselves,
And pour their silver richness
Over us as night compels,
The sea begins to stir,
Dark, heavy, soft it beats,
And murmurs forth its rhythm
Along the path where slumber meets.
Till finally its dance subsides,
Sinking gently into rest,
And in the vast, deep silence,
The dreamer is addressed.
Then in a pale and tender
sky of blue the moon does stand,
With but a single star,
Left alone upon the strand.
Infinitely soft and mild,
The day ascends above the sea,
No breeze stirs in its gleaming shield,
No sound, save bird-song's melody.
Gently breaks the shell
In which the night enclosed us tight,
Until the morning overflows
And bathes us in light.[1]

[1] Several poems written by E.E. Boesch will be presented in this book. These poems originate from a private collection that Boesch personally published for his friends and colleagues (Boesch, E.E. (2005). Reime und Bilder, Privatdruck). The author was granted access to this collection through the kind facilitation of Jaan Valsiner. Copyright was granted kindly by Hedwige Boesch (see author's preface). Parts of the poems will be discussed in each chapter.

E. Freiherr von Fircks, *Culture of Action*, Theory and History in the Human and Social Sciences, https://doi.org/10.1007/978-3-031-88808-3_1

Personal, Scientific, and Historical Background of Boesch's Symbolic Action Theory

E.E. Boesch was a Swiss psychologist deeply invested in establishing the foundation of the discipline of cultural psychology (Valsiner, 1997). Boesch envisioned this emerging field not as an additional subbranch of psychology but as a unifying approach capable of analyzing phenomena integral to everyday life (Boesch, 1980). He critiqued the dominant trends in psychology during his time, which he perceived as overly focused on statistical and quantitative methods that prioritized fragmented analyses of psychological operations (Boesch, 1980). In contrast, cultural psychology was positioned as a holistic approach, enabling the analysis of psychological processes as they develop and transform (Boesch, 1980). Thus, from its inception, Boesch's cultural psychology was inherently developmental in nature.

Boesch, eager to integrate cultural psychology with psychoanalytic principles, became increasingly engaged with studying how culture shapes individual actions and how generalized systems of meaning regulate action tendencies (Boesch, 1977/2006). Central to Boesch's work was the concept of symbolism, which he defined as the idea that persons, events, and situations carry symbolic meanings tied to specific actions and tendencies of conduct. Drawing inspiration from the French psychoanalyst Pierre Janet, Boesch viewed action as the fundamental unit of psychology (Boesch, 1991; Janet, 1919). However, he emphasized that understanding action requires attention to both the ecological and personal determinants of a given situation (Boesch, 1975, 1976). This dynamic interplay between ecological and personal factors formed the basis of what Boesch termed *active symbolism*, a concept that became a recurring theme throughout his academic career (von Fircks, 2023a).

Boesch's ideas were significantly impacted by his experiences in Thailand, where he participated in UNESCO's developmental aid program. This immersion in Thai culture allowed him to observe the everyday lives of the local people (Boesch, 1980). His publications often illustrate the pillars of his symbolic action theory through vivid examples drawn from his time in Thailand, his visits to colleagues and friends, and his observations of people navigating daily activities, such as traveling from one place to another using various means (Boesch, 1980). These experiences were foundational in shaping his theoretical insights, as living as a foreigner and adapting to the cultural patterns of Thailand offered him unique perspectives. His second marriage to *Supanee* (1931–2020), a native of Thailand, further deepened his understanding of Thai cultural life. This dynamic interaction between personal and ecological determinants fostered a nuanced understanding of how symbolism emerges both for the individual and the generalized social context. In this sense, Boesch can be regarded as an early autoethnographer, long before the methodology gained prominence in psychology and other disciplines (Ellis et al., 2010). His writings reveal a seamless integration of personal experience, observation, and theory generation.

As a psychotherapist, albeit without active practice, Boesch was not solely concerned with how individuals structure space and time through generalized meaning-making systems. He also focused on how people can become trapped in particular personal and social situations (Boesch, 1976). He analyzed how pathological conditions manifest in daily life, exploring how individuals may develop a state of disequilibrium that conflicts with their personal needs. Boesch was acutely aware that not all meaning-making patterns support individuals in achieving an I/World balance that satisfies their personal goals and needs (von Fircks, 2022). This sensitivity distinguished him from many other cultural psychologists and underscored his holistic approach to understanding human experience.

Historically, Boesch's symbolic action theory represents one of the first comprehensive scientific systems within cultural psychology. His work laid the groundwork for the holistic analysis of psychological phenomena, preceding the cultural-psychological contributions of scholars like Bruner (1986, 1996, 1997) and Valsiner (1998, 2007). This pioneering role has led some contemporary cultural psychologists to refer to the *Saarbrücken School of Cultural Psychology*, recognizing Boesch's institutionalization of his theory in the European region of Saarbrücken (Valsiner, 1997). Cross-fertilization between Boesch and Valsiner is evident in their respective works, particularly in Valsiner's major contributions to the foundations of cultural psychology (Valsiner, 2014).

Boesch's intellectual courage is evident in his willingness to challenge the dominant trends of his time, which prioritized large-scale data accumulation and advanced multivariate statistical analyses focused on population-based inferences (Boesch, 1980). His advocacy for a holistic approach not only established the foundations of cultural psychology but also proposed the concept of an *Anderswelt*—an alternative world that defies prevailing choices and narratives (Boesch, 1991, 2005). This alternative perspective remains a cornerstone of his enduring legacy within the field.

Action as the Germ Cell of Cultural Psychology

Having examined the personal, social, and historical context of Boesch's contributions, we now turn to the core theoretical underpinnings of Boesch's symbolic action theory. This chapter aims to provide a general overview of his cultural-psychological system while attempting to organize and systematize his body of work. Kurt Lewin, one of the most prominent social psychologists, emphasized the importance of conceptual clarity and the systematic development of theory as crucial yet often underestimated scientific achievements (Lewin, 1982). In the early decades of his work, Lewin concentrated on refining his field theory, aiming to systematize dynamic psychological phenomena. Similarly, Boesch's work has been critiqued for its complexity; various scholars have argued that Boesch's theoretical system and concepts are challenging to comprehend and apply, limiting their adoption in mainstream psychology (Straub, 2021).

While I partially agree with this critique, it is important to highlight that Boesch's focus on the transformation and alteration of psychological phenomena, both personal and social, inherently complicates the task of systematization—especially for students and researchers accustomed to psychology's statistical and reductionist frameworks. In recent years, I have made modest efforts to address this gap, and this book, particularly this chapter, aims to assist psychologists in leveraging Boesch's theory to address not only the challenges facing cultural psychology but also those confronting psychology as a broader discipline. As Valsiner (2021) suggests, laying the groundwork for a unifying framework from which specialized subfields can emerge is an essential, though often neglected, task in psychology today.

Defining Action in Boesch's Symbolic Action Theory

Boesch conceptualized action as the germ cell of his symbolic action theory (Boesch, 1991). But what exactly is an action? For Boesch, an action is a fundamental means by which an individual seeks to change the external world to better align with personal needs. Through action, individuals transform their environment to make it more compatible with their internal states (Boesch, 2021). Simply put, action mediates between a *what-is* condition (the current state of the world) and a *what-should-be* condition (a desired state of the world).

Importantly, an action is not a static phenomenon—it exists within the interplay of past, present, and future. It serves as a bridge between the crystallized structures of the past (*what-is*) and the envisioned structures of the future (*what-should-be*). For instance, consider the simple act of preparing food. This process begins with a need—hunger—that prompts the individual to take steps toward satisfying it. In this context, the act of cooking becomes the action that connects the *what-is* condition of hunger to the *what-should-be* condition of satiation.

Moreover, the act of preparing food alters the environment in multiple ways. For example, the resources used in cooking—such as food from the refrigerator—become unavailable for future use. Similarly, tools like knives may dull through repeated use. This illustrates a fundamental principle of action: it leaves tangible traces in the environment and never occurs in a vacuum (Boesch, 1980, 1991, 2005).

The Transformative and Residual Effects of Action

Boesch's theory underscores that every action consumes resources, both personal and external. Personal resources include the energy and time invested in performing the action, while external resources involve the materials or tools used. For example, the act of cooking not only changes the state of food but also affects the physical state of the instruments (e.g., dulling a knife) and leaves behind environmental traces (e.g., a messy kitchen).

On a deeper level, actions also leave what we call personological traces (von Fircks, 2024a). These traces are evident in how individuals engage with their environment over time. Consider, for example, people walking along a street. Each individual has a unique gait, and over time, this way of walking creates visible wear patterns on their shoes—such as uneven wear on one side. Prolonged repetition of these walking patterns can even lead to long-term orthopedic issues, illustrating how seemingly mundane actions can leave lasting personal and environmental effects.

Through these examples, it becomes evident that action is not merely a momentary event but a process that integrates past, present, and future, transforming both the individual and the world around them. This dynamic, integrative perspective lies at the heart of Boesch's symbolic action theory and represents a critical foundation for understanding psychological phenomena within cultural psychology.

An action, as Boesch emphasizes, fundamentally transforms the world in personal and meaningful ways, including its impact on the individual's body situated within that environment. Consider, for example, the footpaths we encounter in forests. These paths were initially nonexistent but were gradually formed by individuals seeking to cut across specific areas to save time or reach a viewpoint. The construction of such paths over time is the result of repeated actions—walking—driven by personal needs, such as the desire to arrive at a destination sooner or enjoy a panoramic view of the surroundings. In this context, the action mediates the transition from a *what-is* condition (e.g., an untrodden area) to a *what-should-be* condition (e.g., a desired viewpoint or destination).

However, this mediation is not a one-dimensional process. There are often multiple trajectories or actions that can help achieve the same *what-should-be* condition. For instance, one could navigate a forest by creating a shortcut, using a map, or relying on GPS navigation. Similarly, the same action—such as walking—can simultaneously fulfill multiple needs, such as arriving at a destination and experiencing physical exercise or relaxation. This phenomenon, which Boesch refers to as the *polyvalence of action*, underscores the multiplicity of ways in which actions can fulfill a given need structure or how a single action can address multiple needs (Boesch, 1991, 1998).

Actions as Part of an Interconnected System

To fully understand the concept of action within Boesch's symbolic action theory, it is essential to go beyond the germ cell of a single action and consider its interconnectedness with other actions. Actions rarely occur in isolation; instead, they form part of broader networks or systems of actions, which Boesch terms *action systems* (*Handlungssysteme*) (Boesch, 1991). These systems are critical for the closure of a given need structure, as they involve a series of interconnected steps that collectively lead to the fulfillment of the need.

For instance, if I wish to arrive early at a friend's house by taking a shortcut through a forest, my initial action might be walking along an unbeaten path. However, this single action may not suffice to fulfill my need. I might also need to reorient myself periodically by checking my smartphone or GPS to ensure I remain on the correct route. Additionally, I may need to navigate physical obstacles, such as climbing over rocks or avoiding fallen trees. These supplementary actions—navigating, climbing, and checking directions—are interconnected and collectively form an action system that enables me to achieve my goal.

Boesch (2021) highlights that actions within an action system are not merely sequential but dynamically interrelated, adapting to changes in the environment or new resistances that arise. For example, encountering a new obstacle, such as an unexpected river, might require improvising a new action—like finding a bridge or crafting a temporary solution to cross. This dynamic interconnectedness of actions within a system is a critical aspect of Boesch's theory, emphasizing that no action occurs in isolation.

The Action System as a Holistic Framework

One of the most significant contributions of Boesch's symbolic action theory is not only the analysis of individual actions and their ecological consequences but also the understanding of their relationships within a broader action system. These systems, composed of interconnected actions, are essential for the resolution of complex needs and goals. By focusing on the holistic nature of action systems, Boesch offers a framework that extends beyond isolated acts, recognizing the interplay between individual actions and the broader contexts in which they occur.

In conclusion, the polyvalence and interconnectedness of actions highlight their dynamic and transformative nature, as well as their role within larger systems. Boesch's focus on action systems underscores the need to study actions not only as discrete phenomena but also as components of broader, goal-oriented processes. This perspective enriches our understanding of human conduct, emphasizing the complexity and adaptability of actions within ecological and personal contexts.

Fantasms as Personal Regulators of Action

This contemplation leads us to an essential pillar of Boesch's theory: the inclusion of symbolism within scientific analysis. According to Boesch, actions are inherently symbolic, reflecting the individual's underlying needs and motivations. To illustrate this, consider a modern example: individuals choosing to follow a diet. One person might decide to diet to lose weight, another might engage in fasting for spiritual reasons, and yet another might aim to adopt healthier habits. While the external action—being on a diet—appears identical from an ecological perspective, the motivations (meaning) behind it differ significantly based on the personal needs of each individual.

For this reason, Boesch introduces the concept of *fantasms* to explain how actions are personally regulated. In my own interpretation, the action becomes refracted through the individual's personal lens, shaped by their unique hierarchy of needs. In essence, the symbolic meaning of the action is deeply tied to the individual's subjective experience and motivations what Boesch calls fantasms. Regarding fantasms, Boesch explains the following:

> [A] fantasm is here regarded as a kind of overarching goal; it concerns the nature of the anticipated ego-world relationship of the individual, including both hopes and fears. The personal contents the individual gives to a term like "happiness" would constitute a fantasm, as would the subjective meaning given to the term "self-realization" or "order." (Boesch, 1991, p. 124)

With the concept of fantasms, we enter Boesch's interpretation of psychoanalysis (Boesch, 1977/2006). Boesch did not approach psychoanalysis dogmatically, nor did he rigidly adhere to its principles. Instead, he incorporated psychoanalysis into the foundation of his cultural psychology, utilizing it as a means to understand and interpret the personal symbolism underlying human actions. According to Boesch, individuals may act in certain ways without fully understanding the motivations driving their conduct. They might believe that an action satisfies one need, while it actually addresses another, deeper one.

For example, consider an individual who chooses to play soccer. They might claim that their primary motivation is the enjoyment of being part of a team, suggesting a desire to fulfill social needs. However, upon closer observation, a psychologist might find that the individual also uses sport as a way to satisfy other needs, such as enjoying the competitive and aggressive aspects of tackling opponents, thereby addressing a latent need for aggressive expression. Alternatively, the motivation to play soccer could stem from an intricate interplay of needs and fantasms. According to Boesch and other psychologists, such as Stern (1924, 1935), the motivations for an action are often polyvalent, meaning they derive from multiple overlapping needs.

A Nuanced Approach to Psychoanalysis

Although Boesch acknowledged the significance of sexual needs within human conduct (Boesch, 1980), he rejected the notion that all fantasms are necessarily rooted in sexuality. Instead, he argued that symbols and their meanings are far more complex and multifaceted than the sexual interpretations traditionally emphasized in classical psychoanalysis (Boesch, 1991). Boesch's perspective aligns with those of Stern (1935) and Valsiner (2024), who advocate for a flexible and undogmatic application of psychoanalysis. This approach has been a challenge for psychoanalysis since its inception, as it often struggled with rigid interpretations. When discussing fantasms and their connection to individual needs, Boesch identifies four types of symbolism: instrumental, functional, situative, and analogous. Each type requires further explanation to clarify its role in Boesch's framework.

Instrumental symbolism refers to the symbolic meaning of an action as a tool for bridging the gap between the current "what-is" condition and the desired "what-should-be" condition. This symbolism emphasizes the practical and goal-oriented nature of an action, where individuals address specific issues, problems, or goals. To achieve their objectives, individuals often rely on environmental objects or tools that facilitate their actions. For example, someone who wishes to prepare a meal will use a knife and stove as instrumental tools to bridge the gap between their hunger (the "what-is" condition) and the completed meal (the "what-should-be" condition). Instrumental symbolism focuses on the ecological determinants of actions, as these are the tangible resources that individuals employ to fulfill their needs. Understanding the instrumental symbolism of an action requires analyzing not only the actions themselves but also the tools and resources used to carry them out. Boesch refers to these resources as *operators*, although I find the term *tool use* to be more precise and applicable (von Fircks, 2023b). This nuanced understanding of instrumental symbolism illustrates the dynamic interplay between actions, their underlying motivations, and the resources employed to achieve desired outcomes. By integrating these elements, Boesch's framework provides a comprehensive perspective on how actions are symbolically and functionally tied to personal needs and ecological factors.

Functional symbolism also needs to be discussed when undertaking a specific action or relying on a given tool. Here, the person experiences some kind of mastery over herself as well as over the material environment (Boesch, 1975, 1980). In other words, the person feels herself capable of reaching her very own *should-values* within her larger value system, and secondly, the person feels functional pleasure when relying on specific tools while undertaking the action. Thus, we need to differentiate between two kinds of functional symbolism—the first relates to the mastery of one's very own *should-values* (*I should master this exercise*), while the second shows some kind of mastery over the environmental challenges when undertaking an action. Here, we can say that the person can master personal and environmental challenges by herself. Yet, in a similar vein, we need to clarify that these personal and environmental challenges are to be understood in a dynamic fashion. This was especially shown by K. Lewin, who could prove that the person shifts her aspiration level after success and lowers it after continuous failure (Lewin, 2013). This might also concern the environmental challenges the person faces when undertaking a given action. Here, the person might be bored with a specific tool use when wanting to reach a certain state of equilibrium and starts to alter the way she manages environmental challenges. One of the most interesting things I have seen regarding this fact is when I visited a garage and saw an employee not carrying car tires but rolling them with a certain rhythm. I will not forget the ease of the person rolling the car tires in that manner, nor will I forget the pleasure it unfolded for the person as well as for myself when observing this employee. This specific technique of rolling the car tires towards a particular location helped the person to achieve *should-values* of her own while mastering a given way of doing that in a pleasant manner. Experiencing oneself as capable over one's very own aspirations and the environmental challenges denotes the functional symbolism Boesch does not get tired of speaking about (Boesch, 1976).

Analogous symbolism is important when a given action evokes analogous experiences from the past. Boesch uses the example of building a house in Thailand while planting wooden pillars into the fresh ground—here, he refers to the person experiencing the act of sexual penetration while doing so (Boesch, 1980). In other words, we are confronted with symbolism based on actions that appear to be analogous to each other. However, it is important to note that such analogous experiences are not limited to the domain of sexual experience but extend beyond this specific aspect of human experience. As we know from William Stern (1923, 1924), the person is and remains *unitas multiplex*, meaning that the person has many different needs that shift to the foreground depending on the personal value system as well as the environmental demands the person faces at that moment. Referring to analogous experiences solely in terms of sexuality would be to equate Boesch's cultural psychology with a dogmatic interpretation of psychoanalysis. Such scientific equalization would be erroneous according to Boesch's thinking and major works.

Another example not related to sexuality might involve the experience of switching on a light, which could be interpreted as an analogous experience of having an insight (an *Aha-Moment*). Buddhist and Taoist literature are particularly full of analogous experiences; here, nature is interpreted through a specific meaning-making system that originates from Yin and Yang or the Tao (Boesch, 1980; von Fircks, 2023c). Switching on the light might be interpreted by Taoist or Zen practitioners as the natural flow of life, where light is one pole and darkness the other. In this context, the moment of switching on the light could be seen as the analogous experience of being part of the cosmos and connected to the earth. This interpretation is similar to Boesch's early writings, where the Swiss psychologist did not shy away from advocating that any view of nature is already culturally regulated—and that culture is the natural biotope of the human being (Boesch, 1980).

Having elaborated on the first three kinds of symbolism, we now turn to the explanation of what Boesch calls situational symbolism. For example, if a person experiences in free association the climbing of a certain mountain, she might be reminded of a specific situation in which she climbed with her father and where they both had a nice breakfast once they reached the summit (Boesch, 1977/2006). Here, the person experiences the warmth and care of her father as she sees the mountain in front of her. Hence, the person is brought back to a specific event where she felt certain emotions, which she is able to feel once again through her memories. The image of the mountain allows her to relive this experience or memory once again, reminding her of the tender qualities of her father that she enjoyed when she was still a child. For example, when I personally see a tennis court in front of me, I am reminded of how I learned to lose, because I had difficulty dealing with losses when I was a child. Seeing the court reminds me of the journey I needed to undertake in order to learn this essential personal quality. The power of literature and cinematography lies in how they elicit situational symbolism in the reader.

However, when we talk about different kinds of symbolism, we must also mention that different symbols can cross-fertilize each other. In other words, if a person is confronted with a specific image or picture of a mountain—say through free association in therapy—she might be reminded of her past experience with her father, but the mountain might also evoke an analogous symbolism. The mountain might symbolize success, danger, or beauty, among many other potential experiences. Therefore, when confronted with a complex picture, it is often difficult to encounter a single type of symbolism. On the contrary, we are usually confronted with an interconnected set of symbols that also mirror an interconnected set of needs—something we have referred to as the need hierarchy. What the contemplation of symbols has illustrated is the fact that private needs become refracted through fantasms that carry different kinds of symbolic meaning. To gain insight into an individual's need hierarchy, it is crucial to understand the type of fantasm (symbolism) involved and what it reveals about the interconnection of needs (von Fircks, 2023d).

Myths as Social Regulators of Action

However, we need to become aware of the fact that the person is not an individual living an isolated life. On the contrary, the person is always surrounded by a given social environment that helps her to satisfy her needs, shows her new pressing needs, as well as ways to fulfill those needs. In other words, this social environment—family, peers, colleagues, friends, and so forth—is already present in the action field of a given person. Hence, the person is and remains a social being—something which is clear even inside the mother's womb. This social environment consists of different zones (Boesch, 1980) that either promote actions or inhibit actions in specific directions. If we are confronted with an individual having his own room, he might decorate and ornament this room according to his central fantasms. This room is then his private meaning-making arena—a promotion of individuality (Boesch, 1998). Here, the person might act in this room as he wishes to do—without the parents interfering in this process. Yet, the room is part of a larger social environment and atmosphere—the house or apartment. In other words, once the person steps out of his room, he finds himself in other zones that work under other social premises. Thus, the person enters spheres of joint living, penetrates areas of privacy by his parents, and leaves the house to go to school. Hence, what we can say is that the person does not live a simple life according to his very own fantasms. However, granting the person his own space and time for privacy becomes limited to a certain degree by the person being part of the family and joint living spaces in the house. Here, the joint living spaces are organized according to social should-values (Boesch, 1980). These should-values can gain positive or negative value regarding environmental stimuli. For example, the person entering the kitchen is confronted with the value: "You should cook in here, but you must not play." Or the person enters the living room, and his father explains: “You should relax in here, but you must not run. For running, you should go into the garden.”

What we can learn from these everyday examples is that space and time (e.g., “in the morning you should eat breakfast”) are structured according to social should-values that are transmitted via social interactions between different people (Boesch, 1980). Clearly, these should-values might only exist psychologically (e.g., “You should not penetrate your father’s room while he is sleeping”), but they can also manifest themselves materially within the social environment of the person. For example, the father might lock his door when going to sleep so that he will not be disturbed by his children. Or the family will hide a specific location inside the house from their children to avoid them being confronted with particular should-values. If people do not want other people to find out what they did or did not do in the past, they hide things either in the cellar or in the attic so that others might not get the opportunity to discover this or that. This holds especially true for children and their fears of the unknown, which may act as a developmental inhibition to enter the attic or the cellar. In short, we can say that the environment is not only structured personally but also socially, with regard to the importance of social should-values.

When referring to individual fantasms, we have discussed the fact that they are organized in a given hierarchy, meaning that fantasms are not only interconnected but also have different levels of individual importance. This fantasmatic importance is then associated with the importance of needs within the life of a given person, meaning that some needs are more important than others. Surely, the importance of needs—the need hierarchy—might change over time, and we need to operationalize the hierarchy in a dynamic fashion. Yet, this does not change the fact that they find themselves in a given hierarchy. Something similar holds true for social should-values. These are also organized in a hierarchical and dynamic manner according to the most pressing issues within the life of a given community (von Fircks, 2024b).

Let us illustrate this with an example: My parents own a pharmacy; they have a business background because the pharmacy has existed for three generations, and they have a medical, pharmaceutical background, meaning they are natural scientists. My parents work almost 12 h a day, 6 days a week. My father holds a PhD in chemistry. They are highly rational people and do not have a strong preference for the humanities. In other words, they wanted their children to study something practical—arguing that one should not study philosophy, for example, because it is difficult to earn a living from it. Or they encouraged me to study pharmacy, arguing that I should study it because I was very good with medications and their use when I helped out in the pharmacy. Here, we can say that my parents put a special emphasis on two should-values: you should earn a good living with an appropriate job, and you should do something you are good at. Implicitly, I could also hear that they said I should do pharmacy because it was the family’s heritage, as they often complained about the fact that no one in the family was willing to take over the pharmacy, which was associated with negative emotions for them, such as sadness.

What we can see from this rather autoethnographic example is that a family prioritizes certain should-values; yet those should-values are also hierarchized and interconnected—and they pave the way for a certain should-value ground-theme with which an individual person is confronted. What Boesch tries to illustrate with his writings is that these should-values are transmitted in interactions (see my example), but they also manifest themselves in the material environment of a given system or community. In his later works, Boesch speaks of the social over-determination of the environment by the externalization of particular myths and mythemes (Boesch, 2005, 2021). In Boesch's own voice:

> [Myths] exemplify underlying central themes and attitudes toward life, but it is important not to lose sight of the fact that such central themes can be expressed in many different ways. Myths, therefore, are here treated as collectively accepted means of explanation, justification, and exhortation, which might be expressed in the form of myth-stories, but also of mythemes, that is, isolated themes relating to the underlying myth, be they the theme of a story, like little Red Riding Hood, be they in the form of proverbs or other popular exhortations. In other words, a myth is a superordinate pattern of explanation and motivation regulating social action. (1991, p. 124)

In other words, myths refract should-values in peculiar ways. For example, when I was a kid, our house was full of pharmaceutical journals (Apotheken-Zeitschrift) that my parents even read in their free time. Or the conversations were fully flooded with medical and pharmaceutical details—something which I only understood slowly. My siblings and I call my father—Apotheken-father (pharmacy dad) as he is not only our father but also the father of the pharmacy, knowing that he loves to work, and we cannot imagine him as a person without working in a pharmacy. This event went so far that teachers in my school approached me, wanting to know what they can take against a cold, and then they went to buy the product in our pharmacy. What we become aware of within this autoethnographic example is that myths are certain meaning-making patterns related to social should-values. The pharmaceutical myth of providing medication and healing is an important should-value of taking care of the generalized social other. The myth of running your own business and sacrificing yourself for that business is an important should-value of being responsible for yourself and having some duties. Importantly, those social should-values relate to social needs (the need for economic independence, the need for taking care of others, and so forth). Hence, we need to admit that myths, social should-values, and collective needs are in a particular interrelationship, meaning that the environment is overdetermined by specific myths that become crystallized in material manifestations which, in turn, mediate should-values pointing to significant social needs of a given community. Hence, social interactions are not the only pivotal force in transmitting social should-values but also material manifestations such as ornaments, decorations, and so forth, with which the individual necessarily interacts in order to fulfill its basic human requirements.

What we are left with at the end of the paragraph is to show how myths are collective meaning-making patterns that regulate actions. In our autoethnographic illustration, we have seen how the environment and social interactions are structured according to certain should-values and collective needs, including which newspapers to read, which themes to debate during conversation, which books to read, among many other material and social manifestations. What is rather important to say is that those collective meaning-making patterns do not cause action, but they frame or regulate action (Boesch, 1980). In other words, those collective meaning-making patterns set specific social borders within which needs must be fulfilled and how to fulfill them, and how they become refracted through specific materialization, mediating clear messages to the person that should engage with it.

The Interaction of Fantasms and Myths

What we have done so far is a simplification of what actually happens in our daily lives from a cultural-psychological perspective. For reasons of conceptual clarity, we have juxtaposed myths and fantasms one with the other. Yet, this is and remains a simplification. Rather, we are compelled to speak of an interaction between myths and fantasms.

> This process of reciprocal assimilation of myth and fantasm probably lies at the core of interpersonal coordination of actions. Myths and mythemes are collectively shared forms and contents of action, and to the extent that the individual is able to assimilate myths to his or her own fantasmic system on the one hand, and to integrate the fantasms with myths on the other, a process of socialization will take place in which the overarching goals of the individual will be transformed. Thus, action theory construes "individualization" (here: the assimilation of myths) and "socialization" (here: the assimilation and structuring of myths reciprocallly with fantasms) as taking place jointly and leading to the socialized individual. (Boesch, 1991, p. 125)

For instance, if we focus on a person who has been restricted or inhibited in the sexual domain of living, we could imagine the following. Once the sexual functions have matured biologically within the individual person we are speaking of, immediately her parents try to separate the person from different-sex contacts. Here, the parents are rather afraid of the person discovering her sexuality while fearing some negative consequences for her. For instance, they might be afraid of the person becoming a parent; they might be afraid of the person accumulating negative experiences; they might be afraid of their child having a better sex life than they have, among many other potential reasons. Hence, the parents try to inhibit actions in the domain of sexuality. Boesch would argue that those parents would try to restrict the child's access to the action sphere of sexuality (*Handlungsbereich*). Hence, in the action sphere of sexuality, the parents try to inhibit sexual actions performed by the child—in other words, the child's parents feed specific messages into the child's social environment, such as: "You should not act out your sexuality," or "You should be afraid of sexuality." Consequently, the parents try to overdetermine the environment of the child with specific should-values, trying to connotate sexuality negatively, showing the peculiar social need of the family to restrict the living out and experience of one's sexuality.

Concretely, the parents could envision controlling the child's smartphone, checking whether it has access to pornography; they could forbid non-same-sex friendships, being afraid that friendships might turn into romantic partnerships or adventures; they could forbid the child from going to sex education in school—among many other potential inhibited actions from the parent's point of view. In all these instances, we become aware that actions are regulated by overlaying meaning-making patterns incorporating specific should-values promoting or inhibiting actions. While inhibiting actions, Boesch relies on the notion of taboo spheres (*Taboo-Bereiche*). However, despite the overdetermination of a given action sphere with myths and certain should-values, this overdetermination does not work in a deterministic manner. On the contrary, the person might resist this overlaying meaning-making pattern of inhibiting the sexual domain of living and experiencing.

Despite the restrictive access to books and media, the person nonetheless gets a hold of sexuality. For example, a good friend of the person might borrow her a book thematizing sexuality in a rather open fashion. Here, the person discovers through reading that sexuality can also be positively connotated, for example, as the synthesis of two loving partners and not as the social other being the mere platform for living out one's drives. Through reading the book, the person might discover that sexuality can be a good thing and can be something liberating and empowering. Hence, by reading the book, the person has embedded herself in another community, has widened the social field of action in which sexuality is discussed in both a positive and negative manner. Here, the person might have internalized a more positively connotated myth of sexuality, which relates more to her fantasms—individual needs—of dealing with sexuality in a transparent manner and not simply ignoring it. Soon, the person might decide to stop going to church with her parents because this church denies an open appreciation of sexuality; the person might reject the restriction of the parent's attempts to control what media she watches and buys herself a second smartphone through which she listens to sex podcasts and interviews, for example.

The central role of resistance becomes evident in this example. An individual will confront this restriction of the sexual sphere with resistance if they perceive that such a restriction does not align with their purposes and goals. Resistance, in this context, is fueled by certain fantasmatic constructs—and other myths. A person who resists their parents' restrictions concerning sexuality does not typically do so out of mere opposition or a general desire to oppose. On the contrary, their resistance is rooted in their core fantasmatic constructs, such as the fantasm of security and a fantasm of love, which envisions love—whether physical or spiritual-moral—as inherently free. Resistance, therefore, emerges as a fascinating phenomenon—indeed, a compelling action—because it consistently points to fundamental fantasms that initiate and provide the motivational basis for the act of resistance.

What the examples show is that the person rejects the myth-like guidance of her social action field regarding the notion of sexuality, particularly the parent's mediation of a negatively connotated picture of sexuality. For that matter, she redirects her private needs to other channels where she hopes to learn about sexuality in an open and transparent manner—and hence, she decides to listen to podcasts and interviews. In other words, the person chooses a different community that assists her in fulfilling a particular need hierarchy. It comes as no surprise that homosexual or bisexual people often leave restrictive families once they are able to earn a living and move to another city with a more positively connotated picture of homosexuality or bisexuality. To put it differently, a myth can become so suffocating regarding one's very own fantasms that the person feels forced to leave the action field and search for a concurrent action field satisfying her underlying needs. Putting the notion of sexuality aside, I know that close friends of mine have left my hometown (a smaller city) because they wanted to get involved with fashion and design, which was virtually nonexistent in my hometown, and they left for Berlin. This seems to be a rather stark motivation: to find a place that is in deep accordance with one's inner world (interconnected fantasms), so that people feel at ease and understood, or in Boesch's words, a certain I/World-Balance (Boesch, 2021).

The interaction between myths and fantasms offers a dynamic understanding of personal and collective meaning-making in response to societal structures. As mentioned, the adolescent seeking to explore her sexuality in a restrictive environment presents a clear example of how an individual may either try to escape a confining myth or attempt to reshape it. In this case, the adolescent's desire to live out her sexuality contrasts sharply with her parents' restrictive beliefs about it. This creates a direct confrontation with the myth-like structure the parents have set up, leading to intense resistance and conflict. Rather than simply accepting the parents' restrictive messages, the adolescent might engage in provocative actions—such as bringing male friends into the house or locking her door to engage in secretive sexual conduct—aiming to challenge and eventually change the myth-like structures that inhibit her sexual exploration.

The adolescent, in this case, aims to renegotiate the myths around sexuality within her family environment. By engaging in actions that bring new information and perspectives—such as introducing her parents to educational materials like sexuality podcasts or even suggesting therapy—she hopes to impact their perceptions and potentially reshape their attitudes. Psychotherapy, in this sense, serves as a negotiation of personal and collective needs, and in this scenario, it represents a means by which the adolescent could attempt to renegotiate the value systems held by her parents, with the help of an external mediator, such as a therapist. Through these actions, the adolescent seeks to change the existing myth-like structure that negatively connotes sexuality and introduces a more open and positive view.

Similarly, in the case of fashion and design, an individual who feels stifled by the rigid norms of their hometown might also try to renegotiate the prevailing myths about the subject. Rather than leaving the town outright, this individual might create a new action field by starting a fashion and design business that challenges the local community's narrow understanding of fashion. While initially facing resistance from those who are invested in the traditional myths of the town, the person can make their case by explaining that fashion need not be static and can evolve. If successful, this person might gain recognition, perhaps through interviews or public talks, thereby expanding the social action field and shifting the local myths around fashion and design.

In another example, the individual might leave their town for a more open-minded environment, such as Berlin, where they can find a community more in tune with their inner world and fantasms. The experience in Berlin would allow the person to actualize their private needs, particularly around fashion and design, while expanding their understanding of the myth-like structures in that larger city. After their time in Berlin, the individual might choose to return to their hometown, bringing with them a more developed sense of their own fantasms. Now, with a deeper understanding and a more refined hierarchy of needs, they can return and challenge the original myth-like structures of their home, potentially impacting change with greater clarity and self-assurance.

This approach shows that actions are often driven by complex, interconnected needs, and the renegotiation of myths is not always linear. Sometimes, a person may leave an environment to actualize their needs and then return with a transformed perspective. In this way, personal growth and myth transformation can occur through both external and internal shifts. An example of this dynamic can be seen in the world of gastronomy: someone might travel to France, fall in love with the culture of wine drinking, and then return to their hometown to open a wine bar, challenging the local food culture. Similarly, local bakeries inspired by French baking traditions—often not run by French nationals but by locals with a connection to French culture—serve as a testament to how individuals can bring new cultural elements to their communities by negotiating existing myths and introducing new ones.

These examples highlight the dynamic interaction between myths and fantasms and emphasize the importance of understanding their transformation and metamorphosis in a cultural context. Myths and fantasms are not static categories to be defined or confined to rigid frameworks such as questionnaires or scales. Instead, they are fluid, evolving constructs that can shift over time based on personal experiences and societal changes. This perspective, rooted in cultural psychology, underscores the need to study how myths and fantasms transform, interact, and evolve within the broader social and cultural context, reflecting both personal growth and the changing collective needs of society.

Toward a Transparent Systematization of Symbolic Action Theory

In summarizing and systematizing Boesch's *Symbolic Action Theory*, we can observe that the theory revolves around the interplay between the *what-is* and *what-should-be* conditions (see Fig. 1.1) focusing on the transformative nature of actions within a cultural context. The *what-is* condition represents the current state of affairs, while the *what-should-be* condition points toward an ideal or desired state. Through action, an individual attempts to bridge this gap, and once a goal is achieved, the *what-should-be* condition becomes temporarily fixed as a new *what-is* condition, creating a new starting point for future actions.

A key aspect of Boesch's theory is its recognition of the dynamic, cyclical process of transformation: actions lead to changes in structures, which in turn become the basis for new actions. This cycle is foundational to cultural psychology, as it

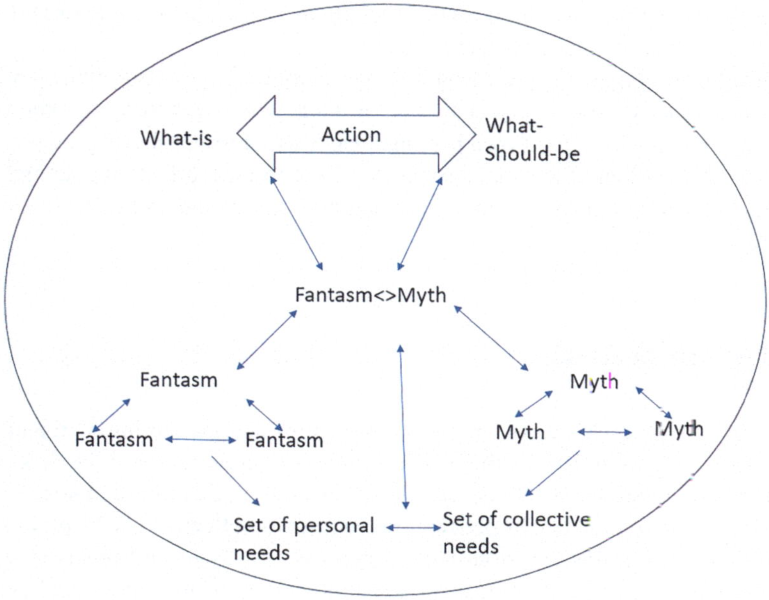

Fig. 1.1 Systematization of Boesch's symbolic action theory based on actions as the germ cell of cultural psychology

helps to understand how culture, personal needs, and collective values are intertwined and evolve over time. The theory posits that actions are not just performed for their own sake but are deeply embedded in the personal and social context in which they occur.

Boesch emphasizes that actions are framed by the interaction between personal and social domains of living. Personal needs motivate actions, but these needs are often unconscious or refracted through *fantasms*—symbolic representations of underlying desires, fears, or needs. Fantasms, therefore, serve as mediators, providing individuals with symbolic signs that influence how they perceive and engage with their world.

However, *fantasms* are not isolated phenomena. They are closely connected to *myths*, which reflect collective values and norms that shape social expectations and guide conduct. Myths, in turn, are part of a broader system of interconnected themes that structure the way people think, feel, and act within their cultures. For example, a myth in a religious community might emphasize the moral duty to help the poor, which becomes a foundational value shaping individual conduct in that context.

The central task of symbolic action theory is to study the interaction between personal fantasms and collective myths. This interaction determines how individuals reconcile their personal desires with the myths and values of the society they inhabit. In some cases, individuals may align their personal needs with collective myths, reinforcing existing social structures. In other cases, individuals may resist, challenge, or even provoke these myths, seeking to renegotiate them in order to better align with their own desires and needs.

This dynamic interaction can lead to the transformation of both myths and fantasms over time. As individuals challenge or adjust the myths they are exposed to, they may also transform their own personal needs and desires, altering the way they perceive the world and their place within it. This, in turn, affects the actions they undertake, creating a feedback loop in which both individual and collective meanings are constantly evolving.

The Genesis of Action

At the heart of Boesch's theory is the study of how actions are initiated and what motivates people to undertake them. The *genesis of action* refers to the process by which an individual decides to act based on personal motivations and social structures. The interaction between personal fantasms and collective myths plays a pivotal role in shaping this decision-making process. As myths and fantasms evolve through their interaction, they reshape the motivations behind individual actions, leading to new actualized conduct and new understandings of what is needed or desired.

In essence, the systematization of Boesch's symbolic action theory highlights the dynamic and evolving nature of human actions within a cultural framework. It stresses the importance of studying the interplay between personal and social meanings, recognizing that individual actions are not just reactions to external pressures but also reflect the ongoing negotiation between personal needs, societal myths, and the broader cultural context. By examining the transformation of myths and fantasms over time, cultural psychology provides insights into how actions are motivated and how they contribute to the shaping of both individual and collective identities.

This perspective is indeed described in the final two stanzas of the opening poem. It is precisely this dynamic perspective—the interplay of day and night, the interweaving of both phenomena (light and darkness) in twilight—that symbolizes, for our purposes, the transformative interaction of myths and fantasms.

Infinitely soft and mild,
The day ascends above the sea,
No breeze stirs in its gleaming shield,
No sound, save bird-song's melody.
Gently breaks the shell
In which the night enclosed us tight,
Until the morning overflows
And bathes us in its light.

Guiding Students and Researchers in Symbolic Action Theory

This chapter is particularly important for novices in cultural psychology who may be uncertain about how to systematize their scientific endeavors. The goal of this section is to provide guidance for both future students and experienced researchers, highlighting essential steps for studying a given phenomenon in a structured and systematic manner—something still lacking in Boeschian cultural psychology. We aim to show that when researching a phenomenon, the following steps should be considered:

A. *Identifying the What-Is and What-Should-Be Conditions*
   The first step is to clearly identify the *what-is* condition of the person or group being studied. This refers to their current state, whether psychological, cultural, or social. It is important to contrast this state with the *what-should-be* condition, which represents the goal or ideal state the individual or group is striving to reach.

B. *Assessing the Actions Taken to Reach the What-Should-Be Condition*
   The next step is to assess how the person or group attempts to achieve the *what-should-be* condition. This involves analyzing the actions taken to reach the goal and evaluating whether those actions are successful in achieving the desired outcome. It is also important to examine how the *what-should-be* condition manifests as a structure and how it differs from the *what-is* condition.

C. *Investigating the Personal Motivation for the Action*
   Researchers should then focus on understanding the individual's motivation for undertaking a given action. This means shedding light on the personal needs that drive the action. However, it is important to recognize that the individual may not be fully aware of the needs motivating them or may underreport certain needs due to social pressures. It is also crucial to acknowledge that needs are often refracted through *fantasms* (symbolic representations of unconscious needs), and researchers may need to unravel this symbolism to clarify the individual's pressing need hierarchy.

D. *Understanding the Relationship Between Fantasms and Myths*
   It is important to recognize that *fantasms* do not exist in isolation; they are always connected to larger cultural myths that point to an interconnected set of social needs. Researchers should assess the predominance of social needs within the action field of the individual or group and explore how these needs are interconnected. Additionally, it is crucial to examine how myths and *fantasms* interact with one another.

E. *Examining the Dynamic Interaction of Myths and Fantasms*
   The interaction between myths and *fantasms* leads to the actualization of both the individual's *fantasms* and the myths they are influenced by. This process can pave the way for the actualization of both personal and social needs. Researchers should recognize that both *fantasms* and myths are dynamic, and they can transform as a result of their interaction. This highlights the importance of studying how these transformations occur. Researchers should examine the

interaction between individuals and their communities, as well as how people interact with their material environment, as myths are often manifested through material culture. This dynamic interaction may give rise to new needs, which in turn may relate to previously formed needs.

I strongly believe that cultural-psychological research projects will benefit from this systematization and the research advice presented here. This framework is especially beneficial for young, aspiring cultural psychologists who are in need of guidance for their important projects. This chapter is a humble attempt to provide such guidance, and we encourage interested scholars to further develop and refine this approach. After all, Boesch argued that symbolic action theory may be better equipped to analyze experience and conduct than activity theory (Boesch, 2021), yet one big advantage of activity theory—especially inspired by the work of Engeström—is its clear transparency and its holistic systematization. With the present chapter (and Fig. 1.1), we lay the foundation for a long-needed and holistic systematization of Boesch's symbolic action theory. The time was ripe long ago.

References

Boesch, E. E. (1975). *Zwischen Angst und Triumph: über das Ich und seine Bestätigungen*. Huber.

Boesch, E. E. (1976). *Psychopathologie des Alltags: zur Ökopsychologie des Handelns und seiner Störungen*. Huber.

Boesch, E. E. (1980). *Kultur und Handlung: Einführung in die Kulturpsychologie*. Huber.

Boesch, E. E. (1991). *Symbolic action theory and cultural psychology*. Springer New York.

Boesch, E. E. (1998). *Sehnsucht: Von der Suche nach Glück und Sinn [Longing: On the search of joy and meaning]* (1st ed.). Huber.

Boesch, E. E. (2005). *Von Kunst bis Terror: Über den Zwiespalt in der Kultur [From art to terror: On the cleavage of culture]*. Vandenhoeck & Ruprecht.

Boesch, E. E. (2006). *Konnotationsanalyse: zur Verwendung der freien Ideen-Assoziation in Diagnostik und Therapie*. Privater Nachdruck. (Original work published in 1977).

Boesch, E. E. (2021). *Musik, Sprache und die Sehnsucht nach dem Paradies*. Psychosozial-Verlag.

Bruner, J. (1986). *Actual minds, possible worlds. The Jerusalem-Harvard lectures*. Harvard University Press.

Bruner, J. (1996). *The culture of education*. Harvard University Press. https://doi.org/10.2307/j.ctv136c601

Bruner, J. S. (1997). *Sinn, Kultur und Ich-Identität: Zur Kulturpsychologie des Sinns*. Carl Auer.

Ellis, C., Adams, T. E., & Bochner. (2010). Autoethnografie. In G. Mey & K. Mruck (Eds.), *Handbuch Qualitative Forschung in der Psychologie* (pp. 345–357). Springer.

Janet, P. (1919). *Les medications psychologiques*. Payot.

Lewin, K. (1982). Formalisierung und Forschritt in der Psychologie. In C. F. Graumann (Ed.), *Kurt Lewin Werkausgabe: Feldtheorie* (Vol. 4, pp. 41–73). Klett-Cotta.

Lewin, K. (2013). *A dynamic theory of personality-selected papers*. Read Books.

Stern, W. (1923). *Person und Sache: Die Menschliche Persönlichkeit* (3rd ed.). J.A. Barth.

Stern, W. (1924). *Person und Sache: System des Kritischen Personalismus (Wertphilosophie)*. J.A. Barth.

Stern, W. (1935). *Allgemeine Psychologie auf personalistischer Grundlage*. Martinus Nijhoff.

Straub, J. (2021). Ernst Boeschs Umwege zum Menschen. Eine Einführung. In J. Straub (Ed.), *Musik, Sprache und die Sehnsucht nach dem Paradies (XI-XXV)*. Psychosozial-Verlag.

Valsiner, J. (1997). The Saarbrucken tradition in cultural psychology, and its legacy. *Culture & Psychology, 3*(3), 243–245.

Valsiner, J. (1998). *The guided mind: A sociogenetic approach to personality*. Harvard University Press.

Valsiner, J. (2007). *Culture in minds and societies*. Sage.

Valsiner, J. (2014). *An invitation to cultural psychology*. SAGE.

Valsiner, J. (2021). *General human psychology*. Springer.

Valsiner, J. (2024). *THE MOONLIGHT DOCTOR: Art and Science of Carl Gustav Carus*. Springer.

von Fircks, E. (2022). Setting the seeds for a normative expansion of Lewinian field theory for cultural-psychological practitioners. *Human Arenas*, 1–21. https://doi.org/10.1007/s42087-022-00307-5

von Fircks, E. (2023a). Eine kurze kulturpsychologische Anleitung für zukünftige Freunde. *cultura & psyché*, 1–15. https://doi.org/10.1007/s43638-022-00039-x

von Fircks, E. (2023b). Geštalt TERAPIJA IZVAN POTREBA: Kako UPOTREBA ALATA UTIČE NA OSOBU I OKRUŽENJE. *Geštalt Zbornik, 10*, 46–59. https://www.ceeol.com/search/article-detail?id=1201024

von Fircks, E. (2023c). Understanding the power of mindfulness from a cultural-psychological perspective. *Culture & Psychology.* Advance online publication. https://doi.org/10.1177/1354067X231201387

von Fircks, E. (2023d). The fully objectified catathymic imagery method (FOCI): Assessing the amorphian inner at work. *Human Arenas*, 1–18. https://doi.org/10.1007/s42087-023-00345-7

von Fircks, E. (2024a). *Learning with William Stern: Personology for future [Forthcoming]*. IAP.

von Fircks, E. (2024b). Follow-up study and its reasons: Shifting the focus more to the ecological assessment of the conservative phenomenon. In *Conservativism: A cultural-psychological exploration* (SpringerBriefs in Psychology) (pp. 49–56). Springer. https://doi.org/10.1007/978-3-031-51205-6_5

Chapter 2
The Action Phase Model in Irreversible Time

Vacation *(E.E. Boesch, Enno von Fircks translation)*
You lie beneath the trees, dreaming in the warmth,
And feel the sun, heavy and full,
Nearby, shimmering swarms of lizard's dart,
And everywhere there's light — even under clouds,
The house's walls still glow with enough light
To dazzle your eyes.
From the window, a piano dances.
First Ravel, a light sparkle,
Yet stiff and strange, so sheer,
Like marionettes in a ballet. Then dark,
Brahms, with heavy, soft strokes
That stir old images within you.
And below, the quiet house where you rest,
Where playfully, and ever new, the sea
Cradles the nights. And far away, brimming with unheard-of delight
A blue like none you've seen in so long —
O veil of grace,
O radiant sun, O lyre of summer's vibrant hues.

Understanding the Need to Study Actions as Being Interrelated

In the previous chapter, we discussed that E.E. Boesch was an action psychologist. In one of his posthumous publications (2021), Boesch argues that symbolic action theory, as an integration of cultural psychology, is better suited to analyzing psychological phenomena than competing action-theoretical frameworks, such as activity theory, inspired by Leontiev (1978) and later developed by Engeström (2000, 2001). Additionally, Boesch expressed dissatisfaction with certain symbolic theories, particularly that of G.H. Mead (1934/2015), critiquing the idea that individuals simply

E. Freiherr von Fircks, *Culture of Action*, Theory and History in the Human and Social Sciences, https://doi.org/10.1007/978-3-031-88808-3_2

internalize social forms. Boesch's critique of both activity theory and Mead's social psychology centered on their overemphasis on the social dimension, or in scientific terms, their monistic orientation. Through his work in psychoanalysis and his focus on the power of fantasms (1991, 1998, 2005), Boesch argued that humans do not passively internalize social forms of gestures and simply apply them to their system of purposes. On the contrary, Boesch was fascinated by how social forms, through internalization, become individuated and personally meaningful (Boesch, 1998; von Fircks, 2023a). The alteration of a social form through fantasmatic refraction was a key concern for Boesch when developing his cultural psychology.

While the previous chapter laid the foundation for Boesch's symbolic action theory by focusing on action as the germ cell of psychology and noticeably cultural psychology, this chapter aims to expand on his theoretical framework, adding complexity to the system. This expansion is built upon the simplification presented in the first chapter, which aimed to familiarize the reader with Boesch's thinking style (see Chap. 1).

According to Boesch, a person attempts to change the external world in a meaningful way by performing specific actions. Each action carries personal symbolic meaning, motivating the individual to move toward a specific goal field—this is the fantasmatic origin that drives a person to perform an action within a given action sphere, situated in a larger action field. However, it is important to recognize that an action is not solely personally driven; it is also socially constrained by barriers and borders (Boesch, 1980; von Fircks, 2023b) that channel actions in particular directions. In this way, an action reflects the fundamental cleavage between the personal and collective culture, or, in Boesch's terms, between fantasms and myths, which we discussed in Chap. 1 in depth.

It is essential to note that a person does not perform an action in isolation. An action is always connected—or interrelated—with other actions (Boesch, 1980, 1991). Therefore, we are called to study actions in their symbiotic intertwinement, meaning we cannot simply examine how a single action helps a person achieve a specific goal or fulfill a particular need. As Boesch (1991) notes, actions are always the result of polyvalence, meaning that multiple interconnected actions serve to fulfill more than one need, contributing to a dynamic structure of needs that must be unraveled by psychologists (von Fircks, 2024a).

To illustrate this interconnectedness of actions, consider the example of preparing food. The overarching activity is the preparation of food, but in order to transition from an undesired present state to a desired future state, the person must break the goal down into manageable steps, performing several related actions. For example, the person first retrieves the raw ingredients from the fridge, then selects and uses a knife, cuts the food, seasons it, and finally cooks it in a pan. This example highlights two important points: first, actions are interconnected, and second, they are sequentially ordered within an overarching activity. When analyzing a system of actions, we must consider both their interconnection and their sequential order.

When examining the multiplicity of actions, we need to investigate the action system—specifically, the embedding of the action in a larger network of actions (Boesch, 1980). A challenge in this process is that actions must not be understood in isolation; they must be viewed in relation to the overall activity and the broader action system. All actions contribute holistically to the overall activity, and detaching a single action from its context would distort our understanding. For instance, if we focus only on the action of cutting food, we lose sight of the interrelated actions, such as retrieving the ingredients from the fridge. If we remove one action central to the system—say, the act of retrieving the ingredients—the entire action system changes, and the goal (preparing food) cannot be completed. Similarly, choosing to cook the food in a casserole rather than a pan alters the overall activity, as the dish will taste different depending on the method of cooking.

From a psychological perspective, it is crucial to understand not only how actions are interconnected but also how altering a single action can change the quality of the overall activity and, by extension, the action system. For example, watching TV—a seemingly simple activity—becomes altered if the remote control breaks. Without the ability to adjust the volume or change channels, the activity of watching TV is no longer relaxing and instead may become an emotional challenge, as the person must manage frustration and negative emotions.

> A lover never stops courting, and we all continuously pursue happiness. Such "endless" actions derive their motivation from superordinate or overarching goals, a concept we shall often meet in the course of these deliberations (…). The stamp collector's purchase of a set of stamps may, then, fulfill all the criteria of a complete action, yet its functional value is not independent. In view of superordinate actions single actions acquire the quality of a sub-action. Actions, therefore, will always have to be considered both from the point of view of their inherent structure, and of their place within superordinate action systems. (Boesch, 1991, p. 45)

However, it is not sufficient to merely examine interconnected actions that are organized within specific action systems by the individual. As mentioned earlier, actions are not only organized spatially, but they are also inherently organized temporally (Boesch, 1991). In this regard, actions are divided into specific, complementary phases that are essential for organizing subsequent actions and for anticipating future actions. Boesch emphasizes the dynamic nature of action—that is, the unfolding process of interrelated actions that are actualized over time, shaped by both personal and environmental constraints (Boesch, 1975). What is crucial in Boesch's symbolic action theory is not merely how an action unfolds, but how it is altered during its unfolding. It is precisely at this juncture that Boesch argues his action theory diverges from other action-theoretical frameworks in psychology, such as the one proposed by Atkinson (Atkinson & Heritage, 1984). So, how are actions organized according to Boesch?

> Wir beschreiben die Handlung, indem wir sie in einzelne Phasen einteilten, die im wesentlichen die situativen Vorläufer, die Perioden der Handlungsvorbereitung, der Handlungsausführung, der Handlungsbeendigung und endlich die inneren wie äußeren Ergebnisse umfassen. (…) Trotz dieser Weite des Handlungsbegriffes liegt sein Nutzen eben gerade darin, die Vielfalt von unterschiedlichen Handlungen nach vergleichbaren Phasen zu beschreiben, in denen jeweilen ja auch analoge Prozesse verlaufen müssen—in der Anfangsphase etwa Entscheidungen zwischen alternativen Zielen, in der Verlaufsphase etwa stabilisierende Regulationen, in der Endphase etwa Evaluationen, um nur wenige zu nennen. (Boesch, 1980, pp. 101–102)

> We describe the action through dividing it in different phases which are essentially the following: The situational precursors, the periods of action preparation, the performance of action and the termination of action which include the inner and outer results of exactly that action. (…) Despite the width of this scientific notion (action), its use is found in uniting different types of actions according to a similar pattern of action phases—which means that there happen to take place analogous processes—in the action preparation phase decisions need to be taken between different goals, in the action process phase stabilizing regulations need to be applied and in the action termination phase the person can make use of evaluations just to name a few psychological mechanisms being implied in the distinct actions. (Boesch, 1980, pp. 101–102, author's translation)

What Boesch aims to achieve in laying the foundation for his symbolic action theory is to revitalize the psychological study of everyday occurrences—a domain that mainstream psychology has largely neglected for decades (von Fircks, 2024b).

The Action Preparation Phase and the Role of Fantasms

In this section, we will explore the distinct phases of action and examine their psychological mechanisms (for an overview, see Boesch, 1980). We begin with the action preparation phase. In this initial phase, motivation is catalyzed by the individual herself. Specifically, the person develops a form of goal anticipation, which, as discussed in the first chapter, revolves around the cleavage between the *what-is* and *what-should-be* conditions in relation to an underlying goal. This cleavage not only reveals the person's goal but also clarifies her current position within the goal pursuit process and identifies the necessary steps (actions) to reach the desired condition. In other words, the experience of this cleavage intensifies the individual's goal anticipation, as she recognizes her present state and understands what actions are required to move toward her goal. The experience of this cleavage triggers basic motivation as well as mental images concerning how the action is likely to unfold. This is crucial for anticipating potential barriers or obstacles that may arise during the execution of the action.

What is particularly interesting in Boesch's conceptualization is that individuals do not always have full awareness of what motivates them (Boesch, 1991; von Fircks, 2023c). As previously noted (Chap. 1), the person's fantasms—mental representations that refract her underlying needs—serve as the central source of motivation for action. However, individuals are not necessarily conscious of these fantasms. They may plan and anticipate actions with considerable force without understanding why these actions hold personal relevance or importance within their psychological system. This suggests that studying the initial action phase—action preparation—presents a complex challenge. It is difficult to assess an individual's fantasms merely by observing the external manifestation of the action (i.e., how the environment is altered). Equally important is understanding the personally meaningful significance of the action.

Boesch argues that individuals may become aware of their underlying fantasms—and their interconnections—through techniques such as free association or other psychoanalytic methods (Boesch, 2006/1977). According to Boesch, this awareness of underlying fantasms leads to a transformation in the fantasms themselves, which in turn alters the actions the person undertakes (Boesch, 1980). This aligns with a central tenet of psychoanalysis, often encapsulated by the imperative *know thyself*. Becoming aware of one's underlying motivations illuminates long-standing life patterns that may have excluded important aspects of experience (von Fircks, 2023c). This process of self-awareness fosters a reactualization of meaning that regulates future actions, a concept supported by existential and Gestalt therapies (Schneider & Krug, 2010). In these therapeutic approaches, individuals are guided (by the therapist) to develop new life patterns or meaning-making frameworks that are more aligned with their underlying fantasms.

In summary, the essence of the action preparation phase lies in the coordination of vague goal imaginations with motivational impulses derived from fantasms. However, awareness of one's underlying action tendencies is not always the norm. If such awareness occurs during the initial phase, it alters the planning and anticipation of action. Therefore, we can expand upon Boesch's symbolic action theory by suggesting that the experience of one's underlying fantasms—or the neglect of certain fantasms—creates a feedback loop that influences the very preparation of action. This reflects a cybernetic or systemic perspective, which is evident in Boesch's framework (see also Boesch, 1976).

The Action Process Phase: The Influence of Ecological Determinants

The action preparation phase is followed by the second phase, known as the action process phase (Boesch, 1980). In this phase, the individual initiates the action, and the action is performed or undertaken, requiring preservation from various environmental resistances, hindrances, and barriers. Once the person has determined the goal she wishes to pursue and identified the necessary steps to move toward it, the action can be personally initiated, thus setting the process of action into motion. This second phase is characterized by specific psychological mechanisms:

> Die Verlaufsphase wird gekennzeichnet durch die progressive Zielannäherung: eine kürzere oder längere Aufeinanderfolge von instrumentellen Handlungen führt zu den notwendigen progressiven Situationsänderungen, die einem dem Ziele näher bringen. Solche Vorgänge können einfacher Art sein—etwa Hingehen, Ergreifen, Öffnen—sie können aber auch in komplexen Koordinationen von sukzessiven Handlungen bestehen. Überdies erfordert diese Phase spezifische Regulationsprozesse: trotz sich ändernder Situationen muss die Motivation konstant bleiben (…). (Boesch, 1980, pp. 103–104)
>
> The action process phase is characterized by progressive approximation towards one's goals; approaching one's goal through interrelated instrumental actions leads to progressive changes in the given ecological situation which brings oneself closer to the goal. Such operations can be simple such as going towards something, grasping or opening something—however they can demand also more complex coordination. Further, this distinct

> action phase demands specific regulatory processes; despite the transformation of the situation, motivation needs to be held constant. (…). (Boesch, 1980, pp. 103–104)

To further understand the action process phase, let us consider a concrete example. Imagine a person who intends to read a book. In this case, the person seeks relaxation by reading a love novel, which is not associated with work or self-development but rather with entertainment and leisure. The goal is to move from the undesired present (feeling stressed) to the desired future state (feeling relaxed). To achieve this, the person plans to read a specific book, prepares tea and cookies, and settles into her favorite chair.

At first, the person enjoys reading, undisturbed. However, her peace is interrupted when she receives a phone call from her mother, who demands her attention. The person hesitates, unsure whether to continue reading or answer the call. Eventually, she decides to accept the call but quickly regrets it, as her mother spends the conversation complaining about a recent meal at a restaurant. Despite the conversation's lack of relevance, the person cannot hang up, as she does not want to offend her mother. After 15 min, the call ends, and the person plans to resume reading.

Before she has a chance to return to her book, she remembers that she needs to check her work emails. After addressing those tasks, she returns to her seat but finds herself too exhausted to focus. She ends up falling asleep while attempting to continue reading.

This example highlights the dynamic nature of Boesch's symbolic action theory. The person initiates a goal (relaxation) by undertaking specific actions (reading the book). However, she is not only embedded in one action system but is simultaneously involved in multiple action spheres. Pierre Janet, one of Boesch's major influences, refers to this as the "multiplicity of the personality" (Janet, 1929), which reflects the environmental conditions that arise when undertaking particular actions and push the individual into competing action systems.

In this case, once the person interrupts her reading to take the call, she finds herself in a competing action system. This shift demands a decision: Should she continue reading, or should she engage with the phone call? As she navigates through these competing action spheres, she must regulate her actions and priorities, deciding which activity (goal) to pursue while managing the interference of others.

The ecological determinants of the action field—the external circumstances—are also at play. The person must address these environmental challenges, deciding whether to prioritize the current activity or to shift focus to another. Even the decision to stop reading and take the call involves regulation, as the person justifies her action by believing she will have time to relax later. These decisions reflect a process of regulating one's actions and adjusting to competing demands.

Boesch's theory suggests that the person's actions are not only impacted by competing action systems but also by personal regulation. An individual will prioritize actions based on fantasms—deeply rooted psychological needs or desires. If a person continues to engage in a particular activity, such as working on Christmas Day, despite competing needs, this suggests that the fantasms related to work are more dominant in the person's hierarchy of needs. Activities that persist despite ecological constraints reflect these high priority fantasms, which guide the person's decisions and actions.

To summarize the action process phase, Boesch calls on psychologists to study not only the interconnectedness of action systems but also the personal regulation of actions. It is crucial to explore how individuals manage competing action systems and how their regulations—whether energetic, affective, or cognitive—guide their actions. While Boesch touches on the role of these regulators, further exploration is needed to fully understand how they influence the dynamics of action in various contexts (Boesch, 1980).

Pierre Janet, alongside Kurt Lewin, was one of the first action theorists to apply a rigorous action-theoretical framework to the study of psychology. Boesch, inheriting these foundational concepts, sought to integrate them within a cultural psychology framework, emphasizing the social and cultural aspects of action regulation. In Boesch's theory, regulators play a central role in understanding how individuals adjust and modify their actions in response to internal and external challenges. Let us examine these regulators more closely, starting with energetic regulators.

Energetic regulators are key mechanisms that can either promote or inhibit action. Fatigue is one of the most significant energetic regulators. If a person feels tired and unable to concentrate, they may resort to different strategies to overcome their exhaustion and maintain their action toward a goal. For example, the person may rush through tasks to reach the goal more quickly, or they may choose to take a break—such as going for a walk—to recharge before continuing the action. Alternatively, they might turn to caffeine or stimulants, like pills, to restore their energy and focus.

In extreme cases, individuals may rely on more drastic measures, such as using drugs to maintain their energy levels and meet pressing deadlines. In contrast, some may turn to more mindful strategies, such as meditation or relaxation techniques, to relieve stress and prepare for further action. These actions serve to regulate the original goal pursuit, ensuring that the person remains energized and motivated to complete their tasks.

Interestingly, the regulation of energetic states may itself require regulation. This phenomenon, referred to as *hyper-generalized regulation* by Valsiner (2014, 2021, 2024), occurs when individuals apply broad regulatory strategies that affect their conduct across different contexts. For example, a person who regularly consumes multiple cups of coffee a day to maintain energy might face comments from colleagues about the health risks of such conduct. The individual may justify this practice as a temporary measure due to the demands of an upcoming project, framing it as an "exception" to their usual routine. In this way, personal regulation can sometimes take on a generalized or exaggerated form, adapted to the needs of the situation.

Leaving the biological regulations aside, we also need to contemplate affective regulations. Some of these regulations have been developed and used in therapy, and over time, they have become more sophisticated. For instance, when people with mood disorders (such as depression) face negative events during their day, therapists can work with them to write down those negative events and then throw them in the bin afterward. This technique aims to help the person detach from those negative emotions mindfully. The negative emotions are felt and expressed but then shifted aside so that other actions can take place. Another therapeutic technique involves encouraging the patient to meet with friends, even when they may not feel like it or have the energy to do so. The therapist hopes that the patient will experience positive emotions during the meeting, ultimately helping them to make a corrective experience—showing the individual that even when they feel drained, social interactions can still be rewarding.

In addition to affective regulation, we also need to discuss cognitive regulation. Cognitive regulation involves the use of thoughts or mental strategies to influence or motivate actions. For example, when a person feels fatigued but needs to finish a project to please their boss, they might remind themselves that completing the job will increase their chances of promotion. The thought of career advancement acts as a motivational cognitive regulator, encouraging the person to continue working despite their fatigue or lack of energy. Cognitive regulation works by linking immediate effort with long-term rewards, making it easier to push through obstacles.

In practice, however, it is often difficult to separate these different types of regulation. Regulation of action typically involves a combination of energetic, affective, and cognitive strategies working together. For example, imagine someone working on a tight deadline. They may feel tired and emotionally drained but still want to finish the project to earn a promotion. In this case, the person might use a cognitive regulator by reminding themselves of the reward (the promotion), encouraging them to keep going. At the same time, they might use energetic regulation by taking short breaks, having a coffee, or getting some fresh air to recharge. Affective regulation could come into play if the person practices mindfulness, acknowledging their tiredness and using deep breathing to reduce stress and refocus. In this way, different types of regulation work in harmony to support the completion of the task.

This interplay between cognitive, affective, and energetic regulation is a central feature of mindfulness-based approaches, which combine cognitive strategies, emotional awareness, and physical relaxation techniques to help individuals manage their actions. From a cultural-psychological perspective, mindfulness allows individuals to address their emotional states (affective regulation), control their thoughts (cognitive regulation), and manage their physical energy (energetic regulation) simultaneously, which promotes a more balanced approach to action.

Both Pierre Janet and Boesch's work laid important foundations for modern therapeutic practices, particularly cognitive behavioral therapy (CBT). CBT focuses on the connections between thoughts, emotions, and conduct, and draws heavily from the regulation strategies outlined by Janet and further developed by Boesch. In CBT, therapists use a combination of cognitive, affective, and behavioral techniques to help clients modify dysfunctional thinking, emotional responses, and conduct. Techniques such as cognitive restructuring, mindfulness practices, and behavioral activation are reflective of Janet's and Boesch's ideas about regulating actions and managing the interplay of these different regulatory mechanisms.[1]

In Boesch's symbolic action theory, regulation is central to understanding how individuals navigate challenges. Energetic, affective, and cognitive regulations each play an important role in guiding action, but in real-world scenarios, these regulators often work in tandem. This understanding of how people regulate their actions provides insight into how they respond to competing demands, environmental challenges, and internal limitations. These concepts are especially relevant in therapeutic settings, where individuals learn to adjust their emotional and cognitive responses to facilitate action.

The Action Termination Phase and the Power of Evaluation

Having contemplated the second phase of action, we are in need to terminate our elaboration about the distinct phases of action. For that purpose, we need to report the last phase of action called action termination phase. Let us illuminate how Boesch understands this last phase within his symbolic action theory.

> Die Endphase der Handlung besteht in den Prozessen des Konsums und der Handlungs-Sistierung. Die Zielerlebnisse bilden natürlich den zentralen Aspekt dieser Phase: man genießt das Erreichte, freut sich seines Erfolges; bei Misserfolg treten die Affekte des Ärgers, der Ermüdung, der Trauer und der Bedrücktheit auf. So wie in jeder Handlungsphase der aktuelle Moment mit den Antizipationen verglichen wird, finden auch hier Evaluierungen statt: man konfrontiert den erlebten mit dem erwarteten Genuss, das Ergebnis mit dem Aufwand an Zeit, Anstrengung oder Mühe, aber auch mit der Leichtigkeit oder Verlaufslust des Handelns (…). Endlich muss aber auch die Handlung beendet werden; man kann im Genuss nicht dauernd verweilen, aber man darf bei Misserfolg auch nicht mit immer wieder erneuten Versuchen ansetzen; irgendwann fällt—normalerweise—die Entscheidung, dass es nun zu Ende sei. An diesem Punkt werden meist Beendigungshandlungen eingeschaltet: man entspannt sich, dehnt sich, geht ein paar Schritte, trinkt eine Tasse Kaffee oder plaudert ein paar Worte mit einem Kollegen. Diese Beendigungshandlungen dienen der inneren Umstellung [.] (Boesch, 1980, p. 104)
>
> The action termination phase consists of the processes of consumption and action suspension. The results of reaching the goal build surely the central core of this phase; one is enjoying the fruits of actions; success is on its way; when failure takes place affects of anger might occur, tiredness, sadness and feelings of depression. Insofar as in any phase of action the present moment is compared to anticipation, evaluations also take place in this phase; one is confronting the experienced pleasure with the anticipated pleasure; the results with the efforts invested but which also includes the ease of action or the pleasure process of action (…). Finally, action needs to be terminated; one cannot enjoy the fruits of action endlessly; however, in failure one is also appealed to not try and try again: sometimes the decision is made to end the action. Once this happens it goes together with actions of termination; one relaxes, stretches, walks around or drinks a cup of coffee or talks to one's col-

[1] Heim & Bühler have done an interesting work figuring out the connection between the ideas of Pierre Janet and the central tenets of CBT (2003).

> leagues. These termination actions serve the goal of adjusting oneself mentally or internally. (Boesch, 1980, p. 104, author's translation)

Boesch reminds us that once an action reaches its conclusion, it must not only be terminated but also consumed and evaluated. This means that the person needs to assess whether their action has helped fulfill the underlying goal and brought closure to the needs that were initially identified. Essentially, the individual must evaluate if they have achieved the "what-should-be" condition they envisioned at the start of the action. If the desired outcome has not been reached, the person may be compelled to adjust their actions and make another attempt to achieve their goal. However, it is crucial that the person symbolically terminates the action, giving a sense of closure. For instance, a businessman may celebrate the completion of a successful negotiation by sharing a cigar with his client, symbolizing the end of the deal. Similarly, when dining at a restaurant, the decision to have dessert is not driven solely by hunger but by the need to conclude the outing with a satisfying final gesture, thus symbolically completing the social activity of dining out with friends.

The concept of action termination extends to therapy as well. For example, solution-oriented therapists often begin the therapeutic process by discussing when and how therapy will come to an end, setting clear criteria for its termination based on the client's perspective (Murphy, 2024). This open dialogue helps the client autonomously conclude the therapeutic journey. In this context, therapy becomes a space where both therapist and client express gratitude and reflect on their experiences, often in moments of silence, allowing the emotions associated with the process to settle. In Gestalt therapy, for instance, clients may express a desire to physically embrace the therapist or send a message of appreciation, marking the end of the therapeutic relationship. These acts help both the client and therapist symbolically and emotionally close the action system of therapy. Notably, this process involves no repression or reappraisal; instead, the individual consciously acknowledges and feels the impact of the action on both a spiritual and bodily level.

A similar pattern occurs in nontherapeutic settings. After a soccer match, for instance, players often gather to share a drink and reflect on the game. This post-match gathering serves as a symbolic closure to the game itself. In cases where the match was particularly significant, such as a win or a stellar performance, the termination phase is extended over time and space. Players might continue their celebrations in the shower, having a beer, chatting about the match, and listening to music on the way home. These secondary actions, in the terms of Pierre Janet, serve to prolong the primary action and deepen the experience of closure.

However, the termination phase can take a different form when an action is unsuccessful. If the desired goal has not been met, the action termination phase may be rushed through, allowing the individual to quickly transition to a new action or attempt to reorient their efforts. For example, if a student fails to meet their graduation requirements, they may avoid attending the prom to avoid confronting their failure. The emotional impact of the failure overshadows the anticipated pleasure of spending time with friends, and they may prefer to disengage from the celebration altogether. This illustrates how regulatory actions, based on individual fantasms and needs, guide the process of terminating actions.

Boesch's theory underscores the idea that meaning and action are intrinsically linked. Whether the action is completed successfully or unsuccessfully, the meaning attached to it—shaped by underlying fantasms and goals—plays a significant role in how the action is terminated. This dynamic is central to Boesch's cultural psychology, where the meaning of action is not just about achieving an end but also about the emotional and symbolic closure that follows. Ultimately, the action termination phase is crucial for psychological well-being, serving a psycho-hygienic function by providing the person with an opportunity to integrate the action's outcomes, whether they are positive or negative, into their ongoing life narrative.

We are convinced that issues within the action termination phase hold significant clinical importance, as problems with terminating or failing to terminate an action are often intertwined with pathogenesis (Polster & Polster, 1974). This connection becomes clear when considering that if a person does not terminate an action that no longer moves them toward their goal, they become stuck. This stagnation can occur because the individual might fear the failure that would come with terminating the action, as doing so would confront them with their own sense of inadequacy. In some cases, people may resist terminating an action because they are overly invested in their success, believing that they can indefinitely sustain the initial momentum. This is particularly prevalent in the entrepreneurial world, where some CEOs become fixated on one idea or product, convinced it will keep them on a wave of success forever. As a result, they fail to fully terminate their initial action and its corresponding system, delaying necessary reassessment and adaptation to changing environmental challenges.

Pierre Janet emphasizes the importance of terminating an action, arguing that it allows both the body and mind to return to a state of equilibrium (Janet, 1919, 1929, 1930). This balance facilitates the reorganization of psychological and physiological forces, enabling them to shift toward other needs. Janet uses the example of sexual love to illustrate this point, describing the culmination phase (orgasm) as a moment when the body and mind enter equilibrium. At this point, the energy invested in sexual activity is released, and the initial needs are satisfied, allowing other needs to come to the forefront. An example of Blanco (2018) is worth considering:

> Before orgasm there is a tendency to the manifestation of contradictory impulses, as for instance to destroy, and at the same time to embrace; to separate, and at the same time to fuse. These impulses never determine definite images, and so I have great difficulty in giving them names *afterwards*. I could say that there are very many different ones, with the predominant characteristic that to each given impulse there is a contradictory one. Mathematically it would be represented with a minus and a plus sign. (pp. 441–442)

According to Janet, depriving an individual of this equilibrium prevents the reorganization of psychological forces (Janet, 1919), which can hinder the emergence of new needs that require attention. This perspective is echoed by some Gestalt therapists, such as Perls et al. (1951), who emphasize the importance of closure and completion in psychological processes.

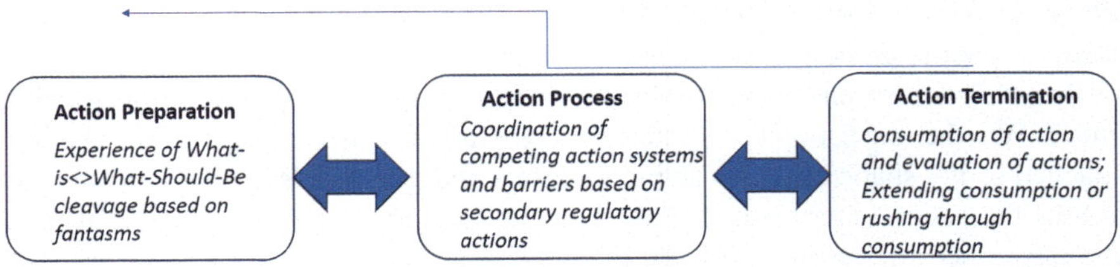

Fig. 2.1 Systematization of Boesch's action phases between Boesch and Janet (with a feedback loop onto the action preparation phase)

From this, we can also elaborate on the role of feedback loops in the action termination phase. When a person terminates an action, particularly after experiencing failure, they can use the insights gained from that experience and the emotional energy generated to reassess the initial phase of action. This creates a dynamic within Boesch's action theory, in which feedback loops allow for reflection and adjustment, facilitating continuous growth and adaptation. This process happens within the context of irreversible time, where actions and their consequences cannot be undone, but can be re-evaluated and adjusted to inform future actions (Valsiner, 2014).

In the following figure (Fig. 2.1), I aim to systematize Boesch's symbolic action theory, outlining the major premises and illustrating how the phases are interconnected. This model should serve as a tool for future researchers, enabling them to apply Boesch's theory in a pragmatic and accessible manner. Through this framework, we can better understand the complex interplay of action phases, the significance of terminating actions, and the feedback loops that shape the ongoing process of personal development and change.

Action-Phase-Feedback-Loops

Before we apply our action phase model (Fig. 2.1), it is important to discuss the concept of feedback loops and how they impact the distinct phases of the action preparation process. Let us consider a person preparing for an exam. In this case, the student has a clear goal: to achieve an A on the exam. What is central to Boeschian cultural psychology is not just understanding what the person wishes to achieve but also why, in terms of their underlying fantasms. For example, the student wants an A because achieving this grade is a step toward studying medicine, which may fulfill deeper needs, such as the fantasm for power, economic freedom, or other personal goals. Therefore, the action has symbolic meaning, linked to the student's motivation for setting and pursuing the goal.

Once the student knows what they want (an A) and how they plan to achieve it (by investing time and effort into studying), they initiate the action phase (the process of studying). In this phase, the student is faced with balancing competing actions and priorities that could interfere with the original plan. For instance, the student might receive an invitation from friends but chooses to decline in favor of studying. However, as time passes, the decision may lead to feelings of frustration, as the student realizes they missed out on social opportunities. Here, we observe the application of cognitive regulatory actions: the student may have told themselves that they needed to study more, and thus declined the invitation, in an effort to stay on track with their goal.

This scenario highlights the difficulty of separating the action phases clearly, as the process is dynamic and intertwined. Even during the action phase, there are intermediate evaluations that shape the process. Once the student completes the exam, this marks the termination phase of the action, which includes the consumption and evaluation of the result. The student may evaluate whether they passed the exam, based on external social standards (i.e., the grading system), and whether they met their own personal standards, such as achieving the desired grade of A. If the student passed the exam but did not meet their personal expectations, this becomes a point of evaluation. The failure to achieve the personal goal can lead to a reassessment of their approach.

In this case, the student might rush through the termination of the action because the outcome did not meet their initial personal expectations. However, this evaluation of failure serves a crucial purpose: it helps the student understand the discrepancy between the "what-is" (the current reality) and the "what-should-be" (the desired outcome). This process of evaluation allows the student to revise their methods, possibly adopting new strategies to improve future performance and bring them closer to their goal.

This dynamic shows that the different phases of action are intertwined and occur in irreversible time. When an initial action phase is completed, it generates new actions, leading to a reconfiguration of the process and the start of a new action phase (as illustrated in Fig. 2.2). Therefore, feedback loops between action phases are essential to understanding how each phase informs the next and how the process of action evolves over time.

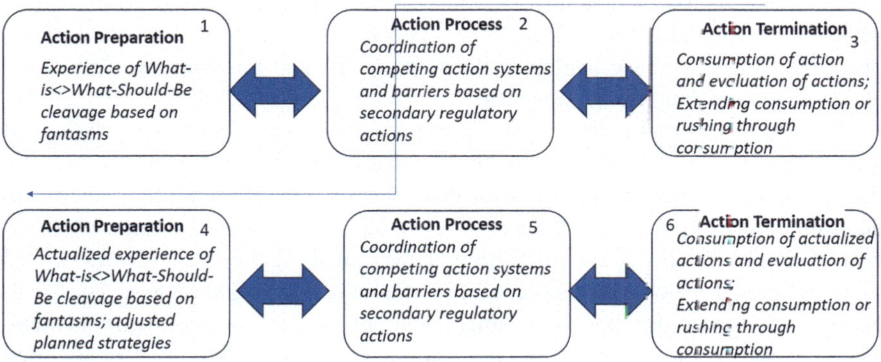

Fig. 2.2 Boesch's action phases in irreversible time

With Boesch, we can argue that action is the germ cell of psychology, as we discussed in Chap. 1. However, actions do not occur in isolation; rather, they are embedded within larger action systems. As we have seen in this chapter, these action systems are structured in distinct phases, which are interconnected through feedback loops. To study actions from a Boeschian perspective, it is crucial to understand not only how actions are organized across these phases but also how the structure and content of these phases evolve over time. This dynamic process is why Valsiner (2014, 2021, 2024) emphasizes that cultural psychology is inherently a developmental psychology. From a Boeschian standpoint, we can fully support this view, particularly when considering the way Fig. 2.2 illustrates the evolving nature of action systems. We see the evolving nature of action well illustrated in the following stanza by E.E. Boesch's poem:

You lie beneath the trees, dreaming in the warmth,
And feel the sun, heavy and full,
Nearby, shimmering swarms of lizards dart,
And everywhere there's light—even under clouds,
The house's walls still glow with enough light
To dazzle your eyes.

Applying Boesch's Extended Action Theoretical Model to Case Vignettes in Psychotherapy

After exploring Boesch's action theory, it is now time to bring this theoretical framework into practice. While Boesch was a psychotherapist, he never formally practiced, but his work has clear clinical implications from the outset. As we delve into his concepts, we can see how they resonate with therapeutic practice. In my gestalt therapeutic work, I have encountered many clients whose struggles can be illuminated by Boesch's extended action theory. One common issue that has emerged over years of practice is that some individuals seek therapy because they have lost touch with what they truly want in life. These clients have often lived their lives for others—partner, family, or friends—sacrificing their own needs for the needs of others (Polster & Polster, 1974). This results in a delayed personal development, where they have neglected their own goals and fantasms, as well as the ability to pursue them in the long term.

This life pattern makes these individuals particularly vulnerable to psychological issues, or in Gestalt terms, they get stuck (Polster & Polster, 1974). In these cases, therapy can help them reconnect with their desires, identify what they need, and determine how to pursue those needs (Perls et al., 1951). What I do in such situations is encourage clients to try something new, something they have always wanted to do but lacked the courage to pursue. For example, they might express a desire to go to a café alone, read a book in solitude, or simply observe the world around them. The goal here is to understand the motivations behind their actions. Clients often say things like, "I would be doing something for myself," indicating a deep yearning for personal fulfillment.

Importantly, I aim for my clients to discover their own motivations, grounded in a phenomenological approach. They need to own their lives and motivations (Perls et al., 1951; von Fircks, 2023d). Once we have identified actions that can fulfill these potential needs, I turn the focus to any barriers that may arise. I ask my clients to anticipate possible fears or obstacles, such as fear of negative judgment, and explore regulatory actions they might use if competing action systems interfere with their chosen actions. This process is crucial for clients to realize that barriers are natural and manageable, and that they can regulate their actions independently.

I encourage clients to commit to their initial actions and to the underlying motivation behind them. If they anticipate challenges and discuss strategies for overcoming them, the negative valence of these barriers—both cognitive and affective—can be reduced (Perls et al., 1951). After taking these steps, I invite clients to reflect on their feelings and the impact of their actions. How do they feel afterward? Do they feel empowered, balanced, or relieved? These feelings are vital in evaluating the success of the action, as they indicate that the client was able to meet their personal demands and expand their potential for future action.

When clients reflect on their successes, they recognize that they have increased their ability to act on their own needs. This positive evaluation of the action provides important feedback, encouraging them to continue prioritizing their self-care and to pay closer attention to how they pursue their goals in the future. The feedback loop in this process helps clients to reaffirm their autonomy and commitment to their personal needs, which ultimately fosters a more active, self-directed life.

With this rather general vignette, we see the benefits of applying Boeschian cultural psychology to the domain of psychotherapy. By examining the core theme of the vignette, we developed an interview scheme that integrates Boeschian tenets into a Gestalt therapeutic setting—particularly in situations where the client has denied living their own life in favor of helping others in their environment satisfy their needs. This is a crucial point because the client needs to rediscover their own needs, which have been suppressed by the demands and expectations of others. In other words, these needs have not disappeared but have become buried under the needs of significant social others. The therapist's role is to assist the client in shifting this need hierarchy.

It is essential to recognize that the therapist's role is assistive, meaning the client needs to make the discovery, not the therapist. This stands in contrast to dynamic or psychoanalytic therapies (Jacobs, 2012). The therapist works in a phenomenological way, helping the client interpret what they need, but not providing interpretations for them. In Table 2.1, I present an interview guide that adheres to this approach. Most of the questions are open-ended, allowing the client to explore their desires, what they are striving for (their "what-should-be" condition), and the personal and meaningful reasons behind these goals. This process helps the client become more aware of the significance of their goals and the underlying needs that motivate them across different situations in life.

Table 2.1 Boeschian Clinical Interview Guide for personal growth through an action-theoretical lens

Action phase	Boeschian premise	Example question
Action preparation phase	Experience of what-is and what-should-be cleavage based on fantasms (private need hierarchy)	What do you want to achieve? What are your short-term and long-term goals? What is so attractive in pursuing those goals? What do you wish to obtain when achieving this goal? Where are you now in relation to your goal? How do you plan to achieve those goals?
Action process phase	Coordination of competing and interfering action systems—development of regulatory (secondary actions); anticipation of barriers and getting used to barriers that might come up when wanting to locomote toward one's goal field	What might be interfering or disturbing events that hinder your goal pursuit? How do you plan to neutralize potential obstacles and hindrances? What are your strategies in dealing with issues that might emerge over the course of your actions? What are your fears when pursuing those actions?
Action termination phase	Consumption of actions and evaluation of actions (whether they fit the goal anticipation)—extended consumption or rushing through consumption	Were you satisfied with the results of your actions? Did you reach your goal? Why not? What happened? What were you thinking when reaching your goal (or when not reaching?) How might your actions be optimized? What do you imagine doing differently next time? Can you imagine locomoting even closer to your goal field if changing some of your actions and strategies? What did you do when reaching your goal? Did you have time to enjoy the results of your action? How did you enjoy? What did it do for you?

Once this is established in the therapeutic setting, the therapist and client can discuss interfering action systems and potential barriers that may arise as the client pursues their goals. The goal is to prepare the client for ecological obstacles, resistances, and barriers, which may be more easily overcome if the client has anticipated them fully. In the final phase of the process, the therapist and client assess whether the action can be terminated according to the client's personal evaluation criteria. If the action has not met these criteria, the two can explore how to adjust the approach.

Importantly, the client is encouraged to reflect not only on whether they met their evaluation criteria but also on whether they took the time to allow the results of their actions to sink in emotionally and energetically. What impact did these actions have on them at an affective and energetic level? If any issues have emerged during the action process phase, the therapist and client may negotiate additional strategies for regulating energy, affect, and cognition to address interfering action systems that prevent the client from reaching their desired goals.

This application of Boesch's action theory to psychotherapy helps the client reconnect with their true desires and personal needs, while equipping them with tools to manage the challenges and feedback loops that occur throughout the action process. By emphasizing self-discovery and active engagement in the therapeutic process, this model fosters autonomy and supports meaningful change in the client's life.

In summary, the application of Boesch's action-theoretical framework has provided valuable insights into how actions are organized and experienced in both personal and clinical contexts, particularly in psychotherapy. The key points of the discussion are as follows:

A: Interconnectedness of actions in action phases and systems

Actions are not isolated but are embedded within larger action systems that are spatially and sequentially organized. These systems are also deeply interconnected with specific action phases, such as the action preparation, action process, and action termination phases. While we separate these phases for scientific clarity, in practice, they remain intertwined, each impacting the others in a dynamic way. These phases come with distinct psychological mechanisms that need to be considered in both research and practice. Understanding how actions unfold across these phases, catalyzed by personal and ecological factors, helps frame actions across time, making them a developmental process within irreversible time.

B: Boesch's dynamic approach to symbolic action theory

Boesch's symbolic action theory is grounded in a dynamic understanding of psychology, which contrasts with psychoanalytic models. He emphasizes the ecological determinants of action, particularly the spontaneous and unexpected events that arise during the process of action. This dynamic quality is most evident in the action process phase, where individuals must maintain motivation while managing competing action systems. These unexpected barriers and hindrances force a transformation in the action process. Boesch's theory encourages a study of how actions change and adapt in response to these ecological challenges, which is essential for understanding the trajectory of personal development.

C: Application to psychotherapy and Gestalt therapy

Boesch's action-theoretical model can be effectively applied to psychotherapy, a field Boesch himself was involved in, even if not in a formal clinical sense. His affinity for psychotherapy can be seen through his intellectual connections, such as his admiration for Pierre Janet, whose work contributed to the foundations of cognitive behavioral therapy (CBT). In this chapter, I have applied Boesch's action theory to Gestalt therapy, providing an interview guide that helps therapists assist clients in exploring their personal needs, motivations, and goals. The guide helps clients navigate the action phases, identify barriers, and evaluate their actions, fostering the development of a personally meaningful future. By circumventing resistance and barriers, clients can engage in actions that bring them closer to their goals, enriching their self-awareness and growth.

In conclusion, this chapter addresses a significant research gap by integrating Boesch's action theory into the context of psychotherapy. By applying and extending Boesch's conceptualization of action theory, we hope to have offered a clear and systematic approach for therapists to use in clinical practice. This not only deepens our understanding of the psychological processes involved in action but also provides practical tools for therapeutic work, helping clients reconnect with their own needs and goals in a meaningful way.

References

Atkinson, J. M., & Heritage, J. (Eds.). (1984). *Structures of social action*. Cambridge University Press.
Blanco, I. M. (2018). *The unconscious as infinite sets: An essay in bi-logic* (2nd ed.). Routledge.
Boesch, E. E. (1975). *Zwischen Angst und Triumph: über das Ich und seine Bestätigungen*. Huber.
Boesch, E. E. (1976). *Psychopathologie des Alltags: zur Ökopsychologie des Handelns und seiner Störungen*. Huber.
Boesch, E. E. (1980). *Kultur und Handlung: Einführung in die Kulturpsychologie*. Huber.
Boesch, E. E. (1991). *Symbolic action theory and cultural psychology*. Springer New York.
Boesch, E. E. (1998). *Sehnsucht: Von der Suche nach Glück und Sinn [Longing: On the search of joy and meaning]* (1st ed.). Huber.
Boesch, E. E. (2005). *Von Kunst bis Terror: Über den Zwiespalt in der Kultur [From art to terror: On the cleavage of culture]*. Vandenhoeck & Ruprecht.
Boesch, E. E. (2006). *Konnotationsanalyse: zur Verwendung der freien Ideen-Assoziation in Diagnostik und Therapie*. Privater Nachdruck. (Original work published in 1977).
Boesch, E. E. (2021). *Musik, Sprache und die Sehnsucht nach dem Paradies*. Psychosozial-Verlag.
Engeström, Y. (2000). Activity theory as a framework for analyzing and redesigning work. *Ergonomics, 43*(7), 960–974. https://doi.org/10.1080/001401300409143
Engeström, Y. (2001). Expansive Learning at Work: Toward an activity theoretical reconceptualization. *Journal of Education and Work, 14*(1), 133–156. https://doi.org/10.1080/13639080020028747
Heim, G., & Bühler, K.-E. (2003). Pierre Janet: Ein Fall für die moderne Verhaltenstherapie? [Pierre janet: A case for modern behaviour therapy?]. *Verhaltenstherapie & Verhaltensmedizin, 24*(2), 205–224.
Hoffmann, N. (2013). *Zwänge und Depressionen: Pierre Janet und die Verhaltenstherapie*. Springer.
Jacobs, L. (2012). Critiquing projection: Supporting dialogue in a post-Cartesian world. In T. B.-Y. Levine (Ed.), *Gestalt therapy: Advances in theory and practice* (pp. 59–71). Routledge.
Janet, P. (1919). *Les medications psychologiques*. Payot.
Janet, P. (1929). L'évolution psychologique de la personnalité (Tome 1). Cahine.
Janet, P. (1930). L'évolution psychologique de la personnalité (Tome 2). Cahine.
Janet, P. (1932). *L'amour et la haine [Love and hate]*. N. Maloine.
Leont'ev, A. (1978). *Activity, consciousness, and personality*. Prentice-Hall.
Mead, G. H. (2015). *Mind, self & society*. The University of Chicago Press. Original work published 1934.
Murphy, J. J. (2024). *Solution-focused therapy. Theories of psychotherapy series*. American Psychological Association.

Perls, F. S., Hefferline, R. F., & Goodman, P. (1951). *Gestalt therapy: Excitement and growth in the human personality*. GST.

Polster, E., & Polster, M. (1974). *Gestalt therapy integrated: Contours of theory and practice* (2. print). Brunner/Mazel.

Schneider, K. J., & Krug, O. T. (2010). *Existential-humanistic therapy. Theories of psychotherapy series*. American Psychological Association.

Valsiner, J. (2014). *An invitation to cultural psychology*. SAGE.

Valsiner, J. (2021). *General human psychology*. Springer.

Valsiner, J. (2024). *THE MOONLIGHT DOCTOR: Art and Science of Carl Gustav Carus [Forthcoming]*. Springer.

von Fircks, E. (2023a). Categorizing personal appropriation processes of cultural givens and investigating their personal and collective functions. *Human Arenas*. Advance online publication. https://doi.org/10.1007/s42087-023-00351-9

von Fircks, E. (2023b). Enhancing and inhibiting: Framing of innovation in educational encounters. *Possibility Studies & Society*, Article 27538699231201014. Advance online publication. https://doi.org/10.1177/27538699231201014

von Fircks, E. (2023c). The fully objectified catathymic imagery method (FOCI): Assessing the amorphian inner at work. *Human Arenas*, 1–18. https://doi.org/10.1007/s42087-023-00345-7

von Fircks, E. (2023d). Understanding the power of mindfulness from a cultural-psychological perspective. *Culture & Psychology*. Advance online publication. https://doi.org/10.1177/1354067X231201387

von Fircks, E. (2023e). Geštalt TERAPIJA IZVAN POTREBA: Kako UPOTREBA ALATA UTIČE NA OSOBU I OKRUŽENJE. *Geštalt Zbornik, 10*, 46–59. https://www.ceeol.com/search/article-detail?id=1201024

von Fircks, E. (2024a). Follow-up study and its reasons: Shifting the focus more to the ecological assessment of the conservative phenomenon. In *Conservativism: A cultural-psychological exploration* (SpringerBriefs in Psychology) (pp. 49–56). Springer. https://doi.org/10.1007/978-3-031-51205-6_5

von Fircks, E. (2024b). Advocating a valid cultural Alltagspsychologie. *Trends in Psychology, 32*(1), 231–250. https://doi.org/10.1007/s43076-022-00231-1

Zwänge, H. N. (1998). *Depressionen. Pierre Janet und die Verhaltenstherapie*. Springer.

Chapter 3
The Action Potential Concept and Its Underlying Mechanisms

Boredom *(E.E. Boesch, Enno von Fircks translation)*
O weary guest of a thousand weary hours,
Where all things merely turn and none endure,
Where every being follows its own allure,
And silently, dull and dark, the seconds sour
And drip away. Distant, remote they lie,
Ruined longings frame the horizon's line,
And neither sunlight nor summer moonlight shines
To glimmer over their somber dreams on high.
Sometimes, a door scrapes open with a sound.
Sometimes, a child laughs on the bathing shore.
Sometimes, as you muse, across the land once more
Swift stars pull their glittering threads around.
Then your heart, startled from its sorrow's power,
Feels in its dryness the scent of coming showers.

When I first entered the research field of cultural psychology and began reading Boesch's foundational works (1991, 1998, 2005), I found myself confused by one of his key terms: "action potential." This confusion was partly due to my prior exposure to the term "action potential" in my undergraduate studies in psychobiology, where it referred to neuronal signal transmission. As a result, I struggled to appropriate the concept and gain a deeper understanding of Boesch's cultural psychology, or what he referred to as symbolic action theory (Boesch, 2021). This confusion stemmed from the English translation of the original German term, *Handlungspotential*, which literally translates to "action potential" in English. My prior experience with the term action potential socially overdetermined from my biology classed did not facilitate my understanding.

Over time, I came to understand the intended meaning of "action potential," but this understanding only came after I grasped Boesch's broader systematic framework of symbolic action theory. However, this journey was not without difficulty. The challenge arose because Boesch used the concept of action potential throughout his career (Boesch, 1975, 1991, 2005), often in varying contexts and interpretations.

E. Freiherr von Fircks, *Culture of Action*, Theory and History in the Human and Social Sciences, https://doi.org/10.1007/978-3-031-88808-3_3

As a result, the key features of the action potential concept and its significance for advocating a holistic general psychology (von Fircks, 2023a) are not easily distilled.

For instance, Boesch connected the notion of action potential to deeply interrelated psychological functions such as identity construction and the development of empathy (Boesch, 1991). In other words, the concept of action potential is central to Boesch's symbolic action theory, which offers a reinterpretation of identity, empathy, and other psychological processes. Given the centrality of the concept, this chapter aims to clarify Boesch's interpretation of action potential and explore its connections not only to identity but also to other psychological functions, such as the formation of personality.

What Is an Action Potential?

Boesch explains the notion of action potential with the relatively vague description of a "readiness and capacity to act":

> We have already encountered this term many times. I used it to designate one of those overarching goals which direct, control, or regulate specific actions, and among those it is probably the most important. I tend, indeed, to consider it to be what one could call the "basic motivation". Indeed, we continuously strive at optimizing our action potential, and this reinforc ing, strengthening, or enlarging of our readiness and capacity to act appears most of the time to be more important than any specific single goal. (Boesch, 1991, p. 105)

Let us illustrate the concept with a concrete example: Consider a classic Boeschian example—the act of wanting to build and construct a house in Thailand (Boesch, 1980). The process of constructing this house occurs within a complex, interrelated network of fantasms and myths (see Chap. 1) and unfolds across specific action phases (Chap. 2). However, the very act of constructing the house in unique ways also influences the person themselves. In the process of building the house—potentially in a foreign country—the individual not only contributes to the external project but also masters their own conduct, expanding their knowledge, skills, and abilities.

For instance, the person might learn how to seal their house against cold and warmth, possibly in collaboration with local people. Additionally, by respecting and engaging in local customs, the person addresses strong fears and anxieties that arise throughout the house construction process. Boesch provides the example of selecting the right type of wood (Boesch, 1980) and manipulating it in a specific manner to avoid attracting bad spirits during the process. In short, the person expands their action potential.

However, this expansion does not imply that the individual entered the process of constructing their house without any prior knowledge or skills. On the contrary, the individual began the process with a set of existing knowledge, skills, and abilities, which were refined during the experience of building the house in Thailand and participating in local rituals, for example. This means that a person acting to achieve change in the outer world already possesses a particular kind of action potential, which evolves over the course of engaging with subsequent developmental tasks. Although the motivation to act occurs within the broader context of the fantasms and myths that overdetermine the action, motivation alone is insufficient without the capacity to act. It is this capacity that enables a meaningful shift in the outer world.

> Optimizing our action potential, then, may mean several things. First, it simply aims at "functional disposability"—the ability to control or to master our functions. We know the pleasures of such functional disposability from playing purposelessly with a new ball-pen, driving a new car around aimlessly, singing for singing's sake, or walking for the pleasure of walking. For similar kinds of pleasures the child will jump, run, climb, shout, or throw stones—feeling able "to function", in any way whatsoever, will not satisfy specific wishes or needs, but, quite independently from such possible anticipations, it will provide the pleasure of exercising the function for its own sake. Whoever sprained an ankle knows what a frustrating functional impairment this can be, but he also remembers the pleasure felt on regaining the lost function. It is true that one may also exercise a function as a prerequisite for attaining a specific goal; yet, the "idle" use of a function offers its own rewards: not the triumph of successfully achieving an aim, but the fulfilling experience of ""possessing one's body". Not rarely, in "vandalism", the search for such functional enjoyment may be more important than the aggressive intentions which are, of course, combined with it. (Boesch, 1991, p. 106)

In this quote, Boesch expresses his concern about understanding the term *action potential* solely in a teleological (goal-oriented) manner. This perspective contradicts Boesch's own premises of a holistic psychology, which emphasizes the importance of artistic and aesthetic experiences for the individual (Boesch, 2000, 2021). In this view, mastering specific biological and cultural operations is inherently pleasurable and a continuous source of positive emotions for the cultural psychologist (Boesch, 2005).

However, there seems to be a second source of positive emotions the mastery of specific personal and ecological demands that bring the individual closer to fulfilling a given need or achieving a particular goal. Returning to the Boeschian example of building a home in Thailand, we see that choosing the right materials for construction, participating in local customs and rituals to cope with fears and anxieties, and seeking help from locals are all actions that demonstrate the individual's mastery of the material environment in distinctive ways (such as building a home from scratch). These actions also reflect the pursuit of symbolic goals, such as protecting the home from demons or bad spirits. Expanding one's action potential, therefore, not only enables the individual to engage with the present moment but also helps them establish a solid self-image, which in turn serves as an anchor for their past or personal history. Hence, action potential is crucial not only for the unfolding of one's future but also for solidifying a foundation for one's past and biography. This process inherently involves the individual's emotional responses which must be managed in order to proceed.

Here, Boesch introduces an intriguing perspective on the psychological study of emotions—not merely from a biological or physiological standpoint, but rather from an action-theoretical framework (see Chap. 6). Expanding one's action potential, in this sense, is a continuous source of positive emotions in two ways: first, through the aesthetic pleasure derived from witnessing, facilitating, and engaging in the actualization of one's bodily and mental functions (analogous to sculpting); and second, by enhancing the individual's capacity to achieve previously unattainable goals.

It is unsurprising, then, that Boesch links the concept of action potential to the interrelated notion of developing competencies for life in general (see Chap. 7). A person who masters the construction of their home amidst the complex meaning field of Thailand does not merely solve a problem or complete a task—although this accomplishment itself is a source of positive affect. The person also expands their general knowledge, skills, and abilities, which may include social skills, particularly when cooperating with locals. Hence, the concept of action potential inextricably linked to the notion of personality.

Action Potential and Transparency

For Boesch, a third crucial function of the notion of *action potential* is its relationship to *ordering* and *transparency* (Boesch, 1998). Enlarging one's action potential enables the individual to anticipate certain situations, including potential problems or challenges they are likely to face. Importantly, the person has the agency to choose how to anticipate not only their own needs but also the demands of their immediate or broader environment (Boesch, 1977/2006).

To illustrate this, let us consider a dramatic example, which remains relevant in some parts of the world (e.g., Africa) and was also true a century ago in the Baltic States. The renowned painters *Julius von Klever* and *Bogdanov Belsky* captured such scenarios with great finesse. These paintings depict children walking to school in the coldest temperatures, crossing dangerous frozen lakes, and then returning home in the dark. In one such example, Fig. 3.1 portrays two schoolchildren who have probably helped their mother with household chores, including purchasing goods from neighboring farmers. This act of purchasing goods is significant from a social perspective—fulfilling an important family duty and adhering to social *should-values*. The children's action is seen as relevant not only to the family's well-being but also as a meaningful contribution to the household's daily life.

However, beyond fulfilling a social obligation, the children may also derive personal pleasure from the task. For instance, they may enjoy operating the sledge and enhancing their motor skills. They might use the sledge for play before completing the errand, enjoy their sense of importance in contributing to the household, appreciate the beauty of the landscape, or interact with school friends on the way back from the farm. All of these actions and their interconnectedness increase the *transparency* of the children's experience. As they engage in this routine, they become familiar with the farmer, who may offer them a surprise next time, learn how to repair the sledge if it breaks, or figure out how to ask for help if the sledge becomes too heavy.

This example illustrates that the simple activity of purchasing goods from a neighboring farmer not only expands the children's ability to act but also provides

Fig. 3.1 *Children in Snow* by Bogdanov Belsky

them with a deeper understanding of their environment. They may gain knowledge about the location of their friends' homes, the residences of important figures in the community, and which people to avoid and why. This all contributes to a broader *transparency* of their surroundings, enabling them to become more familiar with the elements of their action field. In sum, the children's experiences help them expand their action potential while simultaneously enhancing their understanding of the world around them. This is also the reason why it is pedagogically relevant for children to master duties, autonomously.

This is what Boesch means by the notion of *transparency*—the ability to see through things, make sense of them, and integrate experiences into one's existing cognitive framework (Boesch, 1991). For example, the children, having already mastered the use of the sledge, benefit from the new knowledge they acquire about repairing the sledge and knowing when and where to pay attention to avoid breaking it again.

This represents one of Boesch's major contributions: his development of the concept of *action potential*. Action potential is always situated between past and future; it encompasses past experiences—such as knowledge, skills, abilities, and gifts—that are either utilized, expanded upon, or rejected in anticipation of the next present moment. Importantly, a shift in one's action potential through the mastery of new knowledge or skills alters the way a person sees and interprets the world. This change in perspective forms the basis for a subsequent change in action. Such a transformation is not just a matter of quantity but of quality, as the individual alters their worldview and their personality through the activities they engage in.

However, the action potential does not exist in isolation; it is intertwined with action motivation, which is influenced by fantasms and myths. For instance, a child who resists purchasing goods for the family and continues to engage in play demonstrates that they have not fully internalized the social *should-values* and are still more focused on their fantasmal activities. This situation may lead the family to intervene and attempt to motivate the child to participate in household duties. The things we engage with to enhance our capacity to act, as well as the things we avoid because they bother or bore us, offer insights into the complex personality of the individual. Therefore, action potential cannot be analyzed separately from action motivation, as they are deeply interconnected.

Action Potential: An Interpretative Example from Boesch Based on Suhrbier

Be knowledge, skills and abilities transmitted socially and historically, the acceptance, the mastery and its application in various contexts remains a personal endeavor. A rather illustrative example is given by Boesch in the following quote denoting the foldedness of the action potential between past and future.

> Ein beeindruckendes Beispiel fand ich erst kürzlich in einem Forschungsbericht von Birgit Suhrbier. Sie untersuchte bei Indianern im brasilianischen Regenwald die Bedeutungen von Pfeil und Bogen. Dabei fand sie, dass der Bogen nicht einfach eine technisch eingesetzte Jadgwaffe ist, sondern das Heranwachsen des jungen Knaben in rituell festgelegter Weise begleitet: Zuerst werden ihm Bogen aus einem schwächeren Holz gegeben, allmählich größere, doch Bogen aus dem starken Holz erhält der Junge erst nach seiner Initiation, also wenn er selbst erwachsen ist, und erst wenn er heiratet, bekommt er das Recht, selbst einen Bogen anzufertigen, wird ihm also Eigenständigkeit zugeordnet. (Boesch, 2000, p. 35)
>
> Birgit Suhrbier's recent research report offers a notable example of the cultural significance of bow and arrow among indigenous groups in the Brazilian rainforest. Suhrbier's findings reveal that the bow serves not only as a hunting tool but also plays a crucial role in the ritualistic maturation of young boys. Initially, boys are given bows crafted from weaker wood, which are gradually replaced by larger ones. However, it is only after their initiation into adulthood that they receive bows made of stronger woods. Furthermore, the right to

> make their own bows is granted only upon marriage, symbolizing their newfound independence. (Boesch, 2000, p. 35, author's translation)[1]

The quote highlights two key aspects of Boesch's concept of action potential. First, it emphasizes the way in which weaker bows and arrows are replaced by stronger instruments, a change that is essential for expanding the action potential of the boys. This transformation allows them to hunt larger animals and fulfill their role in providing for the tribe. The weaker bows and arrows initially serve to familiarize the boys with the instruments, enabling them to test and refine their skills, which is necessary for achieving the goal of hunting for food. Notably, the action potential evolves over time within a social environment—here, the tribe—facilitating a gradual transmission of skills from older to younger members. As the boys are provided with better tools, their action potential expands, enabling them to meet the growing demands of the tribe.

Second, the concept of *polyvalence* within the action potential is introduced. The action potential is not limited to fulfilling a singular goal but serves multiple functions simultaneously. In this case, the bows and arrows, while essential for hunting, also symbolize maturation and the transition into adulthood within the tribe. This highlights the polyvalence of action potential: as the boys master their tools, their action potential extends beyond material goals (hunting) to social dimensions (becoming respected members of the tribe). The stronger instruments not only enhance their ability to hunt but also elevate their social standing and participation in the tribe's rituals and feasts, thus bringing with it new duties and responsibilities.

This brings us to the broader point that psychologists and researchers must not only analyze how action potential transforms over time but also how the actualization of this potential reshapes the person's relationship with their material and social environment. The expansion of action potential is both a personal and social process, with the individual's capacity to act influencing their social role and identity. Researchers must therefore consider the manifold ways in which action potential can be used, which are crucial for both the individual and their community.

Returning to the Belsky example (see Fig. 3.1), the children's action of helping their family purchase goods illustrates how one action can fulfill multiple needs. It not only enables the children to contribute to the family's well-being, solidifying their place within the community, but it also satisfies personal desires (e.g., helping because of love for their parents). This process expands their action potential, making new elements of their environment more transparent and helping them transition into more responsible members of their social group. Thus, the action is polyvalent, serving both personal and social needs while enhancing the children's overall capacity to act. Boesch encourages us to study the multilayered nature of action potential,

[1] Originally, the ethnographic report comes from Suhrbier (1998)—but Boesch came up with significant new interpretations of the original report—which is the reason why we kept the Boeschian quote.

recognizing its complexity and importance for the individual and their social environment.

Personal and Social Should-Values Informing One's Action Potential

In a previous discussion, we explained that action potential arises from the interplay of personal and ecological demands that evolve within the context of general activities. This concept warrants closer examination from a Boeschian perspective. According to Boesch, action potential can be understood as comprising *subjective* and *objective* dimensions, which he frames in terms of personal and social should-values (Boesch, 1976, pp. 27–29). These should-values act as overarching parameters that guide how individuals anticipate and act in the present and near future.

At the core of this concept is the recognition that personal motivations—rooted in individual goals, aspirations, and internalized values—serve as catalysts for action. For instance, one student may strive to excel in a particular subject in school due to personal ambitions, while another may feel indifferent about the same subject. This contrast exemplifies personal should-values: the internal priorities and standards that shape individual actions.

However, personal motivations do not operate in isolation. As Boesch notes, individuals are also embedded in broader social contexts, where they encounter social should-values—expectations and norms derived from their cultural, familial, and community environments (Boesch, 1991). For example, in a performance-driven family, parents might place great emphasis on academic success, with accomplishments, tangible results, and high marks forming the foundation of their expectations. In this scenario, the child's personal should-values—such as the desire to achieve high grades to pursue a chosen field of study—may align with the family's social should-values. This alignment creates a *performance-driven ground theme* for the individual's action potential, whereby personal aspirations and social expectations mutually reinforce each other.

Boesch's framework also allows us to explore less performance-oriented examples, such as those found in more collectivistic or traditional societies. In his discussion of Suhrbier's ethnographic work on a Brazilian tribe (1998), Boesch highlights how action potential is informed by social should-values specific to the tribe, such as becoming a man, an adult, or a caregiver (Boesch, 2000, pp. 34–57). These values interact with personal motivations in shaping action. For instance, a young boy in the tribe might aspire to become a full member of the community, not only to fulfill social expectations but also due to fantasms, such as impressing and eventually marrying a specific girl.

> Als (…) Hauptfunktion des Bewusstseins entdecken wir das Evaluieren von Handlungseffekten als Indikatoren unseres subjektiven Handlungspotentials. Wir erkannten diese Funktion vor allem am bewussten Erleben positiver Verlaufsphasen, doch gilt das Gleiche selbstverständlich auch für Handlungsschwierigkeiten: jedes Handlungsergebnis

> hat immer sowohl eine sachliche wie eine persönliche Bedeutung (…) [.] (…) Die Bewertung von Effekten als Hinweise auf unser Handlungspotential muss ja Kriterien folgen, die wir für die Selbstbeurteilung als wesentlich erachten (…). (Boesch, 1980, p. 168)
>
> The primary function of consciousness, we discover, is the evaluation of action effects as indicators of our subjective potential for action. This function is particularly evident in the conscious experience of positive phases of progress, but it naturally applies to difficulties in action as well: every action outcome invariably holds both an objective and a personal significance. The assessment of effects as indicators of our potential for action must adhere to criteria that we deem essential for self-evaluation. (Boesch, 1980, p. 168)

The Tension Between Fantasy and Reality as Important Catalyst for One's Action Potential

Boesch emphasizes that the development of one's action potential must account not only for mastery over knowledge, skills, and abilities but also for the interplay between the real and the imaginary (Boesch, 1980, 2000). This tension, as he describes, becomes a driving force that shapes the individual's conduct, motivations, and ability to navigate and adapt to their environment. A particularly vivid example of this dynamic is the psychological mechanisms at play during carnival (Boesch, 1998). During this time, individuals disguise themselves, entering into a dialogue between their actual self and an imagined, potential self. This temporary "dialogue" allows individuals to explore neglected aspects of their identity, surface hidden needs, and recalibrate their relationship with reality. This process highlights the dialogical nature of self-development, where personal identity and action potential are informed by the interplay of self and non-self (Boesch, 1998, pp. 129–134). In this sense, the imaginary serves not only as a tool for creative problem-solving but as a foundation for personal growth and the redefinition of one's goals and aspirations.

The imaginary becomes particularly significant in helping individuals update their action potential. For example, Boesch's Brazilian boy interpretation illustrates how the boy's envy of his father's superior bow and arrows could spark imaginative fantasies of adulthood, tribal membership, and romantic fulfillment. These fantasies, rooted in personal and social should-values, provide a powerful motivation for action. However, these fantasies do not necessarily align perfectly with the realities of life. The boy's idealized imagination (or hypothetical extension of Suhrbier's ethnographic analysis 1998)—such as marrying his childhood friend—may face external constraints, such as another friend marrying the girl first. This discrepancy creates a tensive relationship between fantasy and reality that necessitates a psychological resolution.

Boesch's dynamic model of cultural psychology posits that action potential must continuously adapt to evolving ecological determinants—external forces such as war, disease, or social conflicts (Boesch, 2000). These external realities often challenge the individual's personal and social should-values, compelling them to reassess and realign their motivations and actions. This process creates what Boesch calls the I/World Balance—a dynamic equilibrium between the person's internal needs (e.g., fantasms, aspirations) and the external constraints of the social and material world (Boesch, 2021). Maintaining this balance is an existential condition and crucial for effective action.

In cases where reality does not align with personal fantasy, as in the hypothetical example of the Brazilian boy failing to marry his childhood love, the resulting tension can lead to two possible outcomes: The boy may revise his personal aspirations, de-emphasizing his desire to marry the girl by rationalizing that another partner would suffice. This shift involves accepting external reality as it is and altering the fantasm to maintain harmony with the broader action field. Conversely, the boy might resist external constraints, holding firm to his original personal should-values and striving to achieve them despite potential social repercussions. For example, he may fight against social norms to win the girl's affection, regardless of the consequences for himself or others. Hence, he might enter into negotiation with the gatekeepers of the tribe trying to renegotiate the mythical overdetermination of action.

The tension between fantasy and reality is central to the development of action potential because it forces individuals into a dialogical relationship with themselves and their environment. In the boy's case, his failure to achieve his dream presents an opportunity for psychological growth. By either adapting to reality or challenging it, the boy's action potential evolves: If the boy adapts to reality, his action potential undergoes transformation through the modification of his personal should-values. This adaptation allows him to develop a new outlook on life and adjust his goals accordingly. If the boy resists reality, his persistence in pursuing his original fantasm may lead to an expansion of his action potential through increased determination, skill development, and the reconfiguration of his social environment to meet his needs.

In either scenario, the boy's actions are informed by an ongoing dialogue between the imaginary (aspirations, dreams, fantasies) and the real (external constraints, societal norms, ecological challenges). This dialogue highlights the polyvalence of action potential, as it enables individuals to respond flexibly and creatively to their evolving circumstances while striving for a dynamic balance between their inner and the outer.

Boesch's framework reveals the transformative power of the imaginary in shaping action potential. By acknowledging the tension between fantasy and reality, individuals can confront the constraints of their environment while reimagining their possibilities for action. This tension acts as a catalyst for growth, encouraging individuals to adapt, innovate, and redefine their relationship with the world. In this sense, fantasy is not merely escapism but an integral part of how individuals navigate, expand, and sustain their action potential over time.

Boesch emphasizes that the concept of action potential encompasses more than just the temporal axis of past and future; it also includes the dynamic tension between real and imaginary experiences. This tension is a fundamental aspect of human existence and decision-making, as illustrated in Enlightenment literature, such as Schiller's *Kabale und Liebe*. In these works, characters are often confronted with a clash between personal should-values and social should-values. For instance, lovers of differing social backgrounds must decide whether to rebel against societal conventions (social should-values) to live out their personal fantasies or to conform to societal expectations at the expense of their personal desires. Boesch (2005, pp. 96–107) asserts that this existential conflict—between accepting reality and altering imagination or rejecting reality to pursue one's imagination—captures the core of human experience. Lewin (1982) similarly highlights this interplay between imagination and ecological reality as a defining condition of human life.

Imagination, in Boeschian terms, is as integral to a person's action potential as ecological determinants are to the present moment. It is the interplay between the real and the imaginary that shapes the complexity of human living. By examining this dynamic, we gain insight into the personality of a given individual—how past experiences and personal values impact their future-oriented goals and actions. Imagination, therefore, is not merely an abstract concept but a tangible force that structures a person's subjective universe and action field.

The dynamic nature of action potential becomes evident when we consider Boesch's assertion that individuals seek order, consistency, and transparency not in relation to an objective reality, but within their personal worlds. This personal world reflects a person's overarching action orientations, which guide their interests and activities, such as the books they read, the topics they discuss, or the news they consume (Boesch, 1991, p. 110). Thus, the action potential is deeply personal and reflects the individual's psychological framework for engaging with the world.

Boesch's approach challenges the prevailing static conceptualization of personality, as seen in contemporary psychological research emphasizing the relative stability of personality traits (Costa & McCrae, 2008). Instead, Boesch argues for a developmental and dynamic view of personality, in which past experiences are continuously transformed into future possibilities, and real experiences are reshaped by imagination—and vice versa. This dynamic operationalization of personality allows us to understand how individuals restructure their action potential in response to shifting personal and social should-values.

Social constraints, embodied in social should-values, often act as barriers to the fulfillment of personal should-values. When an individual is unable to realize their personal goals due to these external constraints, their personality development can become inhibited, resulting in negative emotional experiences and frustration. This inhibition illustrates the interdependence of various psychological functions, such as emotions, competencies, aesthetics, anticipation, and mastery. Boesch's notion of action potential offers a holistic perspective on the person, rejecting atomistic approaches that dissect the individual into isolated traits or behaviors. Instead, it allows us to explore the complex, interrelated dimensions of human experience and personality development, providing a rich framework for future research.

The Essential Role of Barriers to Actualize an Action Potential

Before we are turning our gaze to the notion of identity, let us shed some light onto the way how we become aware of our action potential and noticeably about its respective limitations. According to Boesch, we need to contemplate the notion of barriers for this scientific endeavor:

> A barrier is an area that is difficult to overcome, and may need particular interposed actions for the passage; once the barrier crossed, however, the action can proceed more or less as before. (Boesch, 1991, p. 113)

Much of what occurs in the development, expansion, and transformation of one's action potential takes place unconsciously. This aligns with Stern's critical personalism, which posits that individuals must economize their psychological and physiological functions without consciously reflecting on them at every moment (Stern, 1906, 1923, 1924). According to Stern, it is nearly impossible for a person to continuously monitor all their functions, as they are always embedded in social systems and must meet the demands of their environment. Stern argues that it is only during moments of rupture—such as crises, barriers, or resistance—that individuals adopt a meta-position to consciously actualize and realign their psychological or physiological functions. Once this realignment occurs, the functions typically return to a more habituated state.

This can be illustrated through everyday examples. Consider navigating to a train station in an unfamiliar city. The person lacks the geographical knowledge and experience necessary for orientation in this new environment. Consequently, their action potential in this domain—navigation—requires support. In this situation, the individual draws upon what Boesch calls *fundamental action potentials* (Boesch, 1991). For instance, the person might initiate a cooperation process by asking a stranger for directions, using nonverbal cues such as a smile to encourage helpfulness. Alternatively, the person might rely on technological tools, such as a smartphone's navigation app. Here, the lack of a specific action potential (geographical navigation) prompts the activation of a more fundamental action potential (cooperation or tool use). The individual compensates for their limitations by mobilizing these broader capacities, allowing them to overcome the barrier and reach their goal.

Returning to an earlier example, consider the confusion I experienced when first encountering Boesch's concept of action potential. Coming from a psychobiological background, I initially understood the term within a medical or biological context. Encountering its use in Cultural Psychology, however, created a barrier—I was unable to reconcile its meaning in this new theoretical framework. In response, I sought to expand my psychological action potential by drawing upon my fundamental *learning action potential*. I immersed myself in relevant literature, wrote exploratory essays, and engaged in reflective practices to test and refine my understanding. Through this process, I not only overcame the immediate barrier but also enriched my psychological action potential, incorporating Boesch's symbolic action theory into my broader intellectual repertoire.

This process of overcoming barriers and strengthening action potentials is central to Boesch's educational theory (Kölbl, 2020). Boesch emphasizes that "by setting limits, an individual may tend to strengthen his action potential through concentration on specific goals" (Boesch, 1991, p. 115). *Barriers are therefore constructive in nature, as they compel individuals to refine and optimize specific action potentials.* However, from an educational perspective, simply creating artificial barriers is not sufficient. Students also require guidance and support in building their fundamental action potentials, such as learning strategies, to effectively address specific challenges.

In Boesch's framework, the interaction between specific and fundamental action potential plays a vital role in fulfilling one's needs and achieving goals. For educators, therapists, and others in helping professions, barriers present opportunities to facilitate growth by encouraging individuals to draw upon their existing strengths to develop new capabilities. These limitations, when coupled with appropriate support, allow individuals to optimize their action potentials, preparing them to face future challenges. The key is that individuals should not face barriers alone but must be supported in illuminating and strengthening the action potentials required to overcome these obstacles.

Action Potential and the Construction of One's Identity

This contemplation brings us to the last factor in the present chapter regarding our illumination of what Boesch meant with the notion of action potential—that is, the construction of one's identity. The notion of identity is surely a controversial topic within the wider literature of psychology and the social sciences, in general—both relying on a multitude of approaches to explain what constitutes identity (Schwartz et al., 2012). These debates about what constitutes identity are sometimes led in an aggressive manner, denigrating the validity of the social other's perspective. It was only recently that some cultural psychologists such as Valsiner (2018) and Gamsakhurdia (2020) sought to defend a more pragmatic perspective on the notion of identity from a semiotic standpoint. Boesch himself moves in a similar direction, attempting to illuminate the term through his integrated system called symbolic action theory (Boesch, 1991).

For Boesch, it is sufficient to observe a person's actions to gain insight into their identity construction. However, he is aware of the dangers of such an operationalization of identity and cautions researchers and practitioners against adopting an overly static perspective on the phenomenon. As such, Boesch recognizes that actions are not eternally fixed in time but are subject to change—a change that incorporates elements of previous actions rather than their complete abandonment. Grounded in a dynamic perspective, Boesch (1980) explains that a person is continually confronted with new ecological determinants of action that begin to relate to previous determinants and form something novel and higher toward which the person must react.

Consider the following example: A person decides to watch a football game. He purchases some tickets, goes to the stadium, watches the match, cheers for his favorite team, and then goes home. On the surface, this might seem like the person is holding himself constant over time—all his actions are directed toward facilitating the watching and enjoying of the respective football game. However, this perspective is insufficient to fully understand the construction of the person's identity, where the emphasis lies on the notion of construction.

In other words, the scenario is likely to be more complex. While the person decides to purchase tickets for the game, he may be reminded that there is a competing event—the birthday of a good friend. Despite this, he purchases the ticket and justifies it by reasoning that he can visit his friend at another time. During the match, as he grabs a beer to enjoy the game, he meets some colleagues from work who want to initiate small talk. Reluctantly, he engages briefly but quickly excuses himself to return to his seat, unconcerned about appearing impolite. He tells himself that he wants to enjoy the match alone and avoids the risk of being invited to watch the game with them. Additionally, he does not want to discuss work matters during his free time. Later, when he is back at his seat, he receives a phone call from his mother. However, he decides not to answer and continues focusing on the match.

This example illustrates how the person must regulate his actions to maintain consistency over time. In this case, we observe that the person is committed to acting in a particular way—watching the football game. However, various ecological determinants arise, demanding his attention and potentially derailing his plans. He resists these distractions by rejecting small talk, ignoring his phone, and postponing his friend's birthday visit—all to pursue his initial goal of watching the game.

This example demonstrates that the person acts in accordance with his previously planned action, resisting deviation despite emerging ecological constraints. Put more simply, the individual is confronted with a wave of chaos—a crucial threat to inhibiting his planned activity. Nevertheless, he successfully circumvents these challenges and remains committed to his initial action despite external pressures. In Boesch's words, we can depict the following:

> It seems, then, that to integrate specific actions consistently into situational networks and overarching goal orientations will often be difficult, thus jeopardizing the formation of self-identity. Yet, we might also conclude the opposite: could not the formation of self-identity constitute a kind of unifying instance capable of "ironing out" the discrepancies in one's concrete action experience? Levi-Strauss' famous bricoleur (handyman) is surrounded by bits and pieces and, by selecting some, rejecting others, the handyman manages to produce some kind of artful construction: he creates unity within disorder, and this, of course, not only as an external result, but as much as an inner structure. Picasso liked flea-markets or even garbage heaps, and his famous "head of a bull", made from the saddle and the handle of a discarded bicycle, demonstrates the unifying inventiveness of his mind. (…) Yet, these structures will have to be formed in spite of the inconsistencies in the individual's action field we just referred to. (1991, pp. 308–309)

Hence, the construction of one's identity happens against the backdrop of bringing unity to one's environment through a concrete teleological activity by the person. This teleological activity does not occur in a vacuum shielded from ecological determinants; rather, it takes place while actively circumventing these determinants that arise from competing action spheres. In short, identity is always about bringing order to one's environment and to one's self.

Applying this perspective to our football example, we gain a clearer understanding of the identity construction of the individual in question. Watching a football game to relax and cheer for his favorite team plays a significant role in the life of this person, shaping how he navigates and circumvents barriers and resistances that might distract him from his initial activity. In other words, it is not merely the act of watching football that plays a role in the identity formation of this individual, but rather watching football as a means of relaxing and enjoying time alone—for example, as a way to detach from work.

To put it differently, the analysis highlights a particular interconnected identity construction that unfolds over time. Football itself, especially the act of watching it, holds importance for the person, particularly as a way to relax, detach from work, and engage in meaningful time spent alone. Here, we observe interconnected psychological functions that help define the identity of this individual—his general unifying teleological activity. In the current example, this teleological activity involves detaching, enjoying, spending time alone, and engaging in an activity for himself. These actions collectively form a particular personal identity network.

This is precisely the interconnected perspective referenced in the introductory paragraphs of this chapter. It would be erroneous to argue that football alone is the primary feature of this individual's identity construction within this hypothetical example. Instead, identity construction relies on interconnected actions motivated by specific personal should-values embedded in fantasms. These actions and values form parts of a whole that constitutes the individual's identity within this context.

This analysis demonstrates that, methodologically, in order to gain a deeper understanding of the identity construction of a given person, we must not only analyze their personal should-values, the underlying actions, and the interconnectedness of these actions, but also broaden our perspective to include ecological determinants. It is equally important to examine how the individual reacts to these determinants and the regulatory mechanisms they employ.

> These processes will, of course, be based on selective perception of situations and one's actions which, in addition, will be appropriately interpreted. Such interpretations may consist of a justification of motives, an attribution of responsibility to the action of others or to compelling situations, and, where reinterpretations fail, inconsistent aspects of situations or actions may even be "blotted out" of consciousness. Forgetting, repressing, embellishing, projecting, transferring, belittling or enlarging are such processes used for "consistency cosmetics". All these, indeed, are active processes of structuring our self, and we should not be misled into believing that they occur only in neurotic cases; they belong to one's general identity construction, which goes on all the time. (Boesch, 1991, p. 308)

How does this relate to the notion of action potential that lies at the very core of this chapter? Identity, after all, provides the framework for how an individual positions themselves in a related action field. However, action potential serves as the mechanism that enables the individual to maintain that positioning despite encountering barriers and resistances. Identity is more closely aligned with giving life directionality (von Fircks, 2024), while the concept of action potential equips the person with the skills and abilities needed to move toward that direction or to maintain it consistently over time.

Identity construction and action potential, therefore, represent two complementary psychological functions that are interdependent in allowing the individual to maintain constancy in their sense of self over a given period. This interconnectedness can be valuable in therapeutic contexts, where assisting individuals in their identity construction is often a central goal. Frequently, the challenge is not that the person lacks interests or a sense of identity but rather that they lack the skills or knowledge necessary to defend and protect those interests against external barriers and resistances.

The Action Potential Concept: An Example and a Summary

Let us briefly elucidate the revised action potential concept through a specific example (see Fig. 3.2). The object represented in Fig. 3.2 can be described using the straightforward term "coat hook." The coat hook is employed by an individual to hang their jacket, thereby creating order within their home. This example highlights the teleological nature of the action potential concept. By hanging up their jacket, the individual establishes a sense of order that extends into the future. Once the

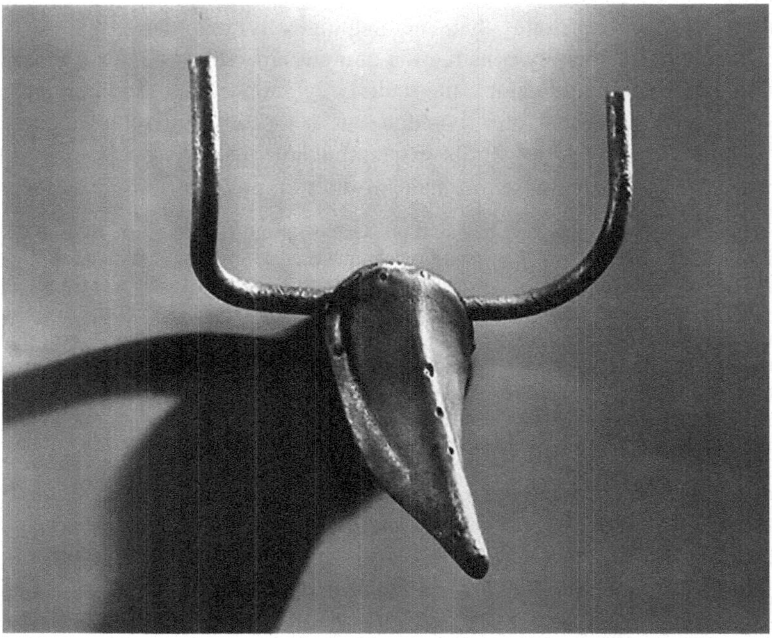

Fig. 3.2 Coat hook and its meaning for the action potential

present moment has passed, the person has constructed order and can thus perceive themselves as orderly and clean—attributes they may integrate into their identity.

Importantly, the need for orderliness—which drives the execution of this action—operates within an interplay between the individual's personal and social "should-values." The need for orderliness arises because the individual has internalized specific personal action-promoters, such as the fantasm of orderliness and cleanliness, which they strive to preserve over time. However, the person may also face barriers, including opposing "should-values." For instance, their partner might criticize them for being excessively neat. In this situation, the individual can anticipate a potential conflict (e.g., "What would happen if I responded to such a comment?") and may choose to disregard the partner's criticism, continuing their action without reacting to the remark.

Moreover, the symbolic meaning of the coat hook itself could interfere with the initial action. The coat hook depicted in Fig. 3.2 might be perceived as resembling the image of a bull, which carries various negative connotations—such as aggression, impulsiveness, or violence—that the individual seeks to exclude from their field of action. In this case, the symbolic meaning functions as an inhibitor of action, as the need for safety and security overrides the need for orderliness and its associated fantasms.

Regardless of whether the symbolic nature of the coat hook is consciously recognized, this example illustrates the complexity of the action potential concept and its significance for the individual as they navigate between past and future.

Having elaborated on the relationship between identity and action potential, we can now move to a temporary summary of what Boesch aims to communicate with the concept of action potential. Here, we can depict the following:

(A) An action potential refers to the capacity of a person to optimize or expand their "readiness or capacity to act" (Boesch, 1991, p. 105). This concept has an experiential quality, meaning that action potential can only be developed and optimized through experience. Expanding one's action potential provides two sources of positive emotions: first, the individual can derive pleasure from completing a particular mental or physical task for its own intrinsic value. Second, completing a psychological or physiological function also aids the individual in fulfilling their needs and achieving their goals. Boesch aims to connect the aesthetics of action—the inherent joy and satisfaction it brings—with its teleological, goal-oriented nature.

(B) Since an action potential is experiential in nature, it exists in the interplay between past and future, with prior experiences being transformed by the individual in the context of specific ecological conditions. To understand a person's action potential, it is necessary to analyze how their skills, knowledge, and abilities evolve over time and how they are shaped by ecological determinants. Thus, when discussing a person's action potential, both the historical development of these capacities and their anticipation of future experiences, which impact the action potential in the present, must be considered.

(C) The expansion of an action potential is guided by particular personal and social motivators that regulate its growth. This process is informed by an interaction of personal and social should-values, which act as regulators in optimizing one's action potential. Boesch emphasizes the polyvalence of actions, which encapsulates the interplay between these personal and social regulators. The task of the psychologist or researcher, then, is to unravel this complex regulatory interaction to understand the process.

(D) Fantasy and imagination are crucial elements in the anticipation of future moments and thus play a significant role in expanding one's action potential. Fantasy is informed by the individual's private and intimate should-values, which serve as a guide throughout their life. Within the realm of fantasy, social constraints are less influential (Boesch, 1977/2006, p. 9), allowing the individual to plan how to integrate or actualize fantasmatic themes into reality in a way that aligns with their personal should-values. However, this alignment may not always succeed. In such cases, the individual must either modify their personal should-values to reduce the negative impact of ecological constraints or actively confront those constraints to shape the environment in accordance with their fantasy. To assess the motivational force of an action potential, psychologists must conduct a careful analysis of the individual's fantasms and imagination, which vividly reveal how they aspire to construct reality in a personally meaningful way.

(E) Barriers are essential to the realization of an individual's action potential. These barriers play a critical role in reshaping the action potential in unique ways, forcing the person to adapt their actions to the demands of their specific ecological context. It is important to distinguish between specific and fundamental action potentials. For instance, a specific action potential, such as the ability to navigate a foreign environment, may reveal gaps in experience that limit the individual's capacity to act. In such cases, the person may rely on more fundamental action potentials, such as a learning action potential, to initiate collaboration. Through cooperation, the individual can acquire the necessary knowledge to actualize the specific action potential, using the broader capabilities of the fundamental action potential as a foundation. Boesch encourages us to examine how fundamental and specific action potentials interact in the context of an overarching activity.

(F) Identity construction arises from the individual's ability to unify their general actions into a cohesive system. To achieve this, it is essential that the person learns how to counteract ecological resistances and barriers, enabling them to maintain a consistent experience over time. This process provides insights into why and how the individual continues to pursue a general activity despite constraints within their action field, revealing their personal should-values, which are diverse and interconnected. This perspective highlights the need for a more comprehensive and integrated approach to identity construction. It also underscores the importance of helping individuals expand their action potentials so they can independently pursue a unified teleological activity. Consequently, the focus of identity construction must extend beyond the thematic aspects of "what" or "what-should-be" (such as interests, goals, and needs). It must also encompass the "how"—that is, how the individual approaches the what-should-be condition by leveraging both specific and fundamental action potentials. If the person not simply becomes aware of their fantasms (what pushes them toward action) but also about the means how to realize that action, then the person has also gained a fundamental insight how to counter boredom, actively—something addressed at the beginning with Boesch's poem about boredom.

Sometimes, a door scrapes open with a sound.
Sometimes, a child laughs on the bathing shore.
Sometimes, as you muse, across the land once more
Swift stars pull their glittering threads around.
Then your heart, startled from its sorrow's power,
Feels in its dryness the scent of coming showers.

Analyzing the Role of the Action Potential in Politics: A Case Study of Friedrich Merz and His Beer Coaster

Analyzing the role of action potential in politics through the lens of Friedrich Merz and his beer coaster proposal reveals the practical applicability of Boesch's symbolic action theory to public and political events.[2] In 2003, Friedrich Merz, a conservative politician who is now the head of the Christian Democratic Union in Germany and Chancellor, proposed simplifying the income tax return process, which was widely perceived as overly complex and difficult for the average citizen to manage correctly (Vereinigte Lohnsteuerhilfe, 2023). Merz suggested a reform in which the tax return could be completed on a single beer coaster—a symbolic and illustrative gesture aimed at conveying the need for simplification in tax regulations. His proposal involved reducing the tax income rules, which then consisted of over 40 specifications, to three major categories determined by income level. This idea quickly attracted both praise and criticism, polarizing public and media opinion. While some acknowledged Merz's intent to streamline the tax process for ordinary citizens, others argued that his suggestion was overly simplistic and potentially detrimental to Germany's federal budget (FocusOnline, 2020).

This proposal became emblematic of Merz's political identity, as the notion of the "beer coaster tax" entered public discourse. Some critics coined the term "beer coaster style" to describe Merz's approach to politics, criticizing it as overly simplistic, polemic, and controversial (Deutschlandfunk, 2023). On the other hand, Merz's supporters praised the accessibility and populist undertones of this style, arguing that it demonstrated his proximity to the concerns of everyday people (Kain,

[2]This political example is derived from the German public context—it may prove difficult to understand outside of the German context—yet the practical implications are valuable for political communication in and outside of Germany.

2024). Notably, Merz himself leaned into this symbolic political branding, referencing the beer coaster idea in subsequent campaigns, even proposing an update to his earlier suggestion by advocating for the use of both sides of the coaster rather than just one.

From an action-theoretical perspective, as articulated by Boesch, this political vignette can be analyzed holistically to understand its deeper psychological and symbolic underpinnings. First, Merz identified a key barrier within the action field of the German populace: the overly bureaucratic and complicated process of completing income tax returns. Recognizing that many people struggled with the demands of this system, he sought to lower these ecological constraints by proposing a simplified system, thereby allowing citizens to approach the task with their existing action potential. In this sense, Merz's proposal can be interpreted as an attempt to modify the ecological determinants of the action field (here, the tax system) to empower individuals to act autonomously without requiring significant additional personal resources.

Merz's symbolic use of the beer coaster played a crucial role in this strategy. The coaster, an everyday object associated with social settings like pubs, became a metaphor for simplification, accessibility, and proximity to ordinary life. This strategic use of symbolism reflects Merz's attempt to align his personal should-values with those of the wider public. The idea for the coaster itself reportedly originated in a casual setting with friends, suggesting that Merz drew on his personal experiences within his social action field to develop a proposal that he believed would resonate with others. His ability to translate this personal experience into a public policy proposition underscores how identity construction and symbolic actions intersect in political contexts.

Merz further utilized the beer coaster as part of his ongoing political identity construction. By proposing similar reforms in the corporate tax law (Der Spiegel, 2020), Merz reinforced his public image as a politician committed to simplifying complex systems. This style of politics has become a defining feature of his political persona, with Merz actively embracing labels such as "simplifying" and "controversial" to distinguish himself from other politicians. Unlike Armin Laschet, a fellow Christian Democrat who failed to secure broader public appeal, Merz has successfully built a sharp political profile centered on these attributes, which he argues are necessary in today's political climate (Hotel Matze, 2024). This ability to integrate past, present, and anticipated future experiences into a cohesive political identity aligns with Boesch's understanding of identity construction as the unification of actions within a larger system of meaning.

From a symbolic action-theoretical standpoint, Merz's beer coaster proposal also illustrates the fundamental mechanism of politics: the modification of ecological determinants within people's action fields. Unlike psychotherapy, which focuses on strengthening individuals' personal resources to expand their action potential, politics often seeks to lower environmental barriers so that individuals can act more effectively within their existing capacities. Merz's proposal exemplifies this dynamic by attempting to reduce the complexity of tax laws, thereby empowering citizens to complete their tax returns autonomously, or at least with reduced effort. While the reform itself never materialized due to Merz's party being out of government at the time, the symbolic and political impact of the proposal endures.

Ultimately, this case study demonstrates the polyvalence of Boesch's theory, which can illuminate not only the identity construction of a politician like Friedrich Merz but also the broader mechanisms by which political propositions are developed, communicated, and perceived. People whose personal should-values reject oversimplification in politics are likely to distance themselves from Merz, whereas those who share his should-values will find his approach appealing. Above all, this analysis highlights how political actions, through the modification of ecological constraints, can enable individuals to mobilize their action potential and address societal challenges more autonomously. By applying Boesch's theory to this case, we gain a richer understanding of how symbolic actions, identity construction, and ecological determinants converge in the realm of politics. This analysis underscores the relevance of action potential not only in psychology but also across disciplines, demonstrating the versatility and depth of Boesch's ideas. The beer coaster is just one example among many. I hope that this chapter offers readers a similar sense of insight and inspiration as they explore these concepts further.

References

Boesch, E. E. (1975). *Zwischen Angst und Triumph: über das Ich und seine Bestätigungen*. Huber.

Boesch, E. E. (1976). *Psychopathologie des Alltags: zur Ökopsychologie des Handelns und seiner Störungen*. Huber.

Boesch, E. E. (1980). *Kultur und Handlung: Einführung in die Kulturpsychologie*. Huber.

Boesch, E. E. (1991). *Symbolic action theory and cultural psychology*. Springer New York.

Boesch, E. E. (1998). *Sehnsucht: Von der Suche nach Glück und Sinn [Longing: On the search of joy and meaning]* (1st ed.). Huber.

Boesch, E. E. (2000). *Das lauernde Chaos. Mythen und Fiktionen im Alltag*. Huber.

Boesch, E. E. (2005). *Von Kunst bis Terror: Über den Zwiespalt in der Kultur [From art to terror: On the cleavage of culture]*. Vandenhoeck & Ruprecht.

Boesch, E. E. (2006). *Konnotationsanalyse: zur Verwendung der freien Ideen-Assoziation in Diagnostik und Therapie*. Privater Nachdruck. (Original work published in 1977).

Boesch, E. E. (2021). *Musik, Sprache und die Sehnsucht nach dem Paradies*. Psychosozial-Verlag.

Costa, P. T., Jr., & McCrae, R. R. (2008). The revised NEO personality inventory (NEO-PI-R). In G. J. Boyle, G. Matthews, & D. H. Saklofske (Eds.), *The SAGE handbook of personality theory and assessment, Vol. 2. Personality measurement and testing* (pp. 179–198). Sage. https://doi.org/10.4135/9781849200479.n9

DerSpiegel. (2020, January 14). a-53502355-b636-45b8-b815-b56f911844bb. *DER SPIEGEL, Hamburg, Germany*. https://www.spiegel.de/wirtschaft/friedrich-merz-fordert-unternehmensteuer-auf-dem-bierdeckel-a-53502355-b636-45b8-b815-b56f911844bb

Deutschlandfunk.de. (2023, September, 30). *Asyldebatte - Kritik an Merz auch aus der CDU: "Viele Mitglieder schämen sich."* Die Nachrichten. https://www.deutschlandfunk.de/kritik-an-merz-auch-aus-der-cdu-viele-mitglieder-schaemen-sich-100.html

Focus Online. (2020, January 14). Friedrich Merz: Bierdeckel feiert Comeback - Forderung nach Reformen. *FOCUS Online*. https://www.focus.de/politik/deutschland/er-holt-alten-vorstoss-aus-der-mottenkiste-comeback-des-bierdeckels-cdu-politiker-merz-fordert-radikale-reformen_id_11552255.html

Gamsakhurdia, V. (2020). *Semiotic construction of the self in multicultural societies: A theory of proculturation*. Routledge.
Greive, M., & Hildebrand, J. (2018, November 11). *Handelsblatt*. https://www.handelsblatt.com/politik/deutschland/kandidat-fuer-cdu-vorsitz-mehr-als-ein-bierdeckel-das-politische-vermaechtnis-des-friedrich-merz/23253958.html
Handelsblatt. (2018, November 02). https://www.handelsblatt.com/politik/deutschland/kandidat-fuer-cdu-vorsitz-mehr-als-ein-bierdeckel-das-politische-vermaechtnis-des-friedrich-merz/23253958.html
Hotel Matze. (2024, May 29). *Friedrich Merz über sein Leben, Angela Merkel, Lobbyismus, Selbstzweifel und die Bundesregierung* [Video]. YouTube. https://www.youtube.com/watch?v=9bKvpBbczG8
Kain, F. (2024, January 29). Immer beliebter!: Darum ist Friedrich Merz plötzlich so stark | Politik. *bild.de*. https://www.bild.de/politik/inland/politik-inland/immer-beliebter-darum-ist-friedrich-merz-ploetzlich-so-stark-86958036.bild.html
Koch, H. (2024, June 20). *Sächsische Zeitung*. Friedrich Merz - das war sein Bierdeckel-Versprechen. *saechsische.de*. https://www.saechsische.de/friedrich-merz-und-sein-bierdeckel-versprechen-5008145.html
Kölbl, C. (2020). *Psychologie der Polyvalenz: Ernst Boeschs Kulturpsychologie in der Diskussion*. Kultur, Gesellschaft, Psyche: Vol. 13. Bochumer Universitätsverlag Westdeutscher Universitätsverlag. https://eref.uni-bayreuth.de/id/eprint/44415
Lewin, K. (1982). Feldtheorie des Lernens. In C. F. Graumann (Ed.), *Kurt Lewin Werkausgabe*: Feldtheorie (Vol. 4, pp. 157–187). Klett-Cotta.
Ramthun, C. (2023, September 5). Friedrich Merz: Statt visionärem Bierdeckel nur mutlose Steuerumschichtung. *WirtschaftsWoche*. https://www.wiwo.de/politik/deutschland/einkommensteuer-der-fall-merz-statt-visionaerem-bierdeckel-nur-noch-mutlose-steuerumschichtung/29372970.html
Schwartz, S. J., Zamboanga, B. L., Meca, A., & Ritchie, R. A. (2012). Identity around the world: An overview. *New directions for child and adolescent development, 138*, 1–18. https://doi.org/10.1002/cad.20019
Suhrbier, B. (1998). *Die Macht der Gegenstände*. Curupira.
Stern, W. (1906). *Person und Sache. System des Kritischen Personalismus System der philosophischen Weltanschauung*. J.A. Barth.
Stern, W. (1923). *Person und Sache: Die Menschliche Persönlichkeit* (3rd ed.). J.A. Barth.
Stern, W. (1924). *Person und Sache: System des Kritischen Personalismus (Wertphilosophie)*. J.A. Barth.
Valsiner, J. (2018). Constructing identity: A theoretical problem for social sciences. In *Beyond the mind: Cultural dynamics of the psyche* (pp. 65–78). Information Age Publishing.
Vereinigte Lohnsteuerhilfe, E. V. (2023, October 18). *Steuererklärung auf dem Bierdeckel – was heißt das?* https://www.vlh.de/wissen-service/steuer-abc/steuererklaerung-auf-dem-bierdeckel-was-heisst-das.html
von Fircks, E. (2023a). Eine kurze kulturpsychologische Anleitung für zukünftige Freunde. *cultura & psyché, 4*(2), 277–291. https://doi.org/10.1007/s43638-022-00039-x
von Fircks, E. (2023b). Understanding the power of mindfulness from a cultural-psychological perspective. *Culture & Psychology*, Article 1354067X231201387. Advance online publication. https://doi.org/10.1177/1354067X231201387
von Fircks, E. (2024). *Learning with William Stern: Personology for future.* [Forthcoming]. IAP.
Vygotskij, L. S. (1996). *Die Lehre von den Emotionen. Eine psychologiehistorische Untersuchung*. (Fortschritte der Psychologie, Bd. 15). Lit.

Chapter 4
The Interplay of Stress, Action Potential, and Cultural Preconditioning: Beyond Symbolic Action Theory

Transience *(E.E. Boesch, Enno von Fircks translation)*
How quickly your strength turns to weakness,
How slowly your sorrow grows into courage,
What changes tides to ice and ice to streams,
What cools each blazing fire into ash's stage?
It's all change, my child. In spring,
The grasses turn to face the sun. Yet in the wind,
Their hope shifts into quiet resignation,
And blossoms, not yet fruits,
Are bound as gifts to love —
Only to wither along the way. In their fragrance,
You've found but a fleeting smile,
And when my lips call out for yours,
Your eyes are already dreaming of the hours
Where distant air holds new meadows to explore.

Culture, action, and symbolism are essential components of Boesch's cultural psychology (Boesch, 1991, 2005). We have explored the relevance of these concepts and their interplay in previous chapters. However, Boesch's work also includes several peripheral topics that play a central role in positioning psychology as a personalistic science. Often overlooked are Boesch's discussions on the action-theoretical significance of stress and affect (Boesch, 1976). In this chapter, we aim to examine and revise Boesch's writings on stress. In doing so, we will not hesitate to consider clinically pathological examples to assess the effectiveness of our stress theory, based on Boesch's work. It remains a peculiarity that Boesch's ideas or conceptualizations have rarely found their way into applied fields of psychology. The fundamental motivation for this chapter is to highlight the practical applicability and utility of these ideas while expanding Boesch's stress concept.

E. Freiherr von Fircks, *Culture of Action*, Theory and History in the Human and Social Sciences, https://doi.org/10.1007/978-3-031-88808-3_4

Stress Emerging Following a Disruption of Action

From an action-theoretical perspective, human beings develop goals (Boesch, 2000). These goals are divided into intermediate objectives, which are organically connected to the overarching goal (Boesch, 1991). Goals are achieved through actions, or actions reduce the distance to the goal (Boesch, 1998). The formation of goals is based on our needs—whether these needs are innate or acquired (Lewin, 1926). In various action fields, the attainment of goals often involves overcoming obstacles and barriers in order for the individual to achieve their respective goal (Boesch, 1980). In this process, the individual requires specific techniques (Boesch, 1976) to overcome these obstacles (or barriers) and continue progressing toward the goal. When the individual encounters resistance, it is clear that we are dealing with a disruption of action. The person is unable to continue the action as intended to reach the goal and is forced to involuntarily slow down their efforts.

This disruption triggers a series of psychological and physiological responses designed to help the individual overcome the resistance and resume the action. The slowing of action, the analysis of the resistance, and the emergence of a new intermediate action (e.g., how can I overcome this resistance?)—along with the development and application of techniques to overcome the resistance—consume synchronously deployed psychophysical energies of the individual.[1] As a result, the person is weakened in their self-assertion and self-development. According to Boesch (1976), stress arises when the disruption of action cannot be overcome independently (whether consciously or unconsciously), and the individual is no longer able to reduce the distance to their goal.

Example: A teenager attempts to wash their clothes independently for the first time. The overall goal is broken down into various actions, each designed to reduce the distance to the desired outcome (an ideal state is pursued). The teenager might gather their clothes, pack them into a laundry basket, and move toward the washing machine. Up to this point, everything proceeds without resistance, and thus without any disruption in action. It is important to note that this goal is driven by the desire to wash clothes independently—a classic fantasm of self-sufficiency, adulthood, etc. The teenager loads the clothes into the washing machine and turns their attention to the control panel. Seeing a variety of buttons and settings, they hesitate, unsure of how to operate the machine. In other words, their action is now being disrupted for the first time.

This disruption in action needs to be analyzed within an interesting field configuration. On the one hand, laundry in this household is done using a machine; on the other hand, the teenager lacks the necessary techniques to operate it. If the teenager

[1] I use the term psychophysical synchronicity to refer to the idea that individuals, in the process of goal attainment, must engage both psychological functions (such as attention, interpretation, and volition) and physical functions of their person. For instance, turning one's head or entire body serves as an illustrative example of how both psychological and physical functions are involved in action. This concept is loosely based on Stern's notion of psychophysical neutrality (Stern, 1906, 1911, 1923).

decided to wash the clothes by hand (a different field configuration), they would not experience the same disruption in relation to the washing technique. As the teenager repeatedly tries to figure out how to set the machine for washing, they begin to experience limitations in their action potential, as well as a depletion of their psychophysical resources. The teenager realizes they are unable to overcome the disruption on their own due to a lack of skills. They may keep trying certain strategies in vain to start the machine, but the disruption persists. This prolonged disruption in action, which they are not yet able to overcome, depletes both their psychological and physical resources.

The combination of failure to overcome the disruption and the consumption of physical and psychological energy leads to the onset of fatigue, which is psychophysically neutral (Stern, 1935). Stress often arises from this combination of an insurmountable disruption in action and the depletion of psychophysical energy; hence, stress is psychophysically neutral.

The Personal Significance of Stress

Conversely, this also means that not every disruption in action leads to stress. In fact, most disruptions are regulated by the individual without causing stress. For instance, if I approach the washing machine to do laundry and realize after a while that it is unplugged, I will likely just plug it back in and continue my action. In this case, I have successfully overcome the disruption using my synchronously deployed psychophysical resources. In most everyday situations, disruptions in action are quickly (and often unconsciously) regulated, preventing stress from arising in the first place.

Thus, stress emerges within a very specific field-theoretical action configuration and is, therefore, highly specific in nature. Stress is not only action-specific but also situation-specific—except in cases like burnout or depression, which we will discuss later in the chapter. The regulation of disruptions in action is a constant feature of everyday life, without the development of stress being observed. In these situations, we are dealing with a well-established and differentiated action potential that enables the individual to carry out these regulatory processes. Now, let us see what Boesch himself has to say on this subject:

> This implies that action-related obstacles must exceed a certain threshold; otherwise, they would be compensated for without the involvement of consciousness. After all, this also suggests that there are continuous gradations between minor and major disruptions in action, and that there must be a more or less broad zone, depending on the situation and the individual, in which the awareness of difficulties in action only vaguely emerges. (Boesch, 1976, p. 304)

This is an interesting illustration by E.E. Boesch. In other words, Boesch also highlights the personal significance of stress. In previous chapters, we have already examined the central theme of this personal significance, situating it within various, yet interconnected, fantasms (Boesch, 1980). These fantasms uniquely reflect the core needs of the individual, which the person perceives as important to their sense of self at any given moment (von Fircks, 2024b). When we apply this complex of fantasms to the genesis of stress, we encounter a significant limitation to what has been elaborated thus far.

Let us return to our initial example: The teenager wants to do laundry but quickly realizes that they lack the necessary techniques to reduce the distance to the goal. They encounter a disruption in action that they cannot overcome with their existing knowledge and skills. In other words, the teenager experiences a persistent disruption in action. As we have seen, this disruption will deplete physiological resources, as various techniques are unsuccessfully attempted to achieve the goal. However, in addition to this, the action also threatens a personally relevant fantasm. For example, we can assume that the act of doing laundry plays a central role in the teenager's potential for action, as successfully doing laundry symbolically represents growing up or increasing independence.

If our individual fails to achieve the objective goal (washing the laundry), they also fail to satisfy their needs for independence and adulthood. In other words, the person experiences not only a factual failure but also a personal failure, as the unwashed laundry is accompanied by the nonfulfillment of the personally significant fantasm. Of course, other personally relevant fantasms could also be imagined in this example, such as maintaining personal cleanliness. Stress will only be experienced by our individual when, in addition to the depletion of physical energy, the failure to achieve the personally meaningful fantasm is also present.

If, for instance, the person has no fantasm that motivates them beyond the objective goal, they may simply abandon the action after two unsuccessful attempts and move on to a different activity through which they can strengthen or consolidate their action potential. The persistence of an action or the change in action to avoid impending failure, and thus the depletion of physical energy, only occurs when a corresponding motivation—that is, a fantasm—is threatened.

Therefore, stress only arises when not only physical energy is depleted, but the fantasm is, in some sense, worn down because it cannot be fulfilled. Stress must thus always be analyzed from a personal-genetic perspective, as it is not solely a matter of objective physiological limits. Viktor Frankl, for example, was able to withstand the physical and physiological burdens of a concentration camp by maintaining a particular method of stress management—a fantasm of mental strength and purpose—which he continuously wielded against the cruelties and barbarities of the Nazi oppressors (Frankl, 1946). Each day that V. Frankl resisted National Socialism was, for him, a reinforcement and consolidation of his action potential—a testament to his complex of fantasms—and thus an essential strategy for enduring the suffering without breaking his entire personal system. Boesch has the following to say on this matter:

> Evidently, it is the external aspects of the situation that make stress unavoidable in this case. However, we note one important point: it is not the situation itself that is threatening, but rather its significance for the individual's action systems. (…) This must necessarily mean that the individual's functional potentiality is drastically limited. (Boesch, 1976, p. 320)

Highlighting the personal significance of stress and how it relates to one's own fantasms is one of Boesch's major contributions to stress research. However, we now find ourselves at a critical juncture. While Boesch's earlier works (Boesch, 1976) already addressed the personal relevance of fantasms, the motivational counterpart with which these fantasms interact is absent from his writings of the 1970s. This missing element is the theme of myths, which mediate certain normative values (Boesch, 1991; von Fircks, 2022). In other words, for a fantasm to take root, it must also be supported by the social environment through certain myths and normative values (Boesch calls those social should values).

The Social Significance of Stress: Cultural Overdetermination

For instance, our teenager (as an example) is likely reinforced in their fantasm if their social environment promotes certain fantasms including rapid maturation and early functional autonomy. The personally relevant fantasm (growing up) may resonate with the myth of early separation from the parental home to achieve economic independence. This myth is repeatedly conveyed to the child or teenager in key interactions, leading many independent actions to be socially over-determined from a moral perspective (Boesch, 2000).

A simple example: A high school student wishes to study medicine. The motivation for studying medicine is likely tied to a personal fantasm of social responsibility, while their environment encourages this path, not only because it can lead to financial independence, but also because it may enable the student to provide services to the family in old age. Thus, the student's fantasm is reinforced from a different symbolic perspective. If, contrary to expectations, the student is unable to study medicine due to insufficient grades, they are confronted not only with the threat to central fantasms but also with increased social pressure from disappointing social should-values.

In addition to the unsuccessful expenditure of physical and mental energy in pursuit of their high school diploma, the fantasm is also worn down. Despite its personal significance, the fantasm is not achieved; the person experiences a diminished action potential that is further undervalued by their environment. Consequently, the action potential is devalued in two ways: first, due to the failure to achieve personally relevant fantasms, and second, due to the failure to meet social normative values. Stress inevitably arises, and the individual must now cope with it appropriately.

In other words, failure to achieve a goal not only weakens one fantasm (here, social responsibility), but also another personally relevant fantasm (the desire to meet family expectations), leading to a dual devaluation of action potential. A person who does not hold both fantasms or who replaces them with other relevant fantasms would likely experience little or no stress in this social-personal situation. This also helps explain the patterns of many teenagers or young adults in relation to preparing for their final exams: a high grade on their exams does not correspond to relevant fantasms—both the motivation and stress are absent—because their social environment has not significantly imparted these values to them through corresponding myths.

An autobiographical example: For me, passing the Abitur (the German university entrance examination) was a personal proof that I could always achieve my own goals and possessed the necessary strengths (fantasm), while my environment (my parents) reinforced this fantasm by focusing less on personal development and more on the achievement of concrete goals. The fact that I wanted to excel in the Abitur in multiple ways indicates that I was confronted with various stressors—rooted in my fantasms—which explains my heightened experience of stress during that time. In short, the experience of stress is polyvalent, as it relates to multiple interconnected fantasms that are also reinforced by the social environment. Stress is not only personally significant for us as psychologists and social scientists, but it also always points to the social overdetermination, that is, the prevailing myths of the environment. Thus, stress becomes a personal-systemic phenomenon, indicating not only the goal-oriented system of the individual but also the symbolic interactions between the person and their respective environment and the values that the environment deems essential. Consequently, stress is not only conceptualized theoretically in terms of action but also operationalized introceptively,[2] that is, from a communal perspective (the internalization of cultural values). Therefore, stress is not only psychophysically neutral, as both components in the experience of stress are difficult to separate, but also personally and socially neutral, as both components play a significant role in the genesis of the experience of stress. Due to the still inadequate foundation of Boesch's cultural psychology, which he later developed comprehensively (Boesch, 1991, 2000), this aspect is not sufficiently articulated in the writings we have prepared for the definition of stress (Boesch, 1976).

It also happens that individuals proactively engage with or willingly accept stressful situations. In other words, stress can not only be reactively processed or integrated by the individual, but it can also be actively sought out by the individual. Here, I do not wish to preempt the literature on eustress (Seaward, 2017) or the humanistic interpretation of stress (Maslow, 1971). My focus is also on integrating this phenomenon of stress into our overall theoretical model of action. We have already stated above that the experience of stress tests or significantly alters our action potential. This means that persistent hindrance to action has a fundamental influence on the metamorphosis of the action potential. This need not only be understood negatively, as the individual experiences the threat to their fantasms through nonachievement of goals or persistent goal distance. It is especially the case in sports that individuals seek situations that allow them to expand their action potential, that is, they want to enlarge their scope of action. This can only occur when the respective athlete enters a temporarily stressful situation and grows from it or improves their action potential.

[2] The term interoception goes back to Stern (1906) with which he explains that people need to unite self-purposes with the purposes of their social environment.

For example, if I want to run a marathon in a certain time—approximately 3 h, compared to the previous marathon I ran in 3 h and 30 min—let us assume I am driven by a specific performance fantasm motivating me to run the marathon in 3 h. Achieving this goal means not only completing the marathon in about 3 h but also fulfilling the performance fantasm and proving to myself that I can achieve my own goals through sufficient willpower. Certain myths of the running community, in which many runners find themselves, also play a consolidating role concerning the performance fantasm. Should our individual achieve their goal, i.e., run the marathon in 3 h, they not only accomplish a concrete objective but also reward their intensified efforts, which involve the expenditure of physical and other resources, thereby expanding their scope of action in relation to their fantasm. It may even be the case that this success allows our individual to lead an elite performance group within the running community. In this context, our individual experiences a double triumph. Not only is the concrete goal achieved, making other, higher-value goals attainable, but the individual also confirms their fantasm, thereby reinforcing and expanding their self-image. Thus, the individual not only enjoys the fulfillment of their goal but also the reinforcement of their self-image based on their action potential (in relation to their fantasies).

This leads to an intensifying sense of triumph, or what I refer to as a double triumph experience. It is precisely the expansion of action potential—based on our highly valued fantasms—that encourages us to voluntarily seek out stressful situations and to actively and successfully cope with them. Successfully managing stressful situations allows for the expansion of action potential and, correspondingly, of our self-image. This also demonstrates why a stress-free life is not desirable—both for adolescents and for adults. A stress-free life would mean stagnation of action potential and thus a standstill of one's self-image and scope of action. This has pedagogical implications: The child or adolescent must be confronted with stressful situations that can be resolved through intensification or actualization of action, thereby integrating them into their self-image. Boesch refers to this type of stress as "should-value stress"to indicate that the stress arises from an inner drive rather than from situational components (Boesch, 1976).

The Personal-Systemic Nature of Stress: The Plea to Study the Interconnectedness of Stress

Regardless of the type of stress, whether it results from internal or external factors, stress also indicates an important social component: it is of a personal-systemic nature—a factor that Lazarus also underestimated (Lazarus, 2006). We particularly want to address this systemic and co-constructive aspect of the experience of stress. Stress, as we have explained with Boesch (Boesch, 1976), results from the experience of internal and external hindrances to action, which we can not only counteract by intensifying action, but which also persist and threaten our action potential.

Consequently, the psychophysically neutral forces of the individual are depleted, and the individual seeks alternative means and methods (techniques) to cope with stress adequately.

Now we turn to the role of these techniques and reflect on their cultural-historical background. Let us say I am experiencing stress at the office: I need to prepare a presentation by tomorrow, for which I will require at least another week. The presentation is important to me because a successful presentation could provide better arguments for my promotion. However, I feel that the remaining time is insufficient to prepare the presentation adequately. I lack the necessary resources to complete it in such a short time. I feel helpless and stressed. What options do I have in such a situation? To reduce the gap to my goal, I could plan the following: I could prepare the presentation with my colleagues; I might rely on artificial intelligence, such as Chat GPT; I could ask my boss for an extension, using rhetorical arguments; I could postpone my meeting with the client for whom the presentation is intended to another date; or I could try to work through the entire night. Many other actions beyond these seem unimaginable in this situation to still achieve the goal. In other words, our individual can neutralize the hindrance to action through a variety of techniques and approach the goal.

However, the choice to apply a specific technique falls within a specific co-determined or co-constructed field of action, which has already been culturally and historically pre-shaped. Techniques of various kinds are thus culturally and historically pre-shaped; one could also say, as Boesch does, that they are socially overdetermined (Boesch, 1998, 2005). The use of a hammer is culturally and historically pre-shaped for different tasks, and it is this cultural-historical shaping that must be internalized when one wishes to drive a nail into a wall. In other words, the techniques and tools we use to achieve our goals already exist before us and have been used prior to our time. Tools and techniques are material crystallizations of culture and when an individual decides to use a tool or technique, they are not only employing the material object but also the knowledge of the cultural-historical community that has imbued this tool with narrative meanings. In short, tools and techniques are pre-determined in their significance and usage (Boesch, 1993).

Another example: knives and forks are essential components of cutlery and serve to eat in a hygienic and portioned manner. A child who seeks to use these tools and techniques is introduced to the correct use of the fork and knife by the gatekeepers of the cultural community, equipped with the reasons why humans should eat with a knife and fork. In other words, the use of tools and techniques always conveys myths within the narrative of meaning, through which the use of techniques and tools is anticipated for future situations. It is not uncommon for a child to ask why they must eat with a knife and fork. Similarly, it is not rare for a child to question why they should sleep in a bed rather than on the floor. We thus grow into a cultural pre-shaping that we must first acquire (internalize) before we can participate in it and contribute to it. Reading and writing are such techniques that are important for a cultural community—not only for the development of a child's personality (including body and psyche)—but also to equip children with essential skills to participate in and contribute to the culture. In a second step, the child or adolescent can employ these techniques for their self-purposes (autotelic). However, the process of growing into or internalizing this cultural pre-shaping remains the prerequisite for the flexible application of tools and techniques to meet their own needs and goals or those of the cultural community.

How does this relate to stress? Let us return to our example: If I, as an adolescent, do not know how to operate the washing machine, it indicates that I have not yet sufficiently grown into the cultural preconditioning required to acquire the use of tools and techniques. In other words, I cannot fend off the experience of stress because I have not yet internalized the techniques through symbolic interactions to neutralize the hindrance to action. When I, as a marathon runner, realize that I cannot achieve my goals—that is, the hindrance to action persists and exceeds my action potential—I initially experience a temporary sense of stress. Based on this, I will attempt to expand my action potential in order to circumvent the hindrance to action. Therefore, I turn to my running community. My running community may offer me an improved training regimen; this time—my colleagues advise—I should incorporate more interval training into my overall regimen. Additionally, I should also adjust my nutrition to extract a few more percentage points of performance. All of this means that through interactions with the members of the running community, I gain additional knowledge and concrete strategies on how to approach my goal from a different perspective. The interactions are, symbolically speaking, functionally valenced because they allow me to incrementally expand my action potential and adapt it to upcoming situations. In other words, I am preparing better for my next marathon—I am anticipating a stressful experience—by leveraging the cultural preconditioning of my running community, both in terms of knowledge and skill. This can only occur through sustained dialogue and my regular participation in this dialogue (here, I must regularly attend the training sessions of the running community). Thus, the experience of stress is, in this sense, adaptive.

Through the feelings of unease that the experience of stress triggers, I must not only endure the negative physiological reactions of my body but also seize the opportunity to realize how this stress experience came about. In short, the experience of stress is adaptive because it indicates to me that my current action potential is still insufficient to neutralize certain hindrances to action. With this knowledge, I can initiate the process of expanding my action potential to anticipate or preemptively neutralize an upcoming experience of stress. This succinctly means that we need to discuss not only positive stress (eustress) but also adaptive stress experiences.

Adaptive, in this context, means that the experience of stress allows us to reflectively realign our action potential and confront the world anew in terms of its challenges. Non-adaptive strategies and techniques are those that we repeatedly apply, even though, in the long term, we cannot resolve the challenges with them. Much of what constitutes chronic stress experiences is centered around this non-adaptive stress coping. A chronically fixed action potential is repeatedly employed, even though it has proven insufficient for coping with stress in various situations. For example, shortly before an important concert performance, a singer takes calming drugs. The drugs may have a short-term effect on reducing the physiological stress response, that is, they decrease the physical changes associated with the experience of stress. Regardless of the fact that the singer will develop a tolerance to the physiological effects of the calming drug, they will also confront certain physical and mental long-term effects of drug use. After the concert, the singer may collapse in exhaustion and needs several weeks to recover from a performance; they may also experience self-doubt about their inability to handle the situation adequately, having only minimized the physiological accompanying circumstances of the experience of stress. In other words, the singer will not only fail to feel a sense of triumph because they could not successfully apply new strategies and techniques for coping with stress in that situation, but their action potential will be fixed in a certain static position that prevents them from growing in response to their challenges. Thus, the singer remains stagnant, and in the long term, they will not only face higher stress levels but also the awareness that they cannot independently confront certain situations. In brief, they will feel powerless in relation to their experience of stress, and in the long term, there is a risk that stress will dominate the individual.

The role of the community is as follows: Rather than feeding the singer increasingly effective drugs (as in the case of Michael Jackson), the goal should be to help the singer develop strategies and techniques aimed at the long-term expansion of their action potential. Only when the action potential is designed in such a way that it is greater than or becomes greater than the situational stress determinants will the individual not only be able to neutralize stress but also grow as a person through the challenges presented by the environment. *When properly interpreted, stress acts as a catalyst for personal growth.* The entire tragedy surrounding Michael Jackson's death (which resulted from his use of propofol for sedation) lies not in the fact that the gifted singer had not developed adequate techniques and strategies to counteract the situational stress determinants. One must particularly consider his undeniably challenging childhood, during which he was unable to sufficiently cultivate his action potential in relation to stress (Fast, 2010). This occurs in many families, partly due to the fact that parents and peers reproduce their nonadaptive stress coping strategies in symbolic interactions, which the child subsequently internalizes and applies to their own situations.

The tragedy in Michael Jackson's case lies in his belonging to a cultural community that did not encourage him to expand his action potential concerning certain stressors but rather sought to minimize the physiological accompanying symptoms of stress through drug use. The medical team accompanying Michael Jackson should not have encouraged him to take ever-increasing doses or use the sedative more frequently; instead, they should have trained him in strategies and techniques to face stress independently. Thus, the singer was negatively preconditioned by two cultural communities regarding the experience of stress and stress management: first, by his family and peers during childhood and adolescence, and second, by the medical community, which aimed to eliminate the physiological symptoms of stress.

The public is not aware that Michael Jackson ever sought the help of a psychotherapist, a person who not only seeks to cure the physiological symptoms of chronic stress but also collaborates with the patient to expand their action potential to confidently confront the causes of stress symptoms. Here, we encounter a significant crossroads in our chapter on stress based on Boesch's symbolic action theory.

We can formulate the following equation for the genesis of stress, but we must, in the next breath, put this in quotation marks and further elaborate on it: (where AP = action potential, Str = stressors, Stre= stress experience)

$$AP < Str = Stre$$

When an individual's action potential is less than the stressors they face, it leads to the experience of stress. We have already provided sufficient examples for this equation and can refrain from introducing more at this point. However, we know that action potential does not arise in a vacuum; rather, it is significantly shaped by the cultural preconditioning of a social community. This is a crucial revision of the equation we have presented here, as the introduction of the concept of cultural preconditioning allows us to avoid assigning blame to the individual for their experience of stress. When a person feels stress, it is not because they are too weak for certain situations—despite what popular belief sometimes suggests—but rather because the individual has not yet been exposed to appropriate stress management mechanisms within their community that they can internalize for various purposes. The experience of stress should not become a symbolic expression of a weak personality, as countless videos and podcasts in popular psychology (Kraemer, 2010) often imply. This contemplation leads us to revise our previously mentioned equation accordingly.

The individual, who possesses both innate and already acquired techniques concerning the experience of stress, does not exist in a vacuum. Instead, they continually receive various strategies and techniques from different cultural communities on how to cope with stress. This process is called introception which is an ongoing process (where tec means resources, techniques, coping strategies).

$$AP\left(Person\left(Tec\right)*Culture\left(Tec\right)\right) < Str = Stre$$

This equation also implies that when experiencing stress, not only must the individual be included in our equation (from a therapeutic perspective), but also the cultural preconditioning of the respective community. It is not surprising that individuals with addiction issues are encouraged to attend certain self-help groups. In these groups, they become part of a community and thus a cultural conditioning that seeks to contain stress not by drowning it, but through various non-substantial coping mechanisms. Therefore, a person lacking stress management strategies is advised not only to develop these strategies in therapeutic individual settings and apply them in their private environment but also to approach certain communities that can help them expand their action potential through various techniques (co-construction of action potential). A person in a therapeutic setting who lacks these coping strategies should thus be viewed by the therapist not only as a manifestation of insufficient personal action potential but also as a lack of co-construction of action potential concerning adaptive communities. This awareness of the co-construction of action potential must guide the therapist in terms of their actions and the direction of the therapy. Only then can therapy be holistically successful.

Now, it is the case that Boesch (1976) distinguishes between two types of stress experiences. The first type of stress we have already encountered—Boesch refers to this as situational stress. The situational determinants exceed our action potential at a given moment, rendering us unable to neutralize the action inhibition. This leads to the experience of stress. This is the basic formula that has run through our entire chapter (so far). However, for Boesch, the Swiss cultural psychologist, there is also a more subjective type of stress, which he calls norm-value stress. I am not entirely satisfied with this dualistic distinction of stress, as there is no pure situational stress or pure norm-value stress. In line with Stern's theory of convergence (Stern, 1911), personal and social-material factors always interact in stress experiences. A dualistic separation seems counterproductive from the standpoint of convergence.

Before we elaborate on our criticism, I would like to briefly present Boesch's concept of norm-value stress: In contrast to ecological stress, norm-value stress relates more to the initiative of the individual or their fantasm system (Boesch, 1976). In other words, the individual imposes certain demands on themselves or their environment based on their needs and fantasms. A person with a pronounced performance fantasm will increasingly seek out situations where they can demonstrate their abilities and strive for higher achievements (Lewin, 1935), in contrast to a person who does not have a pronounced performance fantasm. If the individual evaluates every situation based on this performance fantasm and perceives the world solely from this perspective, we speak of a chronicized fantasm (my terminology). In the hierarchy of needs, the performance need is placed at the top, thus dominating the individual's interactions with their social and material environment. As a result, the individual will expose themselves more to situations in which they can live out their performance fantasm, which are primarily of a stressful nature, as they strive to significantly expand their action potential. In this process, it may occur that the individual possesses this performance directionality but lacks the various skills required to achieve the performance. In short, the individual seeks out situations where they wish to prove themselves, even though they do not yet possess the necessary skills—this is the interoception of norm-value stress according to Boesch (1976), with Boesch not referencing the interaction between direction and scaffolding (Stern, 1935).

As previously indicated, I am not entirely satisfied with this dualistic distinction of stress, as both internal and external factors always play a role in the genesis of psychological functions. First, I miss the differentiation of norm values concerning personal and preconditioned values, which always interact with each other. Second, I lack the interaction of these norm values with the ecological determinants of stress in a secondary sense. It is entirely possible that a person, driven by an increased performance fantasm, engages in simple stress management situations, thereby consolidating their action potential. In this case, there would be no significant conscious stress experience, as the stress would be regulated before reaching conscious awareness. However, there would also be no substantial expansion of action potential, because the ecological determinants of the situation are not challenging enough for the person to assess their action potential and possibly recalibrate it. I want to criticize this dualistic view of stress, which merely presents an artificial (perhaps illustrative) separation of various stress determinants. It is indeed the case that the experience of stress in certain situations is more situationally colored (but not excluding fantasies, as otherwise, the situation would be irrelevant to us) or more fantasm-related in nature (consciously seeking stressful situations that cannot then be managed). Our equation clearly indicates that the person, with their already acquired techniques and strategies, their cultural co-construction or preconditioning, and the situational stressors, plays a decisive role in the genesis of stress. Of course, certain components in this equation can be accentuated, but not at the expense of the others. *In short: The experience of stress and stress management is the interaction of action potential, cultural preconditioning, and co-construction of action potential, tested or challenged by situational determinants (stressors).* Isolating one of these components from our equation would amount to a popular scientific explanation of stress that we, as psychologists and social scientists, cannot allow ourselves to adopt.

Boesch's concept of depression as a global regulation of actions implies that a person experiencing depression perceives their action potential across various areas of life—such as work, family, partnerships, and friendships—as insufficient to meet the demands and stressors of these domains (Boesch, 1975). In other words, the person feels powerless and evaluates their action potential as weaker than the stressors they face. Thus, they perceive all situations as surpassing their psychophysical capacities, leading to a sense of being overwhelmed. This not only depletes their energy but also increases the perceived distance to both concrete and symbolic goals. In psychotherapy, the focus is not only on overcoming the dramatizing or catastrophizing evaluation of these challenges but also on developing skills that enable the person to move closer to their goals. Often, it is necessary for the person to both reevaluate their action potential and update it through the acquisition or expansion of new abilities. Some aspects of this are already addressed in solution-focused brief therapy (Murphy, 2024). In short, it is about the transcendence of the stress experience. I am convinced that this stress transcendence feature is well captured in Boesch's poem at the beginning of this chapter (in particular in the first stanza):

> How quickly your strength turns to weakness,
> How slowly your sorrow grows into courage,
> What changes tides to ice and ice to streams,
> What cools each blazing fire into ash's stage?

Toward a Holistic Definition of Stress

With that, we have reached the end of our chapter and now wish to proceed to a summary:

A. Stress often arises when we encounter action inhibition. However, action inhibition is not a sufficient criterion for the genesis of stress. It is entirely possible that the person exits the field upon encountering action inhibition before a stress reaction occurs. This means that when confronted, the person evaluates that this or that action is not important to them and is thus not relevant for their system of purposes. Therefore, if no needs and fantasms motivate the person to continue the action despite the action inhibition, there will be no experience of stress. Thus, we have contradicted and revised Boesch (based on his own theory), who assumes that every action inhibition is accompanied by an experience of stress (Boesch, 1976). The boy who wants to wash his clothes (because his mother has imposed it upon him) will not experience stress during an action inhibition because the action is not personally relevant to him. The action inhibition does not interact with a valenced fantasm that encourages the person to continue the action under different circumstances. Conversely, this means that stress only occurs where the person is motivated to neutralize the action inhibition but does not possess sufficient action potential (such as knowledge, skills, etc.) to circumvent the action inhibition.

B. The management of stress experience not only allows for the situational handling of obstacles and resistance to achieving a given goal. Stress management also enables the individual to significantly expand their action potential, thus updating skills and competencies that help them master similar or even more stressful situations in the future (Lewin, 1935). Furthermore, the individual also experiences that they are getting closer to their own motivation (their own fantasms) or that they can reduce the distance to these symbolic goals. This leads to a double experience of triumph—once over the factual goal and then over the symbolic goal of the respective individual. Thus, the experience of stress is a fundamental prerequisite for personal growth. From this principle, we can derive important educational guidelines aimed at not wrapping the child or adolescent in cotton wool but rather exposing them to gradually increasing stressors to catalyze personal growth. We have thus established that every stress, in the long term, is adaptive since it helps us reflect on and potentially realign our action potential, as the prevailing orientation is insufficient to neutralize the action inhibition. Perhaps we also need to correct the public image of stress based on this interpretation, which still tends to have a rather negative connotation.

C. However, it is also the case that stress does not only manifest as a personal signal; rather the management of stress depends on the cultural preconditioning of the individual. In other words, the action potential (knowledge of stress, abilities to confront stress) is significantly co-constructed by the respective community or by multiple communities in which the individual is situated. A boy who reacts to stress in relation to exams does not exhibit a personal failure in his stress management strategies. On the contrary, the situation indicates that the boy has not yet had sufficient opportunities to internalize certain cultural stress techniques—either because the cultural preconditioning has been entirely absent or only latently present, and the internalization has not yet been completed. The example of individuals with addiction (e.g., alcohol) has illustrated the validity of this argument: The therapist often recommends not only reacting to stressful situations with newly developed stress management mechanisms (which are tested in the therapeutic setting) but also seeking incentives and stimulation from self-help groups, through which the person turns to a certain cultural conditioning that expands their action potential by approaching problems and challenges with other strategies. Thus, the individual can completely realign their action potential—both from personal motivation and also from social motivation—which can lead to a certain reorientation of the person. A successful therapy that is oriented both personally and systemically often has a cleansing effect on the overall individual and frequently leads to a rebirth of the entire personality. Numerous psychotherapeutic studies (Schneider & Krug, 2010; von Fircks, 2024a) provide evidence of this. Therefore, we have relevantly expanded our perspective on stress, as we discuss the catalysis of growth in relation to stress experience, which not only concerns so-called eustress but is possible in relation to any form of stress experience.

References

Boesch, E. E. (1975). *Zwischen Angst und Triumph: über das Ich und seine Bestätigungen*. Huber.

Boesch, E. E. (1976). *Psychopathologie des Alltags: zur Ökopsychologie des Handelns und seiner Störungen*. Huber.

Boesch, E. E. (1980). *Kultur und Handlung: Einführung in die Kulturpsychologie* Huber.

Boesch, E. E. (1991). *Symbolic action theory and cultural psychology*. Springer.

Boesch, E. E. (1993). The sound of the violin. *The Quarterly Newsletter of the Laboratory of Comparative Human Cognition, 15*(1), 6–16.

Boesch, E. E. (1998). *Sehnsucht: Von der Suche nach Glück und Sinn [Longing: On the search of joy and meaning]* (1st ed.). Huber.

Boesch, E. E. (2000). *Das lauernde Chaos. Mythen und Fiktionen im Alltag*. Huber.

Boesch, E. E. (2005). *Von Kunst bis Terror: Über den Zwiespalt in der Kultur [From art to terror: On the cleavage of culture]*. Vandenhoeck & Ruprecht.

Fast, S. (2010). Difference that exceeded understanding: Remembering Michael Jackson (1958–2009). *Popular Music and Society, 33*(2), 259–266. https://doi.org/10.1080/03007761003640574

Frankl, V. (1946). *Ein Psychologe erlebt das Konzentrationslager*. Verlag für Jugend und Volk.

Kraemer, H. (2010). *Soforthilfe bei Stress und Burn-out: Neue Energie in wenigen Tagen–Coaching mit Neuroimagination–Strategien der Vorbeugung*. Kösel-Verlag.

Lazarus, R. S. (2006). *Stress and emotion: A new synthesis*. Springer.

Lewin, K. (1926). *Vorsatz Wille und Bedürfnis: Mit Vorbemerkungen über die psychischen Kräfte und Energien und die Struktur der Seele*. Springer.

Lewin, K. (1935). *A dynamic theory of personality*. McGraw-Hill.

Maslow, A. H. (1971). *The farther reaches of human nature*. Arkana/Penguin Books.

Murphy, J. J. (2024). *Solution-focused therapy*. American Psychological Association.

Schneider, K. J., & Krug, O. T. (2010). *Existential-humanistic therapy*. American Psychological Association.

Seaward, B. L. (2017). *Managing stress*. Jones & Bartlett Learning.

Stern, W. (1906). *Person und Sache. System des Kritischen Personalismus System der philosophischen Weltanschauung*. J.A. Barth.

Stern, W. (1911). *Die differentielle Psychologie in ihren methodischen Grundlagen*. J.A. Barth.

Stern, W. (1923). *Person und Sache: Die Menschliche Persönlichkeit* (3rd ed.). J.A. Barth.

Stern, W. (1935). *Allgemeine Psychologie auf personalistischer Grundlage*. Martinus Nijhoff.

von Fircks, E. (2022). Eine kurze kulturpsychologische Anleitung für zukünftige Freunde. *Cultura & Psyché*, 1–15. https://doi.org/10.1007/s43638-022-00039-x

von Fircks, E. F. (2024a). *Learning with William Stern: Personology for future*. IAP.

von Fircks, E. F. (2024b). *Conservativism: A cultural-psychological exploration*. Springer.

Chapter 5
The Dynamics of Affect: Symbol Action Theory and Affect Theory Reloaded

Memories *(E.E. Boesch, Enno von Fircks translation)*
You were there. So sudden, so warm,
With a bitter sweetness, a foreign scent.
Your breath rose softly upon my arm,
Your smile lingered in the twilight's gent
And mellow air, a shimmer of shared wine.
Then I asked — and in the warmth of your
Words, I first beheld your face,
And saw my own reflected in your grace.
This gives me much to ponder still,
As through the night the hours slowly spill.

Positive and Negative Affects and Their Action-Theoretical Significance

"Affects are regulators of action"(Boesch, 1976, p. 365), Boesch begins his chapter on affect theory—a straightforward statement with profound scientific implications. In the previous chapter, we saw the regulatory role of stress experience, and similar characteristics can be attributed to other aspects of affect experience. Importantly, Boesch analyzes the role of affects differently, meaning they have various consequences for an individual's subsequent actions. This implies that different affects are linked to different regulations of action, suggesting a need to revise Boesch's initial statement. Before highlighting the varied roles affects play in the course of action, let us first discuss the commonalities in affect experience.

Take anger, for example: I am about to start my workday and try to turn on my laptop, but it would not power on. This represents an obstruction to action. In this case (as in the previous), simply experiencing an obstruction does not result in observable anger; instead, the necessary means and ways of acting are slightly

E. Freiherr von Fircks, *Culture of Action*, Theory and History in the Human and Social Sciences, https://doi.org/10.1007/978-3-031-88808-3_5

modified, or a renewed attempt to reach the goal is made. However, the laptop still has not started, and the obstruction persists. I become frustrated because I cannot start the computer or overcome the obstruction. Anger arises, indicating that my action potential (knowledge, skills) is insufficient to clear the obstruction.

Fundamentally, for affect to arise in this situation, the action must be meaningful within the person's fantasmic constructs—that is, relevant to their symbolic or imagined significance—triggering an affect (see Chap. 4). For example, if the laptop was provided by the company, and logging in is tied to a time-tracking system, the individual may attribute the technical failure to the company, thereby deflecting personal responsibility. This attribution would prevent anger from arising and might even lead to joy, as the situation provides an excuse to avoid work for the day. Conversely, if the company has previously implied that ensuring a smooth technical workflow is the employee's responsibility, the act of repairing the technology might be seen as necessary to meet the personal ideal of being a good employee (a private normative value embedded in the fantasm of being a good employee).

Consider a simpler example: My wife's favorite vase breaks. I am not frustrated by my clumsiness because I always disliked the vase, and now it is finally gone (the obstruction = the vase cannot be repaired). The vase, as a symbol of unappealing aesthetics and my wife's dominance in decorating choices, does not evoke a negative affect but rather a positive affect because it undermines the dominance of my wife in terms of decoration aligning with my fantasms. While clumsiness might evoke frustration in another context (e.g., if it displeased my wife), here it triggers no affect or even a positive affect due to the specifics of this object. From this, we can conclude the following:

1. Negative affects arise when we encounter persistent action disruptions that we are unable to (yet) overcome using our action potential. The disruption must trigger a relevant fantasm or threaten it for the affect to emerge. Thus, affect has a symbolic function, pointing to the person's fantasm system and to a specific object through which the relevance of this fantasm system is triggered. Consequently, a lack of affective reaction to action difficulties also becomes symbolically significant in an action-oriented affect theory. In this view, we must analyze two aspects psychologically: the genesis of affect and the genesis of non-affect. The emergence of non-affect has also a significant meaning in a given situation.

Positive and Negative Affects and Their Action-Theoretical Significance

Boesch not only seeks to examine negative affects and their differing impacts on action (such as grief vs. anger), but he also aims to highlight the significance of positive affects for experience and action. To begin, rather than focusing on more intense affects (like triumph or ecstasy), Boesch examines everyday positive affects, such as *flowing ease*. When analyzed through an action-theoretical lens, flowing ease emerges where a person feels in harmony with both themselves and the world, where personal and external demands resonate. This alignment carries and stimulates actions, enhancing their flow and possibly their outcomes—again affirming that affects are fundamental regulators of our actions (Boesch, 1976).

For example, a person experiencing flowing ease in their work feels tasks are effortless, completes them quickly, and encounters little resistance on the path to their goals. This ease is a special affective regulator, implying the person engages in actions that resonate strongly with their core fantasms, rendering the work personally relevant. In this case, the work is not just a duty to the employer or client but an inherent fulfillment of their deeply valued fantasms. Here, job requirements align with the person's needs, allowing them to take ownership of the work and complete it.

Masterpieces of art and literature have often flowed from internal fantasms, unconstrained by social obligations, thus demonstrating the freedom of fantasm-driven engagement in work. Goethe, for example, secured key patrons and even pursued a political career to gain economic independence. Most of his masterpieces sprang from his inner sources (i.e., fantasms), unencumbered by social expectations, as he was not dependent on meeting external demands. Similarly, many pursue self-employment not only for financial gain but because they wish to be their own boss, engaging in work that reflects their fantasms. In the startup world especially, individuals seek freedom not as an end in itself but as a means to shape work in alignment with their fantasm-driven nature. For them, flowing ease is not just a situational affect but an ongoing state of existence (Bruni et al., 2004; Fernández-Salinero & La Riva, 2014).

The core characteristic of flowing ease (as affect or mood) is not the absence of obstacles but rather the ability to overcome them with minimal effort and energy. This ease reflects not only the person's motivation to pursue an action based on a highly valued goal but also the skills and techniques they possess to quickly address challenges with minimal energy expenditure. An action, as the core of psychology, always points to the interplay between fantasm and technique, through which a goal, which is tied to fantasms, can be achieved. Triumph, as a feeling, signals not only high motivation to achieve a goal but also a person's growth beyond their usual abilities to overcome an initially overwhelming obstacle. Triumph thus signifies an expansion of action potential, arising from the motivation to achieve a goal at all costs (Boesch, 1975, 1976).

2. Positive affects arise where a person can perform an action aligned with their high-value fantasms, without overwhelming obstacles that threaten or limit their action potential. Minor challenges may occur but are manageable given the person's skills, and they are resolved before becoming persistent. Positive affects thus indicate two things: first, that the person can undertake a highly valued action which relaxes central fantasms, and second, the presence of competence to successfully manage the action. Increased positive affects emerge where initially insurmountable challenges are overcome through expansion of action potential, highlighting personal growth and strength.

Thus, action obstacles can generate both positive and negative affects, with their interpretation in relation to action potential being crucial for affect genesis. In this way, affect ultimately points to the structure of *fantasm system <> action obstacle <> action potential*.

With this overview of positive and negative affects' significance for motivation and action potential, we are well-equipped to examine individual, more complex affects and their consequences for our action system. We will next focus on anger, tiredness and grief, to derive educational implications from the action-theoretical significance of these affects. This will, in turn, help develop a self-intervention method to assist individuals in understanding and potentially updating their affects and their relevance for action potential. With this kind of research, we find ourselves within the pragmatic domain of defining and analyzing affects. Pragmatism (James, 2000a, 2000b, 2000c) is an important philosophical discipline with which we can ask not why affects are in the world but what follows from emerging affects and the meaning for our actions regarding the affect emergence.

This embodies the larger goal of Boesch's cultural psychology—not merely to make his ideas accessible but also to develop them for everyday applicability, fueling the practical implications of Boesch's cultural action theory. This final source of my personal-scientific engagement, I consider fundamental, as there still exists a gap between Boesch's theoretical framework and applied psychology. The preceding chapters have already offered some insights in this direction (see Chaps. 3 and 4).

The Genesis of Anger and Its Action-Theoretical Significance

> Specifically, anger signifies three things: first, that we perceive a goal as being threatened (anticipation of failure); second, that we experience our ability to function as impaired or restricted; and third, that we are inclined to compensate for this restriction by intensifying our actions." (Boesch, 1976, p. 370)

Boesch begins his exploration of anger by situating it within the framework of affect-antecedents, emphasizing the interplay among the fantasm system, impediments to action, and action potential. According to Boesch, it is not simply the presence of an obstacle that triggers a negative affect—here, anger—but rather the individual's inability to overcome the obstacle effectively. Consequently, this obstacle endures, underscoring specific limitations within the person's action potential. Anger, as a negative affect, thereby reveals a dual deficit: on one hand, the individual's valued goal cannot be achieved due to the obstacle, leaving central fantasms unfulfilled. On the other hand, the persisting obstacle highlights a deficiency in the individual's action potential, revealing a perceived incapacity to resolve the obstruction independently. Thus, the person experiences a "diminished self" (Boesch, 1976, p. 372) in two respects: first, in terms of the fantasms system, which remains in a state of tension, and second, in the limitations of the action potential as exposed by the obstacle itself.

This concept of a diminished self is crucial to understanding the anger affect, as it is precisely this twofold self-diminishment that incites anger. Take, for instance, a student who feels anger upon receiving a poor grade in a favorite subject, a familiar scenario to many. This example reveals that the favorite subject (e.g., literature) taps into high-value fantasms that drive the student's goal setting and attainment efforts. The poor grade, however, symbolizes an inability to reach that goal, thus exposing deficits in the student's action potential. Consequently, the student experiences a doubly diminished sense of self, relating both to the unrelieved fantasms system and the limitations in action potential highlighted by the impediment (in this case, the class test). Therefore, goals carry an "I-symbolic meaning" (Boesch, 1976, p. 372), and anger signifies "symbols of our impotence" (p. 374).

From a theoretical perspective, anger assumes an interesting adaptive function in shaping the individual's teleology, or future orientation. The experience of anger, along with a sense of impotence, prompts an individual, according to Boesch, to confront two essential areas for self-work. One must examine the importance of the goal and why it matters—requiring reflection on the fantasms system—and, if necessary, reevaluate and reinterpret the value of one's goals. Concurrently, one must address action potential deficiencies, sharpening skills and abilities to confront future obstacles with ease. Thus, anger compels a realignment of both action motivation and action potential, requiring the individual to undertake self-work on multiple fronts. Although this can be a burdensome realization, it is also vital to personal development, as people grow through the experience of anger. Anger inherently points to alternatives for action, distinguishing it fundamentally from grief (Boesch, 1976).

While engaging with anger does not always yield adaptive outcomes, as it is possible for an individual to lose themselves in anger, the focus of one's attention becomes particularly crucial in these instances (Greenberg, 2017). Here, it is important to analyze the focus of a person who entirely succumbs to anger: Does the person focus primarily on the obstacle or on the self-meaning that the anger holds? The question of focus is key to anger's catharsis (Warstat, 2011) and thus to its adaptive resolution.

Later in this chapter, we will address the topic of affect regulation as it pertains to methods of self-intervention. For now, it is essential to recognize:

3. Anger arises when we experience a sense of impotence in our action potential, that is, the persistent feeling of inability to neutralize the barrier to action. This results in a dual, symbolic diminishment of the self: a greater distance from the goal, alongside specific deficits in action potential. Anger thus signals a high-value goal and reveals an inadequacy in the skills and abilities required to independently resolve the anger-inducing situation. However, the anger experienced can potentially lead to a realignment of action motivation and action potential, allowing us to refine our skills while reassessing the significance of the goal—a core theme in personal development. That this does not always succeed highlights the importance of affect regulation.

Furthermore, it is essential to recognize the teleological significance of anger in shaping one's orientation toward action: the energy that anger generates can be productively harnessed, underscoring the need for effective affect regulation.

The Genesis of Fatigue and Its Theoretical Significance for Action

> Fatigue arises from the difficulty of a course of action; it consists in the perception that our means of action are no longer sufficient for achieving a positively valenced goal. The fatigue-related actions that then emerge, as we found, temporarily reduce the goal's valence, often only for a short time, in favor of compensatory, distracting, or restorative actions. (Boesch, 1976, p. 428)

Fatigue (or tiredness) is a complex affect within Boesch's symbolic theory of action (1976). This can be deduced from our everyday experience. It can be interpreted both positively and negatively, meaning it can place a person in either a positive or negative mood. Fatigue as a positive affect often occurs after completing a task that we have successfully finished. A factory worker might interpret his fatigue as a sign of success while having a beer at the end of the workday; he has assembled more products today than in previous days. The heaviness in his legs, the muscle soreness in his shoulders, and the worn-out hands all indicate the physiological component of fatigue. The worker's action potential has been heavily taxed, yet he interprets this exertion, leading to the feeling of positive fatigue, as a success from a productive day and looks back on his workday with pride (further affect). The depletion of his action potential signifies his success that day and leads to a positive interpretation of the physiological symptoms of fatigue. *Here, we see that the genesis of affect is dependent on the interpretation of physiological changes in one's organism.*

However, for this positive experience of fatigue to occur, certain psychological processes must be involved. This means that the worker's action potential was sufficiently prepared to meet the challenges of the job, and he was able to apply his efforts to complete the work properly, given enough motivation. In other words, the worker's action potential or its intensification was not stifled by the external conditions of the work but was instead met by them. He was able to experience a peculiar balance between his internal resources and the external demands, despite the intensification of his action potential. The lack of a sense of triumph is also due to the fact that the worker did not need to realign his action potential and therefore did not have to transcend his existing state. The intensification of his action potential was sufficient to establish a balance between internal and external demands.

The significance of this positive experience of fatigue is multifaceted. If the worker chooses to reflect on this experience, that is, bring it into his self-awareness, he may wonder whether this optimal balance between internal and external demands was just an exception in his working life, or whether he can expect it in the future as well. This would prompt the worker to engage in important internal work that allows him to assess how important this type of work is to him and how it enables him to explore the symbolic meaning of the fatigue experience. If the worker finds that this positive experience of fatigue occurred only once in 30 years, or once a week, the action-theoretical consequences are different. In the first case, the worker might decide to change his job to pursue this experience more often, while in the latter case, the worker might feel more connected to the company he works for and even experience a sense of gratitude. In short, the fatigue experience carries a special symbolic meaning in a given situation for the person. Becoming aware of this symbolic meaning, for example, in his own work environment, has consequences for the alignment of his action motivation and his action potential.

This example clearly demonstrates how affects are dependent on context and interpretation (personally relevant). We may also interpret this as a caution against biological theories of emotion (see Vygotskij, 1996).

Hence, affects are not merely somatic alterations of bodily states that subsequently influence the psychological attribution of the respective affect (e.g., "my heart is racing; this must be fear"). On the contrary, the actions of the body are inherently interpreted through a symbolic code system (or sign system), which generates the affect. There is no affect without its interconnection to such a symbolic sign system—a concept we refer to, following Boesch, as fantasms and myths (Boesch, 1991). To illustrate the importance of this symbolic sign system, let us consider the following example provided by Vygotsky:

> Even the simple feeling of hunger can acquire a religious meaning when it leads to the emergence of a symbolic ritual of mystical communication between humans and God. Consequently, no emotion (…) exists that is, by its nature, independent of or unconnected to the body. When James detaches emotions from the development of the system of ideas and defines them solely as organically structured, this inevitably results in a fatalistic conception of emotions, equally applicable to both animals and humans. Such a theory cannot account for the profound differences in human emotions depending on historical periods and levels of civilization (e.g., the mystical adoration of a medieval knight for his lady versus the gallantry of 17th-century aristocracy). If, instead of focusing on the imaginary psychology of single-celled organisms, one directs attention to the remarkable analyses of novelists or draws upon valuable observational data from people in our surroundings, one can no longer assume that the mere perception of a female silhouette automatically triggers an infinite series of organic reactions culminating in a love like Dante's for Beatrice. Such an understanding must first consider the entire ensemble of religious, political, aesthetic, and scientific ideas that inspired the genius of Alighieri. (Vygotskij, 1996, pp. 185–186, author's translation)

Let us return to the example of negative fatigue, this time drawn from my own experience in school. As a youth, I was particularly fatigued by mathematics classes, to the point where I often had to yawn and eagerly awaited the end of the lesson. This was not necessarily due to the teaching staff, but rather to the subject of mathematics itself. I had had several discussions with my math teacher, who had predicted that I had a talent for math, but he nonetheless tried—unsuccessfully—to motivate me in the subject. In one discussion, I told him that mathematics made me tired because it was too precise for me and often only led to one solution (the correct result, the right formula, etc.), making it unattractive. My teacher did not know how to respond to that. This was one of the reasons I distanced myself from mathematics as much as possible during my youth.

There are two key elements we need to consider in this example: First, the insufficient motivation to engage with the content of mathematics lessons and the limitation of certain fantasms it evoked for me (potential for discussion, openness to solutions, critical engagement). The mathematics class thus threatened relevant fantasms for me, and I was forced to engage with its content despite my low motivation. In other words, the mathematics class was contrary to my relevant fantasm system—and therefore made any action in this area burdensome for me. Additionally, I must admit that, despite my teacher's acknowledgment of my suggested talent, I lacked central abilities and skills needed for mathematics. I lacked the basics to solve higher-level problems. I couldn't even always solve simple equations correctly. This meant that even simple math problems felt like a major effort.

Thus, we can identify two things: First, my insufficient motivation due to the threat to central fantasms, and second, my inadequate action potential for solving simple math problems, which made following the lessons difficult. Even small actions in class had to be carried out with great effort and intensification of action potential. The negative fatigue experience, therefore, seems to revolve around a central combination of insufficient motivation and insufficient action potential. The negative fatigue experience can thus be explained by the interplay of fantasm distance and limited action potential. This can also be observed in other examples.

Why do we feel more tired on some days than others? This is related to this theoretical framework of action. Often, we feel more fatigued on days when we need to do a task that we have been putting off for a long time, and that does not align with our actual action motivation—perhaps our employer has told us that we must now do these tasks. But we must also point out the dynamic of our theoretical framework. Children and adolescents often feel unmotivated in a class because they lack the essential abilities and skills required to keep up and participate. In other words, limited action potential can, in the long run, negatively affect action motivation, which then raises the threshold for engaging with the material. Fatigue can thus become chronic in this process, causing the subject to avoid the math class—an outcome that should be prevented from a school pedagogical perspective.

At this point, I want to revise of Boesch's action theory. This example clearly shows that Boesch's theory, especially his engagement with affects, was too individualistic—a component that he later sought to revise (Boesch, 1991, 1998, 2005). We must argue, following major characterologists such as Busemann (1950), Heiss (1936), and Rohracher (1937), that the structure of personality (here, action motivation and action potential) always refers back to the states of the cultural community in which the person is embedded. This means that personality can only be interpreted within its cultural context (von Fircks, 2024a, 2024b, 2024c), and the interpretation of personality must always include the conditions of the cultural community. A person crying over a bad mark in mathematics reflects a lack of fully internalized skills necessary for excelling in the subject, as well as an underdeveloped capacity to process negative emotions in a personally constructive way. This response signals a gap in the adolescent's cultural preconditioning, as she has not yet cultivated the tools to bridge her emotional response and her desired level of performance. The act of crying reveals that achieving a better mark holds significant personal relevance to her, as it aligns closely with her internalized performance fantasms and aspirations. At the same time, her reaction underscores the pivotal role of the gatekeepers of her community—such as teachers, parents, and mentors—in supporting her developmental trajectory. These gatekeepers bear a particular responsibility to guide her in constructing her sense of self while fostering the acquisition of new skills and emotional coping mechanisms that enable her to navigate challenges more effectively in the future.

Referring to our example, we must ask: Were the basic foundations of the subject adequately conveyed? What was the quality of teaching in the early years of secondary school? Was the teaching staff able to adequately motivate students or connect the lesson content to the student's fantasm system? How did the teaching staff interpret and, if necessary, address deficits in the student's action potential? The responsibility for character formation does not lie solely with the person themselves but also with the gatekeepers of cultural communities, who have the task of not only motivating the individual but also providing them with the necessary skills and tools.

Looking at the fatigue experience solely through an action-theoretical lens is insufficient; we must analyze it both theoretically and systemically (i.e., culturally). The role of the cultural community (its gatekeepers) is also always to elevate the individual from one community and facilitate their transition to a new, more advanced community (Busemann, 1956). The pedagogical challenge today is that teachers often face students who are at different cultural developmental stages and thus have not yet made the leap to the currently relevant cultural community, which is essential for the present curriculum.

The fatigue experience thus points to both its complexity and the prior and current transmission of tool use, which was flawed in some way, as seen in our example of mathematics lessons. The genesis of the fatigue experience and its symbolic significance should therefore not only be interpreted individually but must always also refer to the person's cultural embedding, which shapes the experience of fatigue in the present moment. We will return to this in the section on affect regulation.

Fatigue and Depression

In line with Janet (1919), Boesch argues that fatigue must be included in the pathogenesis of depression, as fatigue experiences can spread across different areas and become chronic. In fact, it is common in depressive disorders that the person loses the sense of purpose (action motivation) in their life, and their action potential is perceived as inadequate for achieving relevant goals. The goal of psychotherapy—especially from a solution-focused perspective (De Shazer et al., 2021; Murphy, 2024)—is to help the person identify their relevant needs and goals, and then, through training of skills, fulfill these needs and achieve those goals. Here comes an illustrative example from Pierre Janet (1923):

> La répétition de cette situation, ces efforts continuels déterminent une fatigue, un épuisement qui jouent un rôle considérable dans les névroses. Un exemple emprunté à la vie normale ou presque normale fera comprendre le mécanisme de cet épuisement. Je viens de recevoir une lettre désagréable qui exige une réponse assez difficile et pénible à écrire. Je pense à faire cette réponse et je la compose presque complètement dans mon imagination, mais je n'ai pas le courage de la rédiger immédiatement et je laisse la lettre reçue sur mon bureau. Dorénavant il m'est impossible de m'asseoir à cette table, de passer devant elle, d'entrer même dans la chambre sans voir cette lettre, sans deviner sa présence et sans recommencer cent fois l'effort de combiner la réponse. Cette réponse au début aurait été écrite en dix minutes; si j'additionne tous les moments dépensés pour la combiner en imagination, toutes les tentatives infructueuses et toutes les émotions qui en sont résultées j'ai dépensé à ne pas faire cette réponse des heures de travail pénible et il ne sera pas surprenant qu'après quelques jours je me déclare très fatigué par cette maudite lettre que je n'ai pas faite (p. 96).

> The repetition of stressful situations and continuous efforts can lead to fatigue and exhaustion, which play a central role in the development of neuroses. Let us consider an example from everyday life that vividly illustrates the mechanism of this mental exhaustion. Suppose I receive an unpleasant letter requiring a complex and difficult response to write. At first, I think about writing the response and almost entirely compose it in my imagination, but I cannot summon the courage to actually write it immediately. So, I leave the letter on my desk, within view. From that moment on, every interaction with that space becomes a source of tension: it becomes impossible for me to sit at the desk, walk past it, or even enter the room without seeing the letter, sensing its presence, and repeatedly making the effort to mentally compose the response. Initially, the letter could have been dealt with in ten minutes. However, if I add up all the moments spent imagining the response, all the failed attempts, and all the emotions that resulted from them, I will have wasted hours of exhausting mental work without writing the response. It is unsurprising that after a few days, I might declare myself utterly fatigued by this cursed letter that I have not even completed. (p. 96, author's translation)

In this example, it is not only about helping the potential client to catalyze sufficient motivation to write the letter and to help them in shaping their skills to do so but also to facilitate the client's awareness why the letter is interpreted so burdensome in relation to the person who has sent the letter and requires an answer. Hence, the person is also appealed to work on the relationship between themselves and the supposed person who has sent the letter potentially actualizing their relationship patterns (what to expect from the other, when to expect an answer, etc.).

The modern cognitive-behavioral psychotherapy approach strives for a holistic focus on restoring psychological balance, addressing both motivation and skills (Busemann, 1956). We see that our systemic focus is confirmed through this cross-connection between psychotherapy and the action-oriented explanation of affects and moods. Restoring psychological balance is thus not only crucial for the pursuit of autotelic goals but also relevant for participation in the cultural community, which can be enriched by the individual (Stern, 1935). Helping a person realize that their action-theoretical structure can enrich the world, and that the world can also enrich them, is a central insight for restoring their sense of meaning. In summary, we can draw the following conclusions about the fatigue experience:

4. The negative fatigue experience arises where we have both low action motivation and a depletion or burdening of action potential. We particularly feel the

heaviness of fatigue in a very specific way when our actions generally threaten our central fantasms while simultaneously draining our energy. We do something because we must, and in doing so, we also lose the strength of our action potential, which could otherwise be used for more fantasm-relevant actions. In this process, we also discover a significant responsibility for the gatekeepers of cultural communities in anticipating or addressing the negative fatigue experience: On one hand, these gatekeepers must awaken the person's interest, that is, show intersections with the person's relevant fantasms, and on the other hand, provide the person with a set of techniques and tools that enable relaxation in the face of internal and external challenges. The positive fatigue experience arises when we intensify our forces for fantasm-relevant actions and then deplete them—with visible success in our action, i.e., the achievement of a goal. The symbolic meaning of both positive and negative fatigue experiences has a therapeutic effect when it is made reflexively accessible, leading to a realignment or consolidation of action motivation and action potential. In this process, the responsibility of the cultural community (its gatekeepers) should not be neglected.

The Genesis of Grief and Its Theoretical Significance in Action Theory

Grief, as a central affect, is closely linked to the concept of renunciation in Boesch's theory (1976). This connection is psychologically plausible, as can be illustrated through the following example: A woman decides to separate from her husband. The man, deeply surprised by this decision, goes through a profound crisis and experiences enduring grief. Despite his attempts to win the woman back, she remains firm in her decision, and the separation becomes final. The renunciation of the relationship becomes irrevocable, and the man is initially confronted with persistent grief. Here, the theoretical significance of grief as an affect—especially when it endures—is already evident. The affect is situation-dependent, whereas the mood is more general in nature. However, if the situation remains unresolved psychologically, it can persist and trigger a mood. Thus, the boundaries between affect and mood are fluid, and a clear distinction is not always possible.

Now let us analyze the affective reactions of both partners to the significant change—the separation. First, we focus on the grief response of the man. The man is taken by surprise by the separation; it was unexpected for him. While he perceived the relationship as still intact, his partner had already recognized significant cracks. In other words, the relationship had still been fulfilling central fantasms for the man, and it had strengthened his potential for action. Through the relationship, he was able to experience not only intimacy, love, and security, but his partner had also helped him to balance certain deficits in his potential for action, such as enhancing his ability to engage in discussions. In this sense, the relationship had a strong symbolic significance for the man, carrying positive connotations, as it symbolized not only growth and development but also the sense of security and self-resonance that the relationship provided.

Now, the abrupt end of the relationship threatens his fantasms, strains them, and brings with it a depletion of his potential for action. The communal and dialogical development that the relationship had provided also comes to a halt. Since the relationship is irreversibly terminated, the man can no longer restore it—no matter how hard he tries or how strongly he is motivated, the relationship remains ended. This results in a persistent distance from highly valued goals that were motivated by his fantasm system, which is experienced as threatening and limiting. The distance to the goal—the relaxation of his own fantasms—is permanent and cannot be bridged by action. This means that the man must devalue the goal or reorient his potential for action toward competing goals. In other words, the man in this example must redirect his potential for action toward new motivations and pursue alternative goals that would relax his fantasm complex. Example: The man might begin dating someone new or take up a new hobby in the hope of moving closer to his symbolic goals. Alternatively, the ex-husband might engage in inner reflection, re-evaluating the relationship as more toxic than nurturing. This process could help him recognize that the relationship had effectively ended long before the actual breakup occurred.

Grief, triggered by renunciation, always involves the challenge of reorienting motivation and potential for action. In the case of this separation, the task of finding new, competing goals that provide relaxation to the fantasm system is complicated. Grief represents not only the loss of an important and fulfilling goal (the relationship), but also the difficulty of mobilizing new sources of motivation to replace what has been lost.

> Grief is always a wavering between two gradients: the original goal orientation and the new one that needs to be found. It necessarily intensifies wherever new goals are difficult to achieve. In the section on fatigue, we already encountered the regret of the elderly person regarding the final limitation of their functional abilities; they are inevitably required to forgo previous capabilities. However, we also found attempts to deny this loss or compensate for it by masking it. Renunciation is perhaps the most difficult action that is imposed upon us. (Boesch, 1976, p. 443)

How to Deal with Grief?

Now, let us turn to the action-theoretical interpretation of the end of the relationship from the perspective of the woman. The woman decided to separate from her husband because she had long felt the need for something new. She realized that her husband could not help her achieve her goals in cooperation, and as a result, her fantasms remained constantly tense. She thus found herself trapped in a certain situation; she had some degree of action motivation—based on her relevant fantasms—but her husband was not cooperating with her to achieve these goals or to work toward them. She now perceives her husband merely as an obstacle to her overall potential for action—a persistent hindrance to her ability to act. In the past, she had tried either to ignore this obstacle or urge her husband to work on himself in order to remove this barrier. All these efforts failed. The distance from her goals, the tense fantasm complex, and the barrier to action had chronicled, and she was convinced that this situation would remain unchanged in the future. In the past, she had already distanced herself from her husband and turned to alternative goals, where she could relax her fantasms and sharpen her action potential. These new pursuits may not have necessarily been an affair, but also could include the development of new hobbies, making new friends, or engaging in new spiritual practice, etc. In doing so, the woman found that she no longer needed her husband's cooperation to achieve her goals, develop and consolidate her abilities, or relax her fantasm complex. She had already done the internal work required to realign her action motivation and action potential—which is one reason why the woman ultimately did not perceive the end of the relationship as a loss and respond with sadness, but rather as a liberation.

The end of a relationship or marriage often comes with this discrepancy in affective responses and their action-theoretical significance. While for one partner, the end represents a symbol of liberation and a new beginning, for the other partner, it represents the permanent renunciation of security, love, safety, and transformation. The different affective reactions to the breakup are often the result of one partner already having internally realigned their action motivation and action potential, while the other partner's room for action remains largely intact. What the grieving partner can do here is to slowly re-evaluate the past relationship (not everything was perfect), realign their action motivation (I can also relax my central needs and fantasms in area XY), and simultaneously work on their action potential (I need to work on my willingness to change; I am doing coaching for this). In order for grief not to lead to a depressive mood, which focuses chronically on renunciation and thus the past, the person must engage in both internal and external work to shift their focus back to the present and future.

I would like to conclude this section by pointing out the co-construction of action potential. Not only is the person responsible for the expansion or realignment of their action potential, but the cultural community is also essential in the co-construction and, where necessary, the correction of the action potential. As psychologists and therapists, we should be cautious not to assign sole responsibility to the person for their crossroads in life. The person who does not want to work on themselves, the person who cannot engage in constructive conflict, or the person who is incapable of emotional work, always points to the conditions of the cultural community in which they grew up. If, for example, the person did not have sufficient opportunities to shape or expand their conflict style, this also indicates that they should consider orienting themselves toward new cultural communities in order to expand their action potential. Deficits in action potential often point to deficiencies in the cultural community or its gatekeepers. In a therapeutic context, the person is often encouraged to distance themselves from certain communities and turn to others that are more significant for the expansion of their action potential. When the action potential becomes strong enough, through self-construction and co-construction, the person can confidently strive to reach their private goals or personal should-values (based on their fantasms). Once again, we must point out the deficits in Boesch's early work, which did not include the actual co-construction of action potential and, as a result, also overlooked deficits in the appropriation of affects and moods.

For that purpose (as done previously in Chap. 4), I would like to come up with an equation on how to define a co-constructed action potential. An action potential is the whole of the person's knowledge, skills and abilities that help them to locomote toward their (symbolic) goals. Yet, this action potential is co-constructed and preconditioned by the multitude of cultural communities the person is embedded in—them having the imperative to help the person to sharpen their action potential or rectify it. Hence, we can speak of a collective action potential (in comparison to Jung's collective unconscious) where the person can sharpen their personal action potential in dialogue with the collective action potential—which is a whole of the different communities and their gatekeepers providing skills, techniques and abilities for individual usage. Both types of action potential inform and update each other and so are in continuous dialogue. This dialogue needs to be made fertile for the person to grow in the community and potentially beyond the community—when the collective action potential has become too rigid. The gatekeepers of the cultural communities do have the peculiar responsibility to open up dialogical possibilities where personal and collective action potential can interact with each other updating the person and then themselves. For that matter we can say that action derives its strength from the interplay between the personal and the collective action potential or we put in the following equation (where Act = action, f = function, PACP = personal action potential, CACP = collective action potential):[1]

$$\mathrm{Act}(\mathrm{f}) = \mathrm{PACP}, \mathrm{CACP}$$

Coming back to our grief example: we can summarize the following:

5. Grief arises where renunciation is irreversible. Irreversible renunciation means that goals can no longer be achieved, and thus the individual's fantasm complex cannot be relaxed. Yet, these unattainable goals still appear highly valuable to the person—regardless of the forces under which they strive to reach their goals, the achievement of these goals remains prohibited. Renouncing an object or another person also means a loss of action potential, as the engagement with the irreversibly lost object or person has always sharpened and thus expanded the person's skills and abilities. Grief, therefore, represents a confrontation with one's own action-theoretical framework, which has lost essential support—support from the collective action potential. The power of grief shines through all of this: the realignment of one's action motivation (where do I try to achieve my goals and relax my fantasms) and the necessary expansion of one's action potential (what skills do I need for this, and which should I develop or train) are visible on the

[1] The nature of the equation will be elaborated in Chap. 6.

horizon of the grief experience and serve as a reminder to not become fixated on renunciation, but to take responsibility for one's own goals and the skills and abilities yet to be developed. As difficult as grief may seem in this or that moment, it is an appeal for the person to realign and thus update their action motivation and action potential. Grief, when properly interpreted, becomes a catalyst for necessary change—certainly not overnight, but rather a process-oriented one.

The Role of Affect Regulation and Affect Transcendence

With the elaborations up to this point, we now turn to the role of affect regulation, ultimately introducing a self-experiential method for the action-oriented processing of affects. The term *affect regulation* is paradoxical in nature. The term implies that the affect should somehow be diminished—perhaps even to the point of unrecognizability. In other words, the term suggests that the affect should somehow be controlled to minimize its effects on action. Additionally, the term *affect regulation* also contains an individualistic component, which we wanted to address critically in this chapter.

If I feel anger, the point is not to somehow minimize the anger or how it affects my action. It is also not about pushing the anger aside and focusing on other things. What should be at the center of affect regulation is the transformation of the affect. As we have learned, affects always arise depending on the situation and interpretation. For the transformation of the affect, it is therefore necessary either to change one's attitude toward the situation (interpretation-dependent) or to change the situation itself in such a way that it allows for another, updated interpretation (situation-dependent)—updating the updater (the collective action potential). Surely, this comes with some inner work for the person as reflected in the beginning poem of Boesch—here again an extract:

Your smile lingered in the twilight's gent
And mellow air, a shimmer of shared wine.
Then I asked—and in the warmth of your
Words, I first beheld your face,
And saw my own reflected in your grace.
This gives me much to ponder still,
As through the night the hours slowly spill.

In this context, the context of affect regulation, I would like to speak of *affect transcendence* (this gives me much to ponder still shows that the lyrical subject yearns for affect transcendence), which should supersede the actual term *affect regulation*. The point, therefore, is either to adapt action motivation and action potential so that transformation leads to an updated affect or mood, or to approach the situation in such a way that its obstacles to action are restructured, allowing the affect to yield to a competing affect. Affect is always a call for personality restructuring or situation restructuring and therefore not merely a simple regulator of action—as Boesch formulated at the beginning of his treatise (1976). The call thus has teleological significance and therefore a horizon. When we speak of affects, we must always take this horizon into account—a phenomenological component of psychology (von Fircks, 2023).

Example: A teacher is angry with a colleague who has not adequately taught the students the basics of mathematics. Here, the teacher has several options to transcend their anger. In other words, they can change their interpretation of the situation by restructuring their fantasm complex (e.g., no longer being enthusiastic about teaching mathematics but focusing only on the essentials) or they can offer a remedial course in which the students catch up on the basics (and hence dialogical opportunities for PACP and CACP.) Additionally, they may talk to colleagues and offer them support for certain classes.

All these patterns are only possible if the teacher constructively deals with their affect (here anger) and considers the following things:

1. The teacher analyzes the barriers to action (i.e., the obstacle) and where it may come from (from the cultural preconditioning of the former teacher), which also points to the co-construction of the students' action potential.
2. They can reflect on the techniques they have at their disposal to reduce the students' distance from their goals.
3. They can recognize that they may need the cooperation of other colleagues for this, that is, also the co-construction of their own action potential.

If all this had been successfully achieved in a process, the anger would have been transformed into a triumph affect; the teacher has grown through the barrier to action and has expanded their action potential. In this way, neither the action field nor the affect is regulated, but both the action field and the affect are transformed through adapted actions based on the teacher's fantasm system. How so? By recognizing that students' inability to master the simplest mathematical operations is not their fault, by understanding that they, as the teacher, possess the skills to teach these operations quickly and effectively, and by seeking assistance from colleagues throughout this process, the situation that initially provoked anger is fundamentally transformed—or actively reframed—by the teacher. Consequently, the anger is intentionally shifted into a sense of triumph. This triumph emerges as the teacher successfully expands their action potential while navigating a challenging task that initially seemed overwhelming.

The grieving husband is also encouraged, in the long term, to chart a path out of his grief and transcend the grief. The weight of the affect encourages confrontation with his own inability and possibly to correct it (i.e., refine his skills and capacities) while also revising his action motivation (what do I expect from my partner). If the partner has managed this in a complex process, for example, accompanied by therapy—then the partner would have grown beyond their previous action motivation and action potential and would have transcended both the action field and the affect.

In every affect, the possibility of affect transcendence shimmers through and shows its teleological symbolism. That Boesch implicitly pointed to this affect transcendence is evident in the following quote:

> The affect easily becomes a disruptive factor as long as we consider it solely as a regulator of ongoing action. Indeed, it often seems that an excess of affect readily throws actions into disarray. The true regulatory function of affects becomes evident only when we also relate them to our personal functional potential. Situations, as we have seen, are experienced as symbols of both our current and fundamental functional capacity. In this ego-psychological sense, affect acquires an overarching signal significance that directly concerns us as agents; beyond its merely situational relevance, it is capable of promoting adaptations of our person to ourselves and our environment. (1976, p. 497)

6. In affect regulation, the focus should fundamentally be on affect transcendence, which always subtly emerges within the primary affect experience. This entails an appeal to the individual to either cognitively restructure their action motivation (their fantasmic system) or to neutralize the situation or the action impediment in such a way that the entire field of action is reorganized. Only through the restructuring of one's own person and action field does the unfolding of affect transformation take place, which always involves work on one's action motivation, action potential, and the co-construction of action potential (collective action potential). Affect transcendence must be understood in this holistic sense.

Affect Transcendence Self-Experience Method

Our reflections on affect transcendence culminate in the *Affect Self-Experience Method*, which we will outline below. This method encourages individuals to answer the questions provided (see Table 5.1). The goal is, first, to distill the affect and connect it to the associated event (i.e., the field of action). Next, attention is drawn to the central needs and the significance of the event within the field of action, as well as to the forces tied to the fantasmic complex involved. Once these have been identified, the focus shifts to the co-construction of the action potential,

Table 5.1 Affect-transcending method

Affect	Triggering event	Fantasm complex	Action potential	Co-construction of action potential	Affect transformation
What was the affect you felt lastly?	What was the event that triggered the affect? What did you feel? What preceded the affect?	Which needs did the event evoke? What is important for you within the event? Does the event have some meaning for you? Which one?	Which personal resources do you see within the event? Which skills and abilities are linked to the event? Which personal forces do come to your mind when thinking about the event?	Did the event have anything to do with strengthening your personal forces and how?	Which hopes do reside within the event? Which future does the event show?

examining how the action potential and its ecological determinants have been shaped. Finally, the exercise directs attention to affect transformation, revealing the horizon embedded in the affect toward which the individual can work, ultimately transcending the initial affect—moving toward its teleological meaning.

The questions that individuals can ask themselves—or that can be posed in a therapeutic context—are outlined in Table 5.1. The following is an example of the affect transcendence self-experience method, drawn from my own life.

Affect Transcendence Self-Experience Method: An Autoethnographic Example

The last intense affect I experienced was grief. I associate this profound grief with the death of a beloved uncle, who always treated me with great love and care and with whom I shared many cherished moments. This uncle accompanied me to my soccer games, my tennis matches, always cheering me on and supporting me. He also taught me how to play chess and various card games. He frequently picked me up from school and drove me home. He was incredibly caring, often bringing me food. I loved him deeply.

When I think of him, I feel a profound sense of security, support, unconditional love, care, fun, engaging discussions, and joy. He undoubtedly showed me the essence of friendship and love—how to care for a loved one, how to sacrifice for others, and how to support someone both mentally and physically.

Coming from a performance-oriented household where my parents focused more on instilling achievement values than providing emotional resources, my uncle taught me how to truly love. Even though I am now deprived of his love and our shared adventures, I know that our time together has shaped me. Through our relationship, I learned how to love—my siblings, my friends, and my partners.

Thus, when I look back on my time with him, I feel not only grief but also gratitude and joy for our connection. My grief has moved to the background, and gratitude has taken center stage. In some way, our relationship lives on.

This gratitude was not consciously apparent to me in this way, even though it was somehow present within me. This conscious gratitude—as I now feel it—relaxes me as a person; it soothes my psychological system.

What remains to be said after this journey through Boesch's affect theory? Affect theories are fundamental to establishing a cultural psychology. Without addressing the questions of affects and emotions, no cultural psychology can take shape—as we human beings cannot live a life without the regulatory role of affects. In this chapter, we examined the role of affects, interpreting them through the lens of action theory with the help of Boesch (1976), partially revising his ideas by highlighting the co-construction of action potential through the collective action potential, and concluding with the affect transcendence method.

The concept of affect transcendence has shown us that the question of affect is not only relevant to cultural psychology or general psychology but also deeply intertwined with our humanity, personal development, and self-realization. What more could we, as psychologists, wish for than to inspire growth in ourselves and others?

Such a perspective was already recognized before psychology gained prominence in the nineteenth and twentieth centuries. Religions of various kinds have long employed techniques of affect transcendence. For instance, when a person feels anger because they have sinned (e.g., speaking ill of their neighbor), they often find relief through prayer or confession. Similarly, it was common for young Christians to experience guilt following self-gratification (masturbation), prompting them to seek solace by turning to God for understanding and strength to resist their instinctual drives.

Non-Abrahamic religions, such as Daoism, also provide avenues for affect transcendence. Adherents often find relief by connecting to something greater, such as nature or the cosmos. For example, they might seek forgiveness for prior disrespectful actions against nature by attuning themselves to its rhythms and asking for pardon.

However, affect transcendence can also exacerbate negative emotions. In the case of self-gratification, feelings of guilt and shame may intensify, escalating into despair, self-hatred, or disgust. This highlights the dual nature of affect transcendence—it can either alleviate or magnify emotional distress. Consequently, the responsibility of gatekeepers within various action fields becomes evident. These individuals play a crucial role in guiding and supporting people, ensuring that affect transcendence does not lead to harm or endanger their person.

References

Boesch, E. E. (1975). *Zwischen Angst und Triumph: über das Ich und seine Bestätigungen*. Huber.

Boesch, E. E. (1976). *Psychopathologie des Alltags: zur Ökopsychologie des Handelns und seiner Störungen*. Huber.

Boesch, E. E. (1991). *Symbolic action theory and cultural psychology*. Springer.

Boesch, E. E. (1998). *Sehnsucht: Von der Suche nach Glück und Sinn [Longing: On the search of joy and meaning]* (1st ed.). Huber.

Boesch, E. E. (2005). *Von Kunst bis Terror: Über den Zwiespalt in der Kultur [From art to terror: On the cleavage of culture]*. Vandenhoeck & Ruprecht.

Bruni, A., Gherardi, S., & Poggio, B. (2004). Entrepreneur-mentality, gender and the study of women entrepreneurs. *Journal of Organizational Change Management, 17*(3), 256–268. https://doi.org/10.1108/09534810410538315

Busemann, A. (1950). *Geborgenheit und Entwurzelung des jungen Menschen: Beiträge zur heilpädagogischen Aufgabe unserer Zeit*. Aloys Henn.

Busemann, A. (1956). Das Selbstständigwerden des jungen Menschen. In A. Busemann (Ed.), *Beiträge zur pädagogischen Milieukunde* (pp. 168–179). Hermann Schroedel. Original work published 1950.

De Shazer, S., Dolan, Y., Korman, H., Trepper, T., McCollum, E., & Berg, I. K. (2021). *More than miracles: The state of the art of solution-focused brief therapy*. Routledge.

Fernández-Salinero, C., & La Riva, B. d. (2014). Entrepreneurial mentality and culture of entrepreneurship. *Procedia - Social and Behavioral Sciences, 139*, 137–143. https://doi.org/10.1016/j.sbspro.2014.08.044

Fircks, E. F. v. (2023). Relating cultural psychology and eco-phenomenology – Setting the seeds for a fertile cooperation. *Human Arenas*, 1–19. https://doi.org/10.1007/s42087-023-00339-5

Fircks, E. (2024a). *Learning with William Stern: Personology for future*. IAP.

Fircks, E. (2024b). Making sense of signs: Readjusting William Stern's personological value theory. *Theory & Psychology*. Advance online publication. https://doi.org/10.1177/09593543241279264

Fircks, E. v. (2024c). Raising questions with lamiell about future research directions of critical personalism: A book review essay. *Integrative Psychological and Behavioral Science*, 1–11. https://doi.org/10.1007/s12124-024-09825-z

Greenberg, L. S. (2017). *Emotion-focused therapy*. American Psychological Association.

Heiss, R. (1936). *Die Lehre vom Charakter*. De Gruyter.

James, W. (2000a). What pragmatism means. In G. Gunn (Ed.), *Pragmatism and other writings* (pp. 24–41). Penguin.

James, W. (2000b). Pragmatism's conception of truth. In G. Gunn (Ed.), *Pragmatism and other writings* (pp. 87–105). Penguin.

James, W. (2000c). Pragmatism and humanism. In G. Gunn (Ed.), *Pragmatism and other writings* (pp. 105–119). Penguin.

Janet, P. (1919). *Les medications psychologiques*. Payot.

Janet, P. (1923). *La médecine psychologique*. Ernest Flammarion.

Murphy, J. J. (2024). *Solution-focused therapy*. American Psychological Association.

Rohracher, H. (1937). *Kleine Einführung in die Charakterkunde*. Teubner.

Stern, W. (1935). *Allgemeine Psychologie auf personalistischer Grundlage [General psychology on the basis of personalism]*. Martinus Nijhoff.

Vygotskij, L. S. (1996). *Die Lehre von den Emotionen. Eine psychologiehistorische Untersuchung* (Fortschritte der Psychologie, Bd. 15). Lit.

Vygotsky, L. S. (1976). *Psychologie der Kunst [Psychology of art]*. Fundus Bücher.

Warstat, M. (2011). Katharsis Heute—von der Abreaktion zur Delegation. In M. Warstat (Ed.), *Krise und Heilung: Wirkungsästhetiken des Theaters* (pp. 91–109). Schöningh and Fink. https://doi.org/10.30965/9783846750223_009

Chapter 6
Collective Action Potential and I/World Balance: The Role of Gatekeepers and Dialogical Isles of Mediation

The Carpet Weaver *(E.E. Boesch, Enno von Fircks translation)*
Do you know what you weave,
weaver?
Where, and to what end,
Does what you weave
will lead?
Do you know to which net
you pull your threads?
Do you sense for which treasures
you toil and tread?
Yellow, blue, and red
Resound in your vibrant designs,
But do you know where laughter,
Where hardship, where death entwines
Through your weaving's spirit,
Guiding the threads you inherit?

The Need for a Collective Action Potential and Its Mechanisms

In previous chapters, we have frequently discussed the co-construction of action potential. In doing so, we encountered a possible modification of the concept of action potential. To emphasize the co-constructive nature of the term, we do not merely highlight the social mediation of skills, techniques, resources, and knowledge. These are constituents of the whole potentiality of action by a person within the socially organized setting. Instead, we propose the concept of *collective action potential*, which the individual can make fruitful for their own purposes. I am convinced that this revision is necessary to make Boesch's symbolic action theory both theoretically and practically productive. At the end of this chapter, this revision is illustrated through dream interpretation. In doing so, we also hope to demonstrate

E. Freiherr von Fircks, *Culture of Action*, Theory and History in the Human and Social Sciences, https://doi.org/10.1007/978-3-031-88808-3_6

the broader utility of Boesch research, extending beyond theoretical discourse (Straub et al., 2020).

The concept of action potential was introduced in Chap. 3, so we will renounce an extended definition here. Suffice to say that action potential represents a readiness for action or, more generally, the capacity to act (Boesch, 1991). The concept of being capable of action holds particular significance for us. Action itself occurs in the here and now (the present). However, the concept of capacity refers to the potential to act and inherently implies the teleological nature of the action potential concept. This makes the action potential crucial in psychology, as it enables the study of a psychological process that subsequently solidifies into a structure.

Assuming that individuals do not exist within a single, isolated field of action but instead operate across multiple action fields, we can conclude that people enrich and nurture their general action potential within the context of various communities. For instance, an adolescent may attend school, belong to a specific class, maintain friendships with a particular group, participate in a family unit, enjoy playing football and tennis, and take pleasure in reading books. All of these activities indicate that the individual is part of numerous action fields. Within these fields, the individual learns to navigate various domain-specific challenges, thereby expanding their general action potential, which enables them to increasingly engage with the world in an independent manner.[1]

Let us consider a specific action field as an example: the tennis action field. Here, the adolescents will not only be taught specific techniques for playing tennis but will also acquire other skills relevant both to tennis and beyond. For example, if an adolescent experiences an emotional outburst during practice, the coach may encourage them to manage their emotions and prevent those emotions from interfering with their gameplay. The coach might employ specific breathing techniques to teach the adolescent emotional regulation. Consequently, the action field—mediated by its gatekeepers—possesses a collective action potential composed of various techniques, resources, and skills that the gatekeepers of the field transmit in certain situations (Lewin, 1951).

Latency and Transparency Within the Action Potential Concept

A crucial aspect of collective action potential is its latent nature. It exists in the background and only manifests through specific situations—or, more precisely, experiences. For example, the collective action potential related to emotional

[1] Each community is characterized by a unique action field that constrains the individual with specific normative values (you should adhere to these or those norms) and promotes or limits particular methods for achieving goals (choose this or that pathway, use this or that tool). Therefore, a community is defined by unraveling its action field, which involves uncovering the action-theoretical determinants that are essential for action.

regulation remains dormant as long as the child or adolescent has not yet faced a frustrating experience on the tennis court. It is only when the child or adolescent struggles with an impending or actual loss that the gatekeeper of the action field (e.g., the coach) may decide to introduce them to emotional regulation techniques. In this way, a concrete need triggers the transmission of elements of the collective action potential, which are then made available to the individual for addressing their specific requirements.

Once again, it is worth emphasizing that the collective action potential of an action field represents a conglomeration of techniques, resources, materials, and knowledge that can assist the individual in solving problems and meeting needs in particular situations. The transmission of these techniques depends on the gatekeeper of the action field and the extent to which the gatekeeper has internalized and is able to convey these techniques.

Another example can be found in the case of schoolchildren who are just starting their education. These children are introduced to various techniques to address specific challenges. Lessons are designed to include tasks that enable schoolchildren to learn, practice, and perfect particular skills. These challenges, in turn, trigger the transmission of collective action potential. For instance, if a child is tasked with cutting and gluing something together, the ability to handle scissors and glue is developed through concrete dialogical interactions with other children and the teacher. The acquisition of this skill—mediated by the gatekeeper of the action field and representative of the collective action potential—not only benefits the child in this particular school setting but also has implications for other action areas and fields. For example, by mastering the skill of safely using scissors and glue, the child can independently tackle challenges in the family action field, such as wrapping a Christmas gift.

The transmission of techniques, skills, and knowledge is essential because it enables the individual to expand their overall action potential, not just in a specific domain. This enhanced action potential allows them to explore previously undiscovered parts of their world, interpret those areas meaningfully, and successfully address the challenges they encounter. Boesch denotes this as the process of transparency—the meaningful appropriation of the world and its elements (Boesch, 1991).

> Saying that a culture offers "possibilities of action" implies that, in analogy to a home, it optimizes conditions for action. Culture, therefore, creates those feelings of familiarity, protection, and security, ease of orientation ("transparency") and of anticipation—which we know best when, such as a refugee, we are deprived of them. This 'feeling at home' constitutes the ground on which develop identifications with one's "belonging"; but also, as we shall see later (…) longings for the "elsewhere" or the "otherness." (p. 36)

The Dialogical Isles of Mediation

The successful transmission of collective techniques is, of course, also contingent upon the particular mediator and the extent to which they represent the collective action potential of a given field. There are undoubtedly individuals who only loosely identify with the techniques and knowledge of a specific action field, which results in the transmission of the collective action potential in a diminished or incomplete form. For instance, a conservative individual tasked with teaching critical theory (because it is part of the curriculum) will likely convey the knowledge and skills associated with that action field in a diluted manner, perhaps even critiquing or rejecting the controversial positions it entails. This example demonstrates that understanding resistances and overcoming barriers is an integral part of the action potential concept.

This example demonstrates that there is a plurality of collective action potentials, corresponding to the various action fields. It follows that different worldviews—mediated through these action fields—can conflict with one another, thereby impacting the transmission of skills and knowledge. Returning to another example, an anti-evolutionist needs to teach an evolution class which is likely to weaken the actual the transmission of associated cultural techniques. Thus, we must acknowledge that the existence of diverse, specific action potentials not only enables individuals to engage with and understand the world more transparently but also fosters antagonisms that may place these potentials in competition with one another. This competition increases the likelihood of both interpersonal and intrapersonal conflicts.[2]

Consider, for instance, the anti-evolutionist individual teaching evolutionary biology. Here, a fundamental conflict arises within their action potential, which reflects broader societal ideological fault lines between anti-evolutionary (e.g., creationist) and evolutionary-theoretical action fields. The danger for the individual lies in possessing and employing a variety of specific action potentials that do not function complementarily but instead stand in opposition, leading to persistent internal conflict.[3]

---

[2] Valsiner raised an important question regarding whether different action potentials can be counted. My response is that they can indeed be counted, as they largely depend on the various communities an individual engages with and grows into. The action potential concept refers to a broad set of skills, techniques, and resources that may or may not yet be developed but can potentially evolve through interaction. However, the number of different action potentials should not be seen as static, as the individual can, throughout their life, choose to leave certain communities or align themselves with new ones. Yet, for a scientific analysis, I think it is more worthwhile to shed light onto the interrelatedness of differing action potentials rather than simply counting them.

[3] We are equally in need of a concurrent perspective. Counterexample: A communist party member being also religious. Both action imperatives—derived from the opposing communities or fields of action—might be integrated into the person without any conflict occurring. A barrier might exclude the ambivalence (Valsiner, personal communication) between the communities as well as the intrapersonal conflict that might emerge as a consequence. As far as our perspective goes, this is a valid argument. Yet, both positions might be harmonized through actions derived from a higher order person uniting previous opposites. I might be a community party member only to the degree of wanting the best education for my children—and I am equally religious because I want the benediction of God not only for myself but for all my family members. In other words, through the higher-order person, both previously antagonistic positions are harmonized and separated onto a lower level (I am a communist because I want my children being able to study afterward); I am religious because I want the mercy of God for my children. Hence, the person is a communist for

If we revisit the Kafka example from other chapters, this analysis is further corroborated. Kafka, who viewed himself as a fractured character (Kafka, 1995), lived in a state of perpetual tension mediated by two sharply opposing and irreconcilable action fields: his bureaucratic action field and his literary action field. Kafka was unable to transfer the techniques and skills learned in one field into the other, nor could he effectively utilize them across fields. As a result, he existed in two almost hermetically sealed worlds, leading a double life.

This underscores the fact that the collective action potential of an action field is always impacted—and potentially fractured—by the gatekeeper, who themselves belong to multiple action fields. The interaction between the various collective action potentials can either amplify or diminish the transmission of a specific collective action potential. This brings us back to the example of the conservative teacher who must convey an ideology they do not believe in (critical theory, in this case). The teacher's internal dissonance weakens the ability to fully transmit the collective action potential of the critical-theoretical field. Thus, the interplay of conflicting worldviews, mediated by individual gatekeepers, profoundly shapes how collective action potentials are transmitted and realized.

The Successful Transmission of the Collective Action Potential

For the successful transmission of collective action potential, certain spaces are required, which we will call *dialogical isles of mediation.* A classroom serves as such a dialogical isle of mediation, where the teacher demonstrates, explains, and imparts techniques in a personalized interaction with the child or adolescent. The teacher may also rely on more advanced students, viewing them as dialogical partners who can transmit certain techniques and knowledge to their peers. Thus, dialogical isles of mediation should always be understood as pluralistic and not dependent on a single gatekeeper of the action field. On the contrary, individuals can draw upon a variety of gatekeepers within the action field and its collective action potential.

On these dialogical isles of mediation, a personal learning process always takes place, through which the individual is introduced to the action field. The goal of this dialogical transmission is the autonomous application of the cultural techniques imparted, allowing the individual to use them without the intervention of the mediator. Hence, the goal is autonomy. Of course, these dialogical isles of mediation must also account for the individual's specific challenges and idiosyncrasies to ensure the successful transfer of cultural techniques. For example, a left-handed child being taught to use scissors will require different guidance from the gatekeeper than a

getting education and a religious person for getting the mercy of God. However, the barrier is set by the higher order person (the family person) through which the permeability or nonpermeability of the border is set up (see below for the definition of higher-order person).

right-handed child. Similarly, a person with a physical disability will require a different introduction to a sport compared to someone without a disability.

The mediator, or gatekeeper, of the collective action potential must also draw upon their own individual action potential and, by extension, collective action potentials, to ensure that the transmission process is effective and personalized. Often, errors in the transmission of cultural techniques—whereby individuals acquire only a rudimentary or incomplete mastery of the techniques—can be traced back to the gatekeeper's insufficient didactic action potential, which prevents them from adapting the transmission process to the individual's needs.

For instance, in my own work as a tennis instructor, I encountered a situation where a child was too physically weak to perform a forehand stroke with one hand. In my earlier professional experience, I resolved this by having the child grip the racket higher up, reducing its weight impact on their stroke. At the time, I was unaware of alternative teaching methods. Only after changing jobs and observing other coaches using a different approach—teaching the forehand with both hands initially—did I realize that alternative methods existed for transmitting this technique to children. This realization led me to adopt the new method in my own training sessions. And it worked quite smoothly with the young students.

In other words, by entering a new action field through my job change, I was exposed to a different collective action potential that enriched my individual action potential for addressing specific cases, such as children struggling to hold the racket properly. This example highlights the incompleteness of individual action potential, which is gradually expanded and refined through engagement with diverse action fields and collective action potentials.

It is also necessary to acknowledge that many techniques and skills remain unconscious and often remain so throughout an individual's life. Only rich experiences across multiple action fields expose the individual to various collective action potentials, thereby enabling the expansion of their general and individual action potential. Consequently, individuals have a responsibility to seek out dialogical isles of mediation to actively advance the growth of their action potential.

In this context, "unconscious" does not imply that these techniques and knowledge already reside latent within the individual, merely awaiting activation. Instead, they are "unconscious" in the purest sense of the word: they have not yet been experienced and, as such, have not been integrated into the individual. It is only within the dialogical isle of mediation that cultural techniques are consciously imparted and, if successfully transmitted, can be integrated into the individual's self. Naturally, the transmission of new cultural techniques builds upon previously acquired and habituated skills that have already been largely integrated into the individual's overall organism and no longer require conscious instruction.

For example, if a child is to manipulate scissors, they must already possess a certain degree of manual dexterity that has been integrated into their organism and no longer necessitates conscious awareness. New cultural techniques, therefore, build upon these habituated, unconscious skills. In this sense, we can speak of the metamorphosis of individual action potential, which grows and flourishes through its interaction with the broader atmosphere of collective action potential.

I intend to illustrate this model in Fig. 6.1, highlighting the dynamic interplay between individual and collective action potential. This illustration requires some explanation. Let us begin with Person A. This individual wishes to improve their presentation and public speaking skills. They are already motivated, arriving with specific anxieties they seek to alleviate. For instance, we can say that they aim to overcome a *fantasm of insecurity* through a learning process that allows them to feel more confident during presentations, avoiding blushing or the appearance of stress marks.

Additionally, they are socially motivated to enhance their presentation abilities, as better skills in this area would increase their chances of successfully marketing their company's products (social should-values). These personal and social motivators serve as drivers for engaging in dialogical isles of mediation and expanding their repertoire of cultural techniques.

We can see that the individual is both personally and socially motivated (intrinsically and extrinsically) to deepen their existing skills, which they have developed over the course of their professional life. This motivation is essential for Person A to take a step toward the dialogical isle of mediation and to approach the gatekeepers of another field of action, in this case, Person B. Person B, for example, a highly experienced colleague in the areas of presentations and public speaking, is characterized by particular abilities, skills, and knowledge in the field of rhetoric (for instance, we might assume that they studied rhetoric in Tübingen). This illustrates that Person B has a surplus of skills and techniques in the field of presentations, which they can bring to the dialogical isle of mediation. However, these skills and knowledge did not arise in a vacuum but can only be understood within the

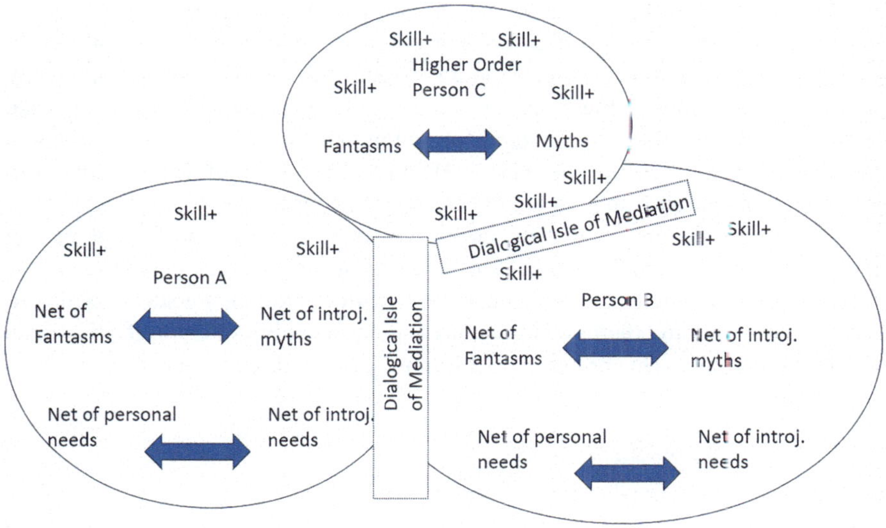

Fig. 6.1 Individual and collective action potential in interaction (introj. = introjected, skill+ = skill needed to give closure to a pressing need)

collective potential for action of their rhetoric studies. In this context, Person B is a gatekeeper of the rhetoric community and can also transmit these skills to others.

We denote Person C as a personoid entity that is hierarchically superior to Person B in the field of rhetoric. Personoid entities exhibit independent personalities—almost as though they were human (Stern, 1906, 1923, 1924). The most essential criterion for attributing person-like qualities to a personoid entity lies in the fact that the entity possesses an autonomous teleology, that is, it operates with certain self-purposes that can be achieved through specific goals (Stern, 1930, 1935). The university program in rhetoric (as it is offered, e.g., in Tübingen) comes with such autonomy, that is, with its own self-purposes. For instance, the program does not merely aim to convey foundational knowledge of German studies but also to enable students to argue persuasively, freely, and plausibly—potentially in preparation for a political career or a management position. This university program brings with it a unique repertoire of foundational knowledge, skills, and resources to achieve this goal. Thus, it possesses a strong, personal potential for action, which the individual Person B perceives as collective potential for action—precisely because they are part of the collective, the higher-order person.

For the individual Person B, the individual potential for action of the higher-order person constitutes a collective potential for action, whereas for Person A, neither applies. However, in our example, Person B can make the collective potential for action, mediated through gatekeepers, accessible to Person A and transmit the learned techniques and resources to Person A on the dialogical isle of mediation. In this way, an unconscious collective potential for action can become tangible and indirectly accessible through the interaction between Person A and Person B, enriching the individual in their already existing skills and thus serving their own purposes. However, Person B may also improve their individual potential for action based on the collective potential for action—for instance, if Person B lacks knowledge on how to provide rhetorical assistance to insecure or nervous individuals. In this case, Person B in our example might consult the materials of the university program, review further literature, or contact colleagues and professors to make further use of the collective potential for action and thus expand their individual potential for action. To do this, Person B must also rely on dialogical spaces to access these techniques and integrate them into their repertoire.

We can thus see clearly that the collective potential for action for one person constitutes the individual potential for action of a higher-order person or a personoid entity. Gradually, through interaction with Person B, Person A will also begin to grow into the rhetoric community—not as deeply as Person B, of course, but at least partially. This brings us to a summary of the discussion so far:

Summarizing the Pillars of the Collective Action Potential Concept

(a) We can distinguish between individual potential for action and collective potential for action. Personal potential for action is enriched by collective potential for action and vice-versa. The interaction between personal and collective potential for action always occurs between individuals and within specific dialogical spaces, which we refer to as dialogical isles of mediation. Here, certain cultural techniques are exchanged, learned, expanded, or replaced, thereby enhancing the individual potential for action.
(b) For the expansion of personal potential for action, a "you"—that is, another person—is always necessary. This other person is part of certain fields of action and, as such, a gatekeeper who possesses more techniques, knowledge, and resources, which can be incorporated into the mediation process. This "you" is part of a higher-order person, a certain cultural community that acts as an autonomous person (or as a personoid entity) and thereby makes its collective potential for action available to the individual. Elements of this collective potential for action can then be integrated into personal potential for action and also passed on to others. In the interaction between two individuals, the collective potentials for action of the respective fields of action are always present and can be utilized for a productive dialogue.
(c) The transmission of cultural techniques always occurs through representatives or gatekeepers of a particular field of action, who are also part of multiple fields of action and thus carry multiple collective potentials for action in the background. This plurality of collective potentials for action may either promote or hinder the transmission of certain techniques, depending on how reconcilable or contradictory the worldviews of the fields of action are. The transmission of certain cultural techniques therefore does not take place against the background of a singular field of action but rather within a plurality of fields of action, which may either disrupt or enrich the process of mediation. A Christian biology teacher, for example, who is required by the curriculum to teach evolution, even though they are a creationist, illustrates the paralyzing effect of negatively interfering fields of action and collective potentials for action. Personal conflicts must thus also be embedded within the framework of larger cultural conflicts—namely, where collective motivations and potentials for action are irreconcilably opposed.

Collective Action Potential and Dream Experience/ Dream Analysis

We now wish to transition to the final part of our chapter on collective action potential—the experience and interpretation of dreams. We are indeed situated within an analytical tradition of dream experience and interpretation. However, we do not wish to view the dream experience solely as a destructive source, that is, we do not wish to interpret dreams solely as a source of repressed needs or in the sense of libido theory, but instead emphasize the constructive nature of dreams. Thus, we align more with Jungian dream experience and interpretation (Winborn, 2024), which seeks to uncover certain complexes in dreams (guidelines in experience and conduct) and relates them to archetypal images, that is, collective experience (Jung, 1916/1948). The engagement with one's own complexes to enrich and amplify them, through a mythical perspective, is the core characteristic of Jungian dream experience and interpretation, which aims to gain a new perspective on certain life paths and then integrate them meaningfully into the self (Jung, 1931/1934).

With our symbolic action theory, impacted by Boesch, we do not wish to simply align ourselves with Jungian dream interpretation, but instead venture to achieve an independent interpretation of dreams. Here, we do not seek to foreground certain complex themes, but rather focus on how mythical and phantasmic components of action interact and point toward a resolution of the relationship between myths and fantasms. Central to the resolution of conflicts is, in fact, the action potential—the acquired knowledge, resources, and techniques through which we bravely address problems and seek to quash conflicts at their root.[4] In this context, however, we aim not only to highlight the individualistic interpretation of action potential but also to trace the collective action potential, the dialogical isles of mediation, and the impact of higher-order persons. The goal of dream interpretation, then, is the dialogical engagement with certain fantasms and myths, or how fantasms can be better relaxed with an updated action potential. This brings us to the teleological significance of dreams, which was also important for Jung and plays a central role in gestalt therapeutic dream experiences, too (Perls et al., 1951).

I would now like to present a dream and then interpret it using our symbolic action theory. Due to gestalt therapeutic research concerning dream experience, I will recount the dream in the present tense (Polster & Polster, 1974).[5]

---

[4] I do not wish to reduce the action potential concept to certain elements, as such a reductionist approach contradicts the principles of symbolic action theory. However, elements are important in defining the action potential concept, as they help differentiate between knowledge and skills, for example, which are crucial for the development of a strong action potential. What I find most important is how knowledge, skills, abilities, and resources are interlinked by the individual in order to strengthen their action potential. Thus, while elements are important in defining an action potential, it is the way they become interconnected through the individual's actions that matters most.

[5] To facilitate the interpretation of the dream, I should add a few autobiographical details. I come from a performance-oriented family and, over the course of my life, I have partially distanced

> I am at home in Germany—in my parents' house. I am talking with my twin sister in the dining room, or at the threshold of the dining room. Suddenly the doorbell rings. My sister leaves the conversation and opens the door. She calls me and says it's for me. When I reach the door, I see my former history teacher, Mrs. XY. She smiles at me and says, "I just wanted to quickly bring you the certificate for the competition you won and congratulate you." And then she adds, "I have two books for you." She then goes through the documents; I see the certificate and realize that it doesn't matter to me at all. I look at the books; I cannot really make out the cover, but I am delighted by both books. I close the door and realize that my history teacher has become completely irrelevant to me.

If we wish to interpret the dream, we must first uncover the strict myths that socially over-determine the dream. The performance myth should be mentioned first, as it has played a major role in my life and upbringing. This is symbolized by the house of my parents and by my history teacher, with whom I never knew whether she would give me a good grade even though I invested a lot of energy in her class. The competition in which I participated (during my school years) also symbolizes this performance myth and brings forth my past thoughts (wanting to make a difference in the world). My sister in the dream, who opens the door for me and has never been as susceptible to this performance-oriented thought as I was, symbolizes love, unconditional appreciation, and security. The books symbolize two things: First, they represent the treasure of knowledge, i.e., they enrich the person with experience and bring new perspectives into their consciousness. Second, the books also symbolize security, as I often sought refuge with my fictional characters during my youth and puberty crisis. Thus, the books represent a personal fantasm of security and knowledge, which is further strengthened by the role of my sister. One could say that the fantasms, the personally relevant meaning patterns, come into contact with no longer relevant myths of the action field. The certificate and congratulations no longer matter to me, as my dream-self realizes. Therefore, the dream indicates that I (the dreamer) have largely outgrown my action field or emancipated myself from it, and now devote myself to relaxing my own fantasms—not in isolation, but in exchange with loved ones, role models, and fictional characters. In another sense, one could say that I am now turning toward a more loving, gentler action field and turning my back on my hard, performance-oriented action field—this process seems to have been completed, as we can gather from the reaction of my dream-self. It is of peculiar importance that those positive emotions I felt during the dream were experienced outside the house and not within it underlying the emancipation from my previous relevant action field. It is also telling that I felt relaxed and at ease after waking up.

But how can we relate the interpretation of this dream to our theory of collective action potential? Referring to collective action potential, we can say that the dream-self in the dialogical isle feels little motivation to address performance or seek recognition for it. In doing so, we specifically highlight the distance from the

---

myself from the strict dominance of this performance narrative and managed to weaken it. To achieve this, I turned to other narratives (mindfulness, meditation, spirituality) that have enriched my original narratives.

performance-oriented action field and its associated collective action potential. The interaction with representatives of a performance-oriented action potential is not rewarding for the individual, and they respond with indifference to these cues. What stands out is the symbolic sense of security, conveyed through the presence of the sister and the books, which is joyfully acknowledged on the dialogical isle. We can thus say that the person is striving for the expansion of the more loving, empathic action potential and wants to integrate this more actively into their self-image—for these reactions are not met with indifference, but with passion. The dream interpretation not only shows the current motivational direction of the individual, to turn toward a loving, empathetic action field, but also suggests the means through which the person seeks to realize this motivational direction—here, through spending time with people in whose presence they feel secure and through role models and fictional characters who understand and validate their needs while also offering new perspectives. This reveals concrete pathways through which the motivational action can be pursued with symbolically significant resources—a crucial factor in relaxing central needs.

In Table 6.1, I present possible questions that can guide the analysis and interpretation of dreams based on our revised symbolic action theory. In this way, we aim not only to stimulate research using our expanded version of symbolic action theory in the field of dream experience and interpretation, but also to encourage psychotherapists to use some of these questions to highlight the responsibility of the community and its role in the genesis of psychopathology.

In order to achieve a balance between the self and the world (Boesch, 2000, 2005, 2021), it is not only the individual who plays a role, but also the gatekeepers of various cultural communities and how they interact and engage with us. To highlight this, we have introduced the concept of collective action potential and demonstrated its practical utility through the analysis of dreams and dream interpretation. As we conclude this chapter, we can say that the dream is not only the via regia for recognizing the power of fantasms and myths but also a means to identify ways in

Table 6.1 Questions for dream analysis and interpretation based on a revision of symbolic action theory

Fantasm theme	Myth theme	Individual action potential	Collective action potential	Collective and individual action potential
Which fantasms are symbolically addressed in the dream??	Which myths appear symbolically in the dream? How do myths and fantasies relate to each other?	How is the individual potential for action interpreted in the dream? Which skills and abilities are articulated through the dream?	Which collectives, dialogical islands, and collective potentials for action are referenced by the dream?	Through which interaction of individual and collective potentials for action is a problem/conflict resolved in the dream (or emerges)? What teleological significance does this solution have?

which myths and fantasms can be reconciled through the interaction of individual and collective action potential. What could be more valuable than promoting reconciliation?

This very topic is also raised as a question in the first stanza of Boesch, as the weaver does not weave only for herself, that is, for her own purposes and goals, but in her action, knowledge and technique crystallize, that is, solidify, as part of the entire collective action potential of the weaving field.

Do you know what you weave,
weaver?
Where, and to what end,
Does what you weave
will lead?

References

Boesch, E. E. (1991). *Symbolic action theory and cultural psychology*. Springer.

Boesch, E. E. (2000). *Das lauernde Chaos. Mythen und Fiktionen im Alltag*. Huber.

Boesch, E. E. (2005). *Von Kunst bis Terror: Über den Zwiespalt in der Kultur [From art to terror: On the cleavage of culture]*. Vandenhoeck & Ruprecht.

Boesch, E. E. (2021). *Musik, Sprache und die Sehnsucht nach dem Paradies*. Psychosozial-Verlag.

Jung, C. G. (1916/1948). General aspects of dream psychology. In *CW 8*. Princeton University Press.

Jung, C. G. (1931/1934). The practical use of dream analysis. In *CW 16*. Princeton University Press.

Kafka. (1995). *Brief an den Vater*. Reclam.

Lewin, K. (1951). *Field theory of social science*. Harper and Brothers.

Perls, F. S., Hefferline, R. F., & Goodman, P. (1951). *Gestalt therapy: Excitement and growth in the human personality*. GST.

Polster, E., & Polster, M. (1974). *Gestalt therapy integrated: Contours of theory and practice*. (2. print). Brunner/Mazel.

Stern, W. (1906). *Person und Sache. System des Kritischen Personalismus System der philosophischen Weltanschauung [System of critical personalism: A worldview]*. J.A Barth.

Stern, W. (1923). *Person und Sache: Die Menschliche Persönlichkeit [The human personality]* (3rd ed.). J.A. Barth.

Stern, W. (1924). *Person und Sache: System des Kritischen Personalismus (Wertphilosophie) [The philosophy of values]*. J.A. Barth.

Stern, W. (1930). *Personalistik als Wissenschaft: Studien zur Personenwissenschaft*. J.A. Barth.

Stern, W. (1935). *Allgemeine Psychologie auf personalistischer Grundlage [General psychology on the basis of personalism]*. Martinus Nijhoff.

Straub, J., Chakkarath, P., & Salzmann, S. (Eds.). (2020). *Psychologie der Polyvalenz: Ernst Boeschs Kulturpsychologie in der Diskussion*. Westdeutscher Universitätsverlag.

Winborn, M. (2024). *Jungian psychoanalysis. A contemporary introduction*. Routledge.

Chapter 7
Dialogues with the Amorphous Inner: An Action-Theoretical Approach to Psychoanalysis

Longing *(E.E. Boesch, Enno von Firck*s *translation)*
Have you kept the dreams,
Those shimmering messengers
Of your most secret world? Their laborious confusion
Is sheltered within the poem. But this — this is home.
No one ever loved the tree solely as a tree,
Not as a tree alone, neither as image
Nor as a verse for its leaves.
Only in secrecy does the brook that sings
Become home; only when the world brings rhyme
Does it become home.
Dark tears of the night
And the bitter serpent of melancholy,
They shine with meaning in the velvet mirror of song.
But love, the silver crown of longing,
And the resonant moon
Above the darkness of sleep —
They are still missing. Stranger, you feel it:
Fulfillment and home are found only
In the fragrant joy of love blooming anew.
Yet trust it not,
Trust the poem instead. This is truth,
And love's enticing red
Becomes a rose and a song
Only in the images of dreams.

The Symbolic Significance of the Amorphous Inner Self

Boesch's reflections on psychoanalysis—admittedly a term he employed with caution—are of profound relevance to cultural psychology. His independent and original considerations regarding the psychoanalytic framework form a rich subject of

E. Freiherr von Fircks, *Culture of Action*, Theory and History in the Human and Social Sciences, https://doi.org/10.1007/978-3-031-88808-3_7

analysis, which will be explored in the following chapter. Before proceeding, one preliminary argument must be presented: Boesch's engagement with psychoanalysis is motivated by a desire to illuminate the human inner world, which remains inherently difficult to access. In this context, Boesch refers to what he terms "inner foreign," a concept representing an internal force that continuously threatens us yet simultaneously serves as a source for the growth of our capacity to act (Boesch, 1998, 2005).

The "inner foreign"refers to elements of our motivational drive and potential for action that are, in certain respects, not necessarily accessible to consciousness—at least temporarily. Nevertheless, we encounter the inner foreign through our varied experiences with the cultural environment, and its elements may surface in specific situations, crossing the threshold of consciousness. This emergence allows us an opportunity to engage with the inner foreign through reflection and examination (Boesch, 2000).

An autobiographical example: After a long hiatus, I decided to return to my beloved football training. Yet something holds me back; I find myself partly afraid to face my teammates again. Before stepping onto the field, I am overwhelmed with anxiety, needing to summon all my willpower to avoid retreating to my car and driving home. This experience forces me to confront the inner foreign. Here, the Inner Foreign is triggered by the symbolic significance of the football field. This compels us to question: what does the football field symbolically represent? Certainly, it embodies camaraderie, as football is a vibrant team sport. It also represents solidarity, as a team can only succeed through collective effort. Furthermore, it symbolizes physical exertion and aggression, given football's nature as a contact sport. Additionally, football signifies the integration of the individual into a larger structure and carries connotations of performance and achievement.[1]

Of these various symbolic meanings, camaraderie resonates with me most strongly. What has always captivated me about football is the convergence of diverse individuals from entirely different backgrounds, united by a common goal. At the same time, I fear that I may have disappointed this sense of camaraderie by staying away from the football field for so long. This leads to another fear: that my relationships with my teammates may have changed irreversibly. Despite these fears, I confronted the football field and resisted the urge to leave. It required significant effort on my part. My initial interactions with my teammates were somewhat awkward, but after some time, the familiar, relaxed atmosphere returned, and we eventually enjoyed a beer together after training. In short, my teammates demonstrated that they are not resentful and that I am always welcome, even after extended absences from training.

In retrospect, this was an extremely reassuring experience, providing me with a sense of safety and belonging. A side note: while football is a sport often associated with aggressive connotations (competition, performance, assertiveness), it also mediates emotional experiences and symbolic associations. It thus fosters a holistic

[1] This aligns with the poem at the beginning of the chapter where Boesch argues poetically that a tree never is just a tree, nor a vase simply remains a vase but carries deeper layer of meanings.

experience for children, adolescents, and adults alike—perhaps a critical factor contributing to its enduring popularity.

The encounter with the amorphous inner self also illustrates how fantasms of threat can be transformed through concrete actions. When the amorphous inner self is triggered and subsequently engaged with, it undergoes a process of reorganization and structuring (Boesch, 1998, 2005). However, Boesch emphasizes that the concept of the inner foreign is only meaningful if we understand it in relation to the interplay between the "strange" and the "non-strange".[2]

> We generally prefer to engage with the external rather than the Inner Foreign. External encounters allow us, at the very least, the option of avoidance. If we choose to confront the external foreign, we can draw on socially recognized arguments and strategies, often garnering approval or even support from others. Moreover, when the external foreign does not pose a threat but instead exerts an allure, it gives shape to our hopes and desires Amorphous longings are transformed into promises—not only of specific pleasures but also of the potential for self-efficacy. Hopes, in this sense, serve as reinforcements of the self. This dynamic suggests that externalization does not merely project an already predefined internal state outward. Inner experiences—be they fear or longing—require concretization; without this, they remain amorphous. Through the process of externalization, we discover in our environment not exact reflections but correspondences to our inner experiences. This interaction allows us to differentiate, transform, and adapt the inner in relation to the external, thereby simultaneously altering its meaning. The communication between inner and outer worlds involves mutual clarifications and transformations. The foreign is connected to the familiar, which enables us to assess and measure our own potential for action. This dynamic interplay allows inner amorphousness to gain structure through interaction with external counterparts, facilitating both an understanding of the self and its capacity to act. (Boesch, 1998, pp. 82–83)

The amorphous inner self, or the inner foreign, inherently reflects longings concealed behind fears or threats. At the same time, this inner foreign serves as a potential source for expanding one's capacity to act. Within the experience of threat, there is often a glimmer of this potential expansion, which can be consolidated through engagement with the contents of the amorphous inner self.

My autoethnographic exploration of aspects of my own amorphous inner self has vividly demonstrated the transformation of threat into self-reinforcement. From the fear and perceived threat of nonacceptance emerged, through lived experience, the recognition of enduring camaraderie—an insight that signified growth beyond the initial situation of threat. External circumstances trigger elements of the amorphous inner self, but only when they bear symbolic relevance to it. For instance, this experience could not have been replicated on a tennis court, as the football field carried unique symbolic significance.

Furthermore, this biographically informed example highlights an essential feature of the amorphous inner self. It is not transformed through intellectual engagement with its uncovered elements—such as by analyzing them on a flip chart or in a journal. Instead, the amorphous inner self is reorganized and transformed

[2] I prefer to equate the Amorphous Inner Self with the Inner Foreign. Boesch (1998) was not able to clearly differentiate between these two terms—and therefore, I use these terms synonymously.

exclusively through concrete experience. It was only through experiencing the resilience and sustainability of camaraderie—even in the face of resistance—that the amorphous inner self, along with its associated fantasms of threat, was restructured. This underscores the primacy of experience as a critical factor in the reorganization of the amorphous inner self, distinguishing it from a purely intellectual engagement with its elements.

This observation carries significant implications for psychotherapeutic practice. A psychotherapist cannot directly resolve the amorphous inner complex but can work with the patient to facilitate encounters with new spheres of experience that trigger transformations of the amorphous inner self.

Thus, we can assert that the amorphous inner self is fueled by fantasms of threat, which can be transformed into positive fantasms if the individual undergoes appropriate mediating experiences. In other words, the amorphous inner self reflects an alternative self, offering a glimpse into deeper layers of the individual (Boesch, 1998). In engaging with the amorphous inner self, the individual continuously operates between fantasms of threat and fantasms of security, navigating the transformation from self-diminishment to self-reinforcement.

Why do I, for instance, enjoy dressing up as a vampire on Halloween? The vampire—often tall, slender, pale, and unassuming, yet possessing a certain elegance—appeals to me in a unique way. In the figure of the vampire, I see aspects of my own self-image brought to life. Behind this elegant façade, however, lies something darker, something hidden that only the vampire's victims or the vampire hunter ever encounter.

The vampire, striking in the shadows and feeding on the blood of others, resonates with symbolic qualities that I can more easily identify with than, for example, the large and clumsy figure of Frankenstein. In embodying the vampire, I experience parts of myself that are otherwise buried in the deeper layers of my self-image—for instance, a sense of *emotional aggression*. A small bite is enough to inflict significant damage. It is not so much the harm itself that resonates with me, but the idea that small actions can have large effects, an idea that also shapes my self-image in connection with my profession as a psychologist.

The vampire, after all, sustains itself on the blood of its victims. Similarly, I might say that I sustain myself on the reinforcement of myself through small, successful actions that I integrate into my self-image. This symbolic meaning is particularly compelling to me in the figure of the vampire, which is why I often find myself drawn to experiencing the "vampire-self" through TV series or films.

In short, the vampire embodies symbolic elements of elegance, hidden depth, and the potential for transformative action—qualities that both reflect and expand my understanding of myself.

The Amorphous Inner Self as a Constructive Source

The amorphous inner self is not a source of chaos or destruction, but rather one of construction and co-construction. In other words, it assists in structuring and expanding the self-image through experiences that reinforce the self. At this point, I would like to touch on the systematic nature of psychoanalysis: I am convinced that it is futile to develop a rigid system of meaning patterns that connects the amorphous inner self to external objects. Such a categorical system is destined to fail because an object like the vampire evokes very different associations—or none at all—in different individuals at different times—depending on their fantasms.

For instance, one person might focus on the vampire's disguise (as they always appear as normal humans), another might emphasize the night as the vampire's essential domain—its habitat—and yet another might identify with the vampire's animalistic nature, given its strong instinct for nourishment. The layers of meaning associated with the vampire are thus polyvalent and inherently ambiguous, interpretable only within the specific context of the individual's fantasms and potential for action. Establishing universal categories of meaning risks erasing the personally relevant fantasms through which the object resonates.

This perspective ties the amorphous inner self closely to the theme of renunciation (Boesch, 1998). For example, when I only dress up as a vampire on Halloween or during Carnival, I am renouncing certain positive fantasms associated with the engagement with fantasms of threat. The fantasm of emotional aggression is only integrated into my self-image on this singular occasion, in a specific situation, and otherwise remains part of an *alternative self*. Unless I actively anchor this fantasm into my everyday self-image through inner work mediated by experience, I am otherwise relinquishing it.

This highlights an important insight: the amorphous inner self represents not only unrealized motivation and potential for action but also possibilities that remain untapped due to this renunciation. By forgoing elements of the amorphous inner self on most days of the year, we fail to utilize the motivation and potential it entails for other purposes. Even alcohol often allows individuals to express aspects of the amorphous inner self, as reduced impulse control can bring these elements to the surface. It is unsurprising, then, that patients in my group therapy sessions have reported feeling more comfortable with themselves when intoxicated than when sober. Alcohol lowers the threshold for embracing these amorphous elements of one's self-image and incorporating them into one's patterns of conduct.

This mechanism reveals the risk of addiction: many individuals believe they are more likable—to themselves and others—or even a "better" version of themselves when regularly consuming alcohol (e.g., funnier, more extroverted, bolder). Similarly, someone who enjoys dressing up as a vampire on Halloween might find themselves stuck in this alternative self, unable to flexibly transition between different self-images. Indeed, entire subcultures exist in which members identify as vampires, drink blood, and hold ceremonies because they believe themselves to truly be vampires.

However, the more therapeutically effective approach lies in recognizing, through lived experience, the elements of the amorphous inner self that are lacking in one's motivation and potential for action. These elements can then be gradually integrated into everyday life. In this way, the amorphous inner self becomes a source of expanded agency rather than an alternative to it.

The amorphous inner self embodies an ambivalent tension. On the one hand, it overwhelms us (there are reasons why these elements reside in the deeper layers of the self). On the other hand, engaging with it experientially can enrich us. It is no coincidence that Boesch refers to the conflict within culture (Boesch, 2005). However, the personally relevant symbolic significance of experiences involving this conflict must always be emphasized.

It is unsurprising, for instance, that French acquaintances of mine reacted differently to the experience of Carnival in Germany. One friend asked me whether it was socially acceptable to engage in infidelity during Carnival, while another focused on the act of dressing up. The first was drawn to sexual fantasms, while the second was captivated by fantasms of metamorphosis. Thus, the amorphous inner self is always concerned with subjectively meaningful content that has yet to be anchored in the self-image—or that one seeks to anchor in a specific way.

As Boesch (1980, 1991, 2021) has long emphasized, individuals do not exist in a social vacuum but are always embedded in a socially over-determined field of action. Within this context, the amorphous inner self is also shaped and constrained by the gatekeepers of this social field, who may either hinder or support the realization of its fantasms. Let us now turn to Boesch's insights in this regard (1998):

> It should be noted that if the amorphous inner self is, to a significant extent, rooted in renunciations and deprivations, it becomes all the more a thorn in one's side the harder it is for us to accept these renunciations. The more reluctantly we have endured these sacrifices, the more likely they are to resurface and burden us again at some point. Thus, the ability to renounce—*Verzichten-Können*—emerges as an important capacity for action. (pp. 86–87)

Here, too, we can rely on everyday observations. Because we live in a certain field of action that promotes specific myths (social norms about how a person should behave) and inhibits others, it is quite possible that a person's fantasms are met with rejection in the field of action and the myths, making it difficult for the person to reconcile the myths with their fantasms. Heiss (1936) rightly views this interplay as the genesis of a broken or contradictory character. Yet, the person always has opportunities to pursue and live out these fantasms in specific moments, for example, when the myths are temporarily suspended, such as during Carnival or Halloween.

But what long-term perspective could be proposed in this example so that aspects of the amorphous inner self can be integrated into the self-image and thus have an ego-strengthening effect? This can only happen by negotiating with the gatekeepers of the field of action—here we could also say: with the censors—the socially accepted fantasms and redefining the boundaries of the taboo zone for expressing certain special needs. How often do we hear educators telling children that they should not be mean? Of course, this advice is very important from an educational perspective to create a harmonious field of action where participants respect and support each other and, thereby, accept the myths. However, in the long term, this myth-strengthening education must also address the fantasms, so they do not get pushed too far into the amorphous inner self, because, in fact, behind these fantasms, there may be abilities and skills that may later become important to the person. The educator must therefore look at the underlying fantasms (behind the actions) and see whether they can be relaxed through other actions, that is, in a socially accepted way that does not distort the field of action. The productive handling of one's amorphous inner self can thus be learned—a process that is often underestimated, as teachers and educators focus more on myth-preserving actions. Behind the action of being mean, many different fantasms may group together, such as feelings of exclusion (threat fantasms in contrast to approach fantasms). It is essential that this feeling of exclusion is productively resolved, as this is fundamental to a humanistic (i.e., one that believes in the potential of the child) education. The child must learn that there are other ways to neutralize threat fantasms, which will help them expand their action potential, such as confronting and addressing the excluders and, if necessary, making peace.

This means that when we work on our alternative self, key abilities and skills are trained, which otherwise cannot be utilized if we let them starve in the deep layers of the person. The tangible confrontation with the amorphous inner self is therefore not only an impulse to make certain parts of our action motivation conscious and actively integrate them into our action planning but also an impulse—through motivation—to gather the necessary techniques/resources to expand the action potential—which then ultimately makes locomotion toward the amorphous parts of the self-image possible. If the reaction to being mean cannot be resolved productively, it is likely that similar situations will trigger the same behavioral patterns, and the person still will not know how to restructure the amorphous inner self. This example clearly shows that it is an educational duty to assist the child early on in how to productively resolve amorphous parts of their self-image—triggered by renunciation—that is, how to pursue them in socially accepted areas. In other words: The alternative self must not become a source of permanent renunciation and thus resignation, but rather the primal source for updating one's motivation and strength. This is not about psychoanalyzing children and adolescents—which, for example, Stern rejected in his productive definition of psychoanalysis (1923, 1924, 1935)—but about accompanying the child on their path in the world with all their personal challenges and the resistances they encounter, helping them resolve them independently and not continuously repressing them through renunciation.

Example: A teenager shows a great talent for questioning things and critically reflecting on them. This is accompanied by specific fantasms of curiosity, so an optimal theoretical framework for action exists, within which the teenager wants to independently impact the world.[3] However, early on, the teenager noticed that they had to hide their attitude and talent because teachers and educators often responded

---

[3] In German, I want to denote the interplay between action motivation and action potential as Handlungsgefüge which could loosely translated to action assemblage.

irritably to their actions. In other words: To avoid delaying the lesson, the teenager's attitude and talent were censored, and parts of this theoretical framework were shifted into the inner foreign. The duty of the educator or teacher is not to inhibit these qualities of the person but to explore with the teenager ways and possibilities for how the action assemblage can be put to use in the service of the lesson. It often happens that certain fantasms are shifted into the inner foreign because they cannot be reconciled with the overall field of action. However, this must always also be an appeal to the field of action to find ways and means for integrating the person's theoretical framework into the field of action—in youth education, this is even an educational duty (note here the importance of the collective action potential). If this does not happen, we are once again confronted with the broken character (Heiss, 1936), which is prone to various psychopathologies.

Another example: A child comes from a nonathletic family and encounters resistance with their sports fantasms and specific potential for action. Often, this results in an inhibition of the theoretical framework for action, and ultimately, renunciation occurs, and the source for independent development is cut off. Education in this case should find ways to harmonize this sport-specific framework with the field of action. On Sundays, you can walk to the bakery to get rolls: stop your time every Sunday and see if you improve. As mentioned elsewhere: *the amorphous inner self is still unshaped power and therefore potential.*

The Role of Dialogue for the Amorphous Inner Self

The question now is how this force can be shaped. One aspect that is not really explicitly developed in Boesch's writings is the role of dialogue in connection with the Amorphous Inner Self. What does this mean? Taking the psychotherapeutic situation as an example, we know that unspoken things find their place here and can be addressed accordingly. A large part of psychoanalytically oriented therapies falls into this area. It is no wonder that many individuals confess something in the psychotherapeutic situation that they would not otherwise confess.

Example: psychotherapists often deal with a coming-out (Evans & Barker, 2010; Hanley-Hackenbruck, 1988). The role of the psychotherapist is to help patients properly categorize their sexuality (in this example) and find an appropriate interpretative level. In this process, the therapist will gently trigger the amorphous inner self (renounced sexuality) in a dialogical manner in order to work through these displaced themes reflexively. It is crucial that the person is given space to find an appropriate interpretation for this complex of themes. The therapist acts in this context as a companion, making individual interpretative suggestions or introducing them to the person. It is essential that the therapist does not shape the amorphous inner self for the person but that the person remains the agent in this process. Otherwise, the person will introject another external interpretation possibility, which may not or not yet align with their own needs. It is of great importance when the therapist helps the patient to shape this amorphous inner self—also in verbal terms. Metaphors may be used (Guenther, 2023, 2024) or objects incorporated into the therapeutic context, which can be actively shaped (von Fircks, 2023), in order to facilitate a reflective engagement with the displaced content. That this happens in this protective space also has significance beyond the therapeutic context. It is not surprising that Rogers declared the human relationship and therefore dialogue as inherently healing (Rogers, 1995), highlighting the importance of corrective experiences that play a role not only in the therapeutic context. Friends, acquaintances, and family can also intervene in this process and help the person shape this inner foreignness and then integrate it into their self-concept. Nothing is more healing than a field of action or its actors that continually offer the person the opportunity to bring the amorphous inner self into dialogue or into the relationship. This is where systemic psychotherapy and community psychotherapy are also significant, showing the social importance in the continuous structuring of the amorphous inner self. The shaping of the amorphous inner self is thus decisively facilitated or inhibited by the field of action or its members, so that in this context, we must also speak of the social component of personality development or personality actualization. The person always needs offers to structure and update the inner foreignness by means of the collective action potential (see previous chapter). The amorphous inner self is thus unshaped power, but in order for this power to be shaped, it requires action-theoretical offers.

In the extreme case, this refers to the psychotherapeutic situation. More often, we are dealing with friends and acquaintances who ask us: "You have something, don't you? Is everything okay with you? Something is going on inside you, right?" All of these are invitations to explore the amorphous inner self and shape it in the ongoing dialogue. Yet, we will reject these invitations if we feel that we do not know the person well enough or do not trust them enough. We fear that our amorphous inner self will be met with rejection by this person—a further reason to hide or flee from the inner foreignness. From these sample questions, it becomes clear that real dialogue is not patronizing but always supportive and expanding. The inner foreignness is therefore not something that we can say, "Today we will go on a discovery journey into it and dive deep within ourselves." This is not how the amorphous inner self works. On the contrary: the person needs an action-theoretical framework that triggers the confrontation with the inner foreignness and thereby reveals it. This is the basic requirement to engage with this inner foreignness. Additionally, genuine dialogue is then needed, which is supportive, expanding, and not patronizing, in order to shape the amorphous inner self and, if necessary, update it. One reason why many people do not engage with their inner foreignness is that they lack the social or action-theoretical framework through which the amorphous inner self is revealed or the person is encouraged to look at and shape this amorphous inner self. If this framework is missing, the amorphous inner self remains amorphous.

Why is it easy for some people to accept their sexuality, while others struggle with it throughout their lives? These questions always depend on the interplay between the person and the field of action—and, of course, we should not only look for reasons for these actions or nonactions in a low score for openness (Kahn, 1991). With our action-theoretical perspective, we would say that the person has not had enough opportunities to become aware of this amorphous inner self and to define or redefine it positively. In doing so, we also want to ward off any naive blame on the person (you just had to be braver) and always seek to consider the interplay of the person and the field of action in terms of the inner foreignness. Hence, the personal example from the soccer field. The amorphous inner self is thus not only updated during Carnival or other annual festivals. On the contrary, the updating should or must become a regular part of the person's daily life. This meta-normative aspect also has ethical and moral implications. We now wish to summarize our chapter:

From Carnival to Everyday Interactions: The Daily Amorphous Inner

(a) The amorphous inner self is a central component of an action-theoretical foundation of psychoanalysis. The inner foreignness always points to certain fantasms of threat and longing of the respective person, which manifest through symbolic interactions or patterns of conduct. My fear of entering the soccer field triggers the amorphous inner self and allows me to update the inner foreignness in concrete experiences and dialogues. The amorphous inner self primarily points to displaced complexes of themes that the person has often renounced in a certain way, but which still hold some significance for their self-concept.

(b) The amorphous inner self cannot be discovered and updated through isolated self-exploration. Here, we cannot speak of a direct, subjective access to the inner foreignness. On the contrary, the amorphous inner self is always revealed in a concrete situation of action (and thus within a certain field of action)—through concrete interpersonal interactions. Only in dialogue can the inner foreignness be disclosed and reflexively processed, and then accordingly updated in the ongoing interaction (which also includes dialogue). The primacy of experience and dialogue are essential mechanisms for the discovery and structuring of the amorphous inner self. Without these two components, the amorphous inner self would not be restructured. The psychotherapeutic space is only one of many fields of action where the amorphous inner self can be redefined and restructured. It is also important to point out other fields of action, such as friendly or familial ones—whether the amorphous inner self or its shaping is hindered or promoted. From a scientific perspective, both processes are of immense significance. Of course, this is also true from a psychotherapeutic perspective.

(c) Thus, the amorphous inner self is not only relevant in exceptional cases—as we have shown—Boesch's examples of masks and carnival are extreme cases that undoubtedly have their validity. Much more important, however, is the everyday amorphous inner self and its shaping in small but meaningful interactions and dialogues with friends, colleagues, and family. In this context, the amorphous inner self is understood as a source of energy for the reupdate of the self-concept, and its importance is unfolded not only for psychopathology but also for personality development. At the end of the day, a person must grow not only through their challenges but also through the significance of their inner foreignness. For this to happen, however, they need dialogical offers, as we have outlined in this chapter.

References

Boesch, E. E. (1980). *Kultur und Handlung: Einführung in die Kulturpsychologie*. Huber.

Boesch, E. E. (1991). *Symbolic action theory and cultural psychology*. Springer.

Boesch, E. E. (1998). *Sehnsucht: Von der Suche nach Glück und Sinn [Longing: On the search of joy and meaning]* (1st ed.). Huber.

Boesch, E. E. (2000). *Das lauernde Chaos. Mythen und Fiktionen im Alltag*. Huber.

Boesch, E. E. (2005). *Von Kunst bis Terror: Über den Zwiespalt in der Kultur [From art to terror: On the cleavage of culture]*. Vandenhoeck & Ruprecht.

Boesch, E. E. (2021). *Musik, Sprache und die Sehnsucht nach dem Paradies*. Psychosozial-Verlag.

Evans, M., & Barker, M. (2010). How do you see me? Coming out in counselling. *British Journal of Guidance & Counselling, 38*(4), 375–391. https://doi.org/10.1080/03069885.2010.503698

Fircks, E. F. v. (2023). The fully objectified catathymic imagery method (FOCI): Assessing the amorphian inner at work. *Human Arenas*, 1–18. https://doi.org/10.1007/s42087-023-00345-7

Guenther, L. P. F. (2023). Allegory analysis from M. Theresa - A case study. *Integrative Psychological and Behavioral Science, 57*(4), 1183–1197. https://doi.org/10.1007/s12124-023-09750-7

Guenther, L. P. F. (2024). The concept of hysteria as mirror of the relation between clinical and cultural psychology. *Culture & Psychology, 30*(3), 538–550. https://doi.org/10.1177/1354067X231185723

Hanley-Hackenbruck, P. (1988). Psychotherapy and the "coming out" process. *Journal of Gay & Lesbian Psychotherapy, 1*(1), 21–40. https://doi.org/10.1300/J236v01n01_05

Heiss, R. (1936). *Die Lehre vom Charakter*. Walter de Gruyter.

Kahn, M. J. (1991). Factors affecting the coming out process for lesbians. *Journal of Homosexuality, 21*(3), 47–70. https://doi.org/10.1300/J082v21n03_03

Rogers, C. (1995). *On becoming a person*. Carl Ransom.

Stern, W. (1923). *Person und Sache: Die Menschliche Persönlichkeit* (3rd ed.). J.A. Barth.

Stern, W. (1924). *Person und Sache: System des Kritischen Personalismus (Wertphilosophie)*. J.A. Barth.

Stern, W. (1935). *Allgemeine Psychologie auf personalistischer Grundlage*. Martinus Nijhoff.

Chapter 8
The Quest for Meaning: Exploring Action-Theoretical Life Philosophy

Farewell *(E.E. Boesch, Enno von Fircks translation)*
Dark green and brown in quiet thought,
The heath lies, wrapped and caught
In the gold of strong suns, richly wrought.
A plowman leads his horse by the rein,
In the distance, on the hill's domain,
The mill turns its sluggish vanes.
Blackberries edge the wide expanse,
Black and red along the road, a chance
Encounter—gypsies in their camp's dance.
Beyond the last hill's distant crest,
The sea still gleams—oh, heart, distressed,
Can farewell bring you joy and rest?

Boesch's Nonphilosophical Background

Boesch was not a philosopher. On the contrary, his works contain little in the way of philosophical foundational thinking. This may stem from two main reasons. First, Boesch aligned himself more with early psychologists such as Pierre Janet, Eduard Claparède, and Jean Piaget, who were also less philosophically inclined. These figures shared a commitment to establishing psychology as a foundation that avoided metaphysical speculation, instead using testable concepts and methods to understand and potentially heal the human mind. This must be seen in the context of psychology's emergence as an independent scientific discipline (Danziger, 1994; Valsiner, 2012), where psychology's autonomy was not to be jeopardized by overly fixating on philosophical thought experiments.

Second, Boesch was a sensitive practitioner. In addition to his early work as a school psychologist in Switzerland, he was involved in UNESCO development work in Thailand (Boesch, 1971), which inspired central aspects of his

E. Freiherr von Fircks, *Culture of Action*, Theory and History in the Human and Social Sciences, https://doi.org/10.1007/978-3-031-88808-3_8

action-theoretical cultural psychology. He linked everyday observations with the personal experiences of individuals, thereby developing an original understanding of the motivations underlying experience and conduct. Moreover, Boesch was also a psychoanalyst (Boesch, 1977/2006). While he did not practice classical psychotherapy, he held group seminars on free association and its symbolism. Boesch approached psychoanalysis not in a strictly biological sense—rooted in Freud's drive theory (Freud, 1923)—but rather through the broader lens of his *fantasm* theory (see Chap. 1).

This distinction is crucial, as we cannot argue that Boesch advocated for grounding psychology on a strictly biological-physical basis—a separate pathway for psychology's independence with historical roots (Funke, 2023). Boesch derived his central ideas for cultural psychology from his work as an everyday scientist and practitioner—a third foundation for establishing psychology (von Fircks, 2024a).

Yet despite Boesch's lack of a philosophical orientation, his work holds significance for philosophically oriented psychology. This chapter focuses on this potential. It situates Boesch's contributions within a long-overlooked tradition of psychology: the psychological *Lebensphilosophie* (philosophy of life). This philosophy of life has significant predecessors and interesting findings about how a successful life can unfold under specific conditions (Klages, 1910, 1926). Connections to characterology must also be noted here (Heiss, 1936), which had a strong philosophical orientation. Ludwig Klages' metaphysics (1910) laid out clear principles for living a cosmically connected life without losing one's ties to nature, the body, and the dynamic flow of life. This foundation also catalyzed the development of ecologically informed psychology (Klages, 1956).

We will use this tradition as a model to analyze Boesch's action-theoretical cultural psychology in light of *Lebensphilosophie* considerations. Ultimately, this approach may yield flexible guidelines for dynamically understood personality development—a hallmark of classical *Lebensphilosophie*.

The Mythical Person

A central feature of Boesch's theory is the interplay between fantasms and myths (see Chap. 1). While fantasms mediate symbolic personal should-values, myths convey social should-values (Boesch, 1991). Social should-values are fundamental for maintaining a particular social structure, that is, the socially meaningful field of action. An example of socially meaningful normative values is restrictive sexual norms. In many parts of the world, where no sexual revolution has occurred, inhibitive norms still constrain individuals' sexual lives. Premarital sex, extramarital relationships, or same-sex relationships are often socially and legally sanctioned in such contexts.

A brief aside is necessary: laws are simply institutionalized should-values. Sharia law serves as a prime example here—a socially overdetermined code of conduct for Muslims that holds legal authority in non-secular states. This illustrates how a worldview shapes institutions and regulates public life. Myths thus act as socially overdetermined regulators of conduct (Boesch, 1991). Many fateful life trajectories result from individuals failing to meet or deliberately disregarding these normative values. For instance, sexual relations driven by passion with a close acquaintance often led to early marriages without intentional commitments of love. Similarly, same-sex love is often met with violence, imprisonment, or even death, leaving the surviving partners socially ostracized.

In aristocratic contexts, marrying someone of unsuitable social standing was historically taboo, often resulting in exclusion from noble society and social isolation (von Fircks, 2025). The rationale is always that maintaining the rules and values of the field of action is necessary to preserve its meaning for the individuals involved.

People who uncritically accept such social should-values without questioning their alignment with personal beliefs can be described as conformist individuals. For the conformist or mythical person, it is always about adapting to the socially relevant normative values without internal or external resistance. This adaptation occurs without reconciling social values with personal ones.

From a psychological perspective, however, caution is warranted: individuals may quietly rebel or feign adaptation while privately resisting, expressing dissent, or addressing the issue continuously. Resistance can take an external form, such as sabotage or boycotting, or remain internal. This internal resistance was vital for certain writers under Hitler's Third Reich, such as Erich Kästner (Hanuschek, 2016). Historians term this phenomenon "inner emigration" (Scholdt, 2011). Externally, a conformist person may appear indistinguishable from a resistant individual. Internally, however, one may experience great anguish over socially relevant normative values, while the other embraces them with optimism. Only actions combined with their motivations can clarify the person's true characterological nature. Consider the example of Theodor Haecker, a German *Kulturphilosoph* who belonged to the group of inner emigrants during World War II. During the final years of the barbaric Nazi regime, Haecker wrote critical essays reflecting on life under Nazi rule, which he published after the regime was dismantled by the Allies. In 1935, Haecker adopted a different strategy: he openly criticized and refused to support Hitler and his followers. However, as the Nazi regime grew more oppressive, Haecker became quieter in public and instead focused on writing secret essays and diaries between 1939 and 1945. For most parts of the day, his conduct was not notably different from that of a person living under the constraints of mythic conformity—except for the clandestine essays and diaries.

With Boesch we can label the conformist individual as the mythical person. The mythical person adapts to shifting social normative values and integrates them into their worldview without passion or inner fire. These values are introjected without personal engagement. Such individuals live formulaic lives, jumping from one trend to another without independent convictions. The Haecker example shows the opposite of a mythical person.

From a Kantian perspective—education toward autonomy—the mythical person does not represent an ideal life path. By suppressing or excluding personal convictions to consolidate socially relevant normative values, the mythical person relinquishes their *fantasms*. Thus, the mythical person does not represent a teleologically meaningful personality development; instead, it serves as a negative example of how life should not be lived and as a cautionary tale for individual development.[1]

However, the mythical person is a societal exception and was historically overestimated in frequency, such as during National Socialism. We will revisit this point later.

The Fantasmatic Person

We now aim to shed light on the opposite personality type through Boesch's action-theoretical cultural psychology. This involves examining private should-values mediated through fantasms. In contrast to the previous terminology, I would like to designate this type as a *fantasmatic person*. What characterizes the fantasmatic person? Unlike the mythic person, the fantasmatic person lives predominantly from within, with their psychological and physical functions often revolving around themselves. Frequently, this inward-directed way of living occurs with partial or total disregard for socially relevant should-values. The fantasmatic person is solely driven by their internal passions and forces. Every opportunity is seized by this individual to express themselves in the world and leave their personal mark on it. Outsiders might label this type of conduct as egotism—but this would be a misclassification.

The fantasmatic person experiences relatively strong fantasms, which mediate private normative values, and they are often overwhelmed by the power of these fantasms. In short: the fantasmatic person lives solely for the sake of their fantasms and experiences a continuous urge to reshape the world according to their own desires. For social coexistence, such individuals are difficult to accommodate and frequently trigger interpersonal conflicts.

Example: A fantasmatic person is called by their family to dinner. During the meal, they speak only about themselves, hardly listen to others, and respond little to objections. They are primarily focused on expressing their own thoughts. Once the fantasmatic person has finished eating and communicated everything they intended to, they leave to continue their pursuits. They will not remain at the table until others have finished eating, nor will they participate in the conversation. They will show little to no interest in the lives of others. Furthermore, they will not assist in clearing the table, as this would cost them time that could otherwise be devoted to their own concerns. In summary, the fantasmatic person lives in their own world, caring little about how their personal world intersects with those of others or how such intersections could lead to meaningful connections. The fantasmatic person exists in a hermetically sealed world of their own making.

---

[1] The neo-Kantian philosopher and psychologist Jonas Cohn described such phenomenon as the bounded personality (Cohn, 1914).

For their social environment, such a personality presents significant challenges, as their approach not only fails to promote shared living but actively hinders it.

Historical Examples

Historically, there are fascinating cases of fantasmatic individuals who exemplify this extreme type. Napoleon Bonaparte, for instance, can be seen as such a figure. Through his imperialistic fantasms, he sought to leave his mark on Europe. Napoleon's aim was less about achieving dominance through alliances or coalitions (i.e., aligning fantasms with myths) and more about enforcing his visions of France and Europe through violence and aggression. His Code Civil has still some remnants in European law.

Similarly, J. Robert Oppenheimer (Thorpe, 2019) displayed traits of a fantasmatic individual in his younger years, using science as a means to realize his independent visions and compelling fantasms. Perhaps the most striking example of this type is the solitary scientist, living both from within and for themselves, who locks themselves away for days to solve an equation, formulate a theory, or advance an experiment. Such a fantasmatic person may entirely neglect other needs, such as physical well-being or the pressing social normative values of family and friendship.

Another example is Franz Kafka. This prominent writer twice broke off his engagements for deeply personal, fantasmatic reasons. In his *Letter to His Father* (Kafka, 1995) and his diaries (Kafka, 1948/2016), Kafka portrays marriage and its mythic obligations as potential barriers to the realization of his central fantasms (his writing). The potential threat or limitation posed to his fantasms by marital and familial duties was sufficient reason for Kafka to end his engagements, despite the positive prospects of the marriages. Kafka's entire body of work is imbued with the tension between the external world, with its impenetrable rules (*The Trial*, *The Castle*), and the inner self, which is constrained and rendered absurd by these externalities. Kafka described himself as a broken character (Kafka, 1995), as he was unable to reconcile private normative values with social ones and, due to his health, often had to prioritize his own needs over those of others. We want to provide the reader with a lengthy quote in order to illustrate the nature of the fantasmatic person; for that, we rely on Kafka, again.

> Warum also habe ich nicht geheiratet? Es gab einzelne Hindernisse, wie überall, aber im Nehmen solcher Hindernisse besteht ja das Leben. Das wesentliche, vom einzelnen Fall leider unabhängige Hindernis war aber, dass ich offenbar geistig unfähig bin zu heiraten. Das äußert sich darin, dass ich von dem Augenblick an, in dem ich mich entschließe, zu heiraten, nicht mehr schlafen kann, der Kopf glüht bei Tag und Nacht, es ist kein Leben mehr, ich schwanke verzweifelt herum. (…) Es ist der allgemeine Druck der Angst, der Schwäche, der Selbstmissachtung. (…) Die Heirat ist gewiss die Bürgschaft für die schärfste Selbstbefreiung und Unabhängigkeit. Ich hätte eine Familie, das Höchste, was man meiner Meinung nach erreichen kann, also auch das Höchste, das Du erreicht hast, ich wäre Dir ebenbürtig, alle alte und ewig neue Schande und Tyrannei wäre bloß noch Geschichte. (…) Wenn ich [aber] in dem besonderen Unglücksverhältnis, in welchem ich zu Dir stehe,

selbstständig werden will, muss ich etwas tun, was möglichst gar keine Beziehung zu Dir hat; das Heiraten ist zwar das Größte und gibt die ehrenvollste Selbstständigkeit, aber es ist auch gleichzeitig in engster Beziehung zu dir. (Kafka, 1995, p. 53)

Why, then, have I not married? There were individual obstacles, as there are everywhere, but overcoming such obstacles is what life consists of. The essential obstacle, however, one that unfortunately is independent of any particular case, was that I am apparently mentally incapable of marrying. This manifests itself in the fact that from the moment I decide to marry, I can no longer sleep, my head burns day and night, it is no longer a life, and I wander about in despair. (…) It is the general pressure of fear, weakness, and self-contempt. (…) Marriage is certainly the guarantee of the most profound self-liberation and independence. I would have a family, the highest achievement in my opinion, and thus the highest you have achieved as well; I would be your equal. All the old and eternally renewed shame and tyranny would become mere history. (…) [But] in the particular unhappy relationship I have with you, if I want to become independent, I must do something that has as little connection to you as possible; marriage is indeed the greatest and grants the most honorable independence, but at the same time, it is in the closest relation to you. (Kafka, 1995, p. 53, author's translation)

The Limits of the Fantasmatic Person

The fantasmatic person does not represent a teleologically significant form of personality development. Living solely from within and for oneself, as Kafka's case illustrates, often leads to isolation. Furthermore, a fantasmatic person cannot update or develop their own fantasms when they completely withdraw from the social world and fail to harmonize their personal needs with societal demands. For instance, Kafka's writing might have undergone intriguing developments had he married Felice Bauer, as this marriage could have updated his system of fantasms. In his *Letter to His Father* (Kafka, 1995), he even acknowledges that marriage might have unlocked a certain strength and energy within him. However, the risk of his fantasms being constrained rather than enriched was ultimately too great, and he chose not to marry Felice Bauer.

The fantasmatic person thus represents an extreme case, serving as a cautionary example for personality development. With this understanding, we have laid the groundwork for examining the concept of the *complete person*, who continuously harmonizes their fantasms and myths.

The Yet-to-Be-Completed Person

A central aspect of action-theoretical cultural psychology is the integration of fantasms and myths (Boesch, 1980, 1991, 1998, 2000). This integration of social and private should-values represents a lifelong task and is thus fundamentally important for our philosophical reflections on life, based on Boesch's cultural psychology. It must always be taken into account that this balance is, at best, a dynamic equilibrium. The world is ever-changing. This means that we are constantly confronted with new biological and value-laden entities in our environment (e.g., natural disasters in the biological domain; new value creations, such as tools or narratives, in the value domain) (von Fircks, 2024d). New individuals entering our lives also count as value-laden objects that penetrate our field of action and necessitate a realignment of our own system of fantasms. The harmonization of a person's fantasms and myths must, therefore, be perpetually renewed and should never be viewed as something permanently achieved. This is the reason why we characterize this type of person as the "yet-to-be-completed person." Completion must be repeatedly pursued because a person does not stagnate. A person is and remains in a state of development, just as the biological and social world continues to evolve.[2]

To illustrate, let us consider the example of professional football players, who, during their careers, often experience a pronounced sense of equilibrium. Through their action-theoretical framework, they are able to significantly shape their social field (the football action-space) and are, in turn, shaped by the key gatekeepers of this field (teammates, coaches, etc.). Players experience immense gratitude and a richness of experiences that enrich them as individuals and enable them to reciprocally enrich their field of action. However, what happens when a football career comes to an end (e.g., due to age, injury, etc.)? Here, we are faced with an example of a person becoming alienated from their field of action. They can no longer exert influence on their cherished world through their action-theoretical framework. A support system in their life collapses, along with sources of recognition, security, and love.

In this scenario, the person finds themselves disconnected from their action framework, unsure of where to apply it or in which new field of action they might find resonance. Unsurprisingly, some former professional athletes report experiencing a profound emptiness, depression, and anxiety (Hagemann et al., 2007), with some even struggling with psychopathologies after their careers (e.g., Gerd Müller). Conversely, there are football players who successfully adapt their action framework to new fields of action, refining it and applying it in other contexts. One prominent example is that of former players transitioning into coaching roles, where they apply their action framework from a coaching perspective and learn from and with the players, they mentor. Notable examples include Jürgen Klinsmann, Xavi (Barcelona), and Miroslav Klose.

Another example is Lukas Podolski, who, after retiring from professional football, became an entrepreneur and restaurateur. For several years now, the former

---

[2] A cross reference by William Stern may be allowed (Stern, 1923, 1924, 1935). The personal world of the person needs to be gained and regained within the social tension the person is confronted with. Hence, the social and material world borders the individual meaning making of the individual person. By times, these borders can be relocated or shifted by means of personal efforts and by means of negotiation (von Fircks, 2024a, 2024b, 2024c). However, not only the individual meaning making patterns evolve but also the social meaning making patterns develop over a time span—relocating the borders in which the individual may find meaningfulness for their life.

German national player has successfully operated a kebab chain. Certain qualities that were crucial to him during his football career (e.g., conscientiousness, discipline, work ethic, and determination) have also been applied to his new field of action, enriching it through his personality while simultaneously enriching himself through this new field (socially = new competencies, and economically = monetary gains). In contrast, consider the case of Adriano from Brazil, who, after his football career, returned to his favela and struggled with alcohol problems. These three examples vividly illustrate our everyday psychological foundation for life-philosophical reflections based on Boesch's cultural psychology. Moreover, they demonstrate that the reconciliation of fantasms and myths is a continuous, never-ending process that must be repeatedly pursued to bring the person and the world into equilibrium. This process always involves effort and consumes certain resources of the individual. This dynamic equilibrium is not something that falls into a person's lap—it must be actively worked on. The person must assume responsibility for their dynamic equilibrium.

Transitions from Working Age to Retirement

This perspective is not limited to football examples but can also be applied to the developmentally significant transition from working age to retirement. Let us examine these processes more closely. A person's action framework is primarily shaped by their field of work, where they can significantly impact the working world through their actions. Of course, this only occurs under the condition that the work allows the person to further develop their action framework, that is, to refine their motivations and abilities while simultaneously providing opportunities for the person to exert a meaningful impact on the working world.

When a person enters retirement, a significant field of action disappears, along with the strengthening fantasmatic and mythical components of their experiences and conduct. Unsurprisingly, many individuals respond to this transition phase with psychopathologies (primarily depression) and display a disruption of equilibrium (Callaghan et al., 2013). In recent years, partial retirement models have been introduced for such cases, allowing individuals to test retirement by working part-time during the final 2 years of their employment (Wanger, 2010).

During this period, it is crucial for individuals to discover new fields of action, test their action framework (action motivation and action potential) within these fields, refine it, adapt it to the new action domain, and assess whether the value creations achieved through these updated actions are meaningful. Many people entering retirement, for example, engage in volunteer work or take on part-time jobs. Others pursue university studies in fields they have always been interested in. Still, others dedicate themselves to new hobbies (e.g., gardening, painting, writing) and seek to apply and further develop their already established action framework within these areas.

As scholars, we must avoid claiming that individuals entirely devoid of prior experience dedicate themselves to a new field of action as if beginning a new life. This is not about a rebirth but rather about updating the action framework. A businessman who, before retirement, had some interaction with art dealers, auctioneers, and artists (e.g., as an investment advisor) will likely draw on these contacts to facilitate the reorientation of his action framework. A painter who wishes to engage with science once again might ask one of his clients (a psychologist) about their studies and profession and subsequently attend some psychology courses as an auditor.

An illustrative example can be found in the life and work of Hermann Hesse, who, in later life, turned to gardening and botany. He was able to express his artistic, lyrical, and creative action framework in this field and, in turn, received decisive inspiration for his magnum opus *The Glass Bead Game* from this gardening activity. Hesse himself reports that the idea for the novel came to him while gardening. Hesse's biography is considered particularly significant in terms of lived completion (Ball, 2006).

The yet-to-be-completed person thus represents an essential criterion for teleologically meaningful personality development and personality actualization. It is crucial that this person continuously works through pluralistic experiences (i.e., experiences in various fields of action).

A Negative Example of Fantasms and Myths Integration

At this point, we must also discuss a negative example that involves the integration of myths and fantasms, which played a significant role in the political power of National Socialism. National Socialism was a polyvalent worldview. In other words, it presented many open myths (social normative values), which individuals could use for orientation but also into which they could integrate their fantasms. The political credo of National Socialism, "from the cradle to the grave," meant that the Nazis claimed to accompany the individual throughout their entire life. The leaders of National Socialism aimed not only to provide the individual with orientation but also to control essential aspects of private life. It is no surprise that many organizations with National Socialist ideology were founded (e.g., *Kraft durch Freude*, Hitler Youth, etc.) in order to carry out this endeavor sustainably. Also, numerous professional associations (such as the National Socialist Pharmacists' Association, the National Socialist Medical Association, etc.) must be mentioned in this context, showing to what extent the NS ideology sought to cover all areas of life (Haffner, 1978).

Returning to the myths, if we examine the historical context, the NS ideology, as an overarching myth, provided individuals with ample space to relax their fantasms. It was especially the threatening fantasms that the NS mythology addressed, thus providing individuals with an ideological foundation to express their personal needs and their associated symbolism. The high unemployment rate, inflation, the wounded pride caused by the 1918 Treaty of Versailles, and political instability due to the numerous governments that arose in the Weimar Republic, all show that individuals were confronted with many threats (both subtle and overt) before Hitler's rise to power, and these needed to be meaningfully framed. Take, for instance, an individual who lost their job in the 1929 stock market crash, is insecure about the political situation, and is uncertain about the future of themselves and their family. This person is confronted with various but connected threatening fantasms (e.g., the "failure" fantasy, the "uncertainty" fantasy, etc.). The comprehensive NS ideology allowed this individual to imbue these threatening fantasms with meaning, thereby relaxing them. This means that, psychologically, the individual could fend off the threatening fantasms by adopting the Nazi ideology, which framed Jews and Communists as scapegoats, thus absolving the person of responsibility for themselves and their own system of fantasms.

In the next step, political dissidents, Jews, and homosexuals were expelled from their jobs or socially ostracized, and new positions were created through numerous Nazi organizations. This shows that not only could the threat be psychologically averted (through an appropriate interpretation), but also that the individual's fantasms of security could be addressed and realized. The widespread support for the NS ideology might be rooted in these two reasons: first, an internal action was updated by reinterpreting the threatening fantasms through the NS mythology, and second, through external actions (new jobs, new roles in private life, political engagement), which in turn could relax the fantasms of security (an internal action).

The entire barbarity of National Socialism ultimately reveals that the integration of myths and fantasms can be abused. From a life-philosophical perspective, this example teaches us that the integration of myths and fantasms must not destroy other fields of action. A way must be found to fend off symbolic threats and approach security without destroying or fundamentally restricting the action fields of other individuals or groups. The NS ideology allowed for the reconciliation of myths and fantasms, but this reconciliation of private and social normative values came at the cost of murder and destruction. History should, in this regard, always serve as a cautionary tale for us as cultural psychologists.

Transcendence: The Ultimate Reconciliation with God

We live in a godless world (May, 1953, 1991). The general belief in deities scarcely present, and often we can only speak of the cultural roots of religions, which occasionally manifest in people's lives. Yet, the human person yearns for God or dialogue with deities[3]—yearns for spirituality (Picione & Marsico, 2024; Zeigan,

[3] From now on, God and deity are used as synonyms. By that we wish to utter the need to be less Eurocentric when speaking solely of God as an Abrahamitic Father. I prefer the term God as the term comes from the indogermanic language tree denoting something like to call somebody (ghuto). Be that as it may, both terms (deities and God) are used with the same credibility.

2025). Without a larger context of meaning, the person cannot exist (May, 1983). The individual requires a connection to a higher power. Even in a godless world, the person cries out for God—however this God may be conceived. This is evident, for example, in the boom of alternative spiritual offerings such as mindfulness, neo-Buddhism, silent retreats, and more.

Let us examine the cry for God more closely. Every religion comes with a specific system of normative values. The Torah or the Bible are extreme examples (the Ten Commandments) of a long-traditioned mythical system designed to provide the individual with order and security. The extent to which belief impacts daily life is still evident today, for example, when businesses are closed on Sundays in some countries. In Jewish faith, for example, the Sabbath includes many rules that the individual must follow to show respect and reverence to the deity. Belief always carries a fixed, social system of normative values, which is conveyed through various institutions and gatekeepers. Rituals are frequently used to ensure that these normative values are continually adhered to. In rituals, the normative values are consistently reproduced, thereby mediating a certain sense of meaning. We all remember our childhood questions about why we should pray, why we should follow the commandments, and what belief actually brings. The answers to these questions are derived through specific inner and outer actions, and ritual plays a major role in aligning inner and outer actions. For instance, I pray to God (outer action) to ask for forgiveness for my sins (inner action).

Jesus' message, "Love your neighbor," which was meant to replace the Ten Commandments, serves as an illustrative example of the integration of inner and outer action. In this, the individual is to adhere to a humanized God, find a role model, and not only internalize the declaration, but also act in accordance with it. God is always an impetus for external action. This is a fundamental characteristic of belief, meaning that the mediation of meaning in rituals is not the only decisive factor. On the contrary, the idea that God is always an impetus means that God is, a priori, always approachable. Regardless of the individual's life circumstances, location, or level of engagement, the dialogue with God is always possible—even if only for a few seconds. Thus, God is always approachable. We humans may avoid the dialogue, turn away, and deny it, but the dialogue with God is inescapable. God is always present, and because He is always present, He remains accessible in every situation. This approachability allows the human psyche to always remain connected, never completely isolated. God is the first and last source of security—through His a priori accessibility.

However, the role of God extends beyond this a priori accessibility: "Love your neighbor" comes with polyvalent connotations. This social normative value is an open value that the individual can fulfill through a variety of actions. The open nature of the normative value itself allows the person to choose their own actions to satisfy it. The person is thus free to adopt and shape this divine should-value according to their own personality. In this private shaping of the social normative value, the individual is again in dialogue with God. For example, "Love your neighbor" may be fulfilled by the individual through an open attitude toward their neighbors, by caring for elderly parents, by offering a loving hug to someone in need or distress, etc. According to their personality, the individual may and should interpret the normative values in accordance with their own action-theoretical framework. Thus, they not only harmonize private and social values, but also embed their inner and outer actions into a larger context of meaning. The individual is therefore not only in a suspended state of balance that appears particularly graceful, but in an ultimate balance, as they experience themselves as part of a greater whole and actively contribute to it. The common dialogue—and also the dialogue with God—often culminates in fruitful cooperation, resulting in a joint work (such as a decision, for example), for which both parties are responsible. From a life-philosophical perspective, the dialogue with God is recommended, as it always results in shared fruits that shape and enrich the further course of one's potential for action and motivation for action.

We can also turn to a Taoist example, which has a similar effect as a Christian one. The Taoist seeks to connect with the cosmic force and allow themselves to be carried by it through life. In doing so, they accept setbacks and perceive them as part of a larger whole (suffering is as much a part of life as joy). The Taoist repeatedly connects with the Tao in more obvious or subtle ways and attempts to cultivate this connection, even if only temporarily, by opening themselves to it and letting the Tao flow through them. In one form or another, the Taoist also communicates and enters into dialogue with the universe and the world in all its transformations. Practices like Tai Chi or Qi Gong are interesting dialogical gestures aimed at aligning the body and psyche with the rhythms of the world (von Fircks, 2025). Furthermore, the most dialogical form of Taoism—the oracle of the I Ching—is another such method for communicating with the world in its flux, integrating this dialogue into oneself, and updating it whenever necessary.

This example illustrates that dialogue does not require a personified deity—although, of course, dialogue with a person is easier than with an ever-present force that sustains and unfolds life.

We may live in a godless world, but we should take our cry for God—often reduced to a mere whisper—seriously and stand by it. God enriches our field of action. Religion, therefore, is not an opiate for the masses, but a catalyst for dialogical thinking par excellence.

With that, we have reached the end of our life-philosophical chapter of action-theoretical cultural psychology. We now aim to summarize the chapter and outline guidelines for action based on action-theoretical cultural psychology. These guidelines should not be considered exhaustive, but rather as an impetus for dialogue about how one's life can continually succeed. The transformative power of numinous meaning is visible in Boesch's poem accompanying this chapter—and is especially vivid in the first stanza:

> Dark green and brown in quiet thought
> The heath lies, wrapped and caught
> In the gold of strong suns, richly wrought.

Philosophical Foundations of Action-Theoretical Cultural Psychology

I. Characterologically, we have identified three different types of life philosophies of the individual. These are the mythical, the fantasmatical, and the person striving for completion. All three life philosophies involve divergences in both inner and outer actions, fundamentally shaping the life path. The mythical person merely internalizes the social normative values, suppressing their own private normative values. In doing so, they completely conform to the field of action, forgoing enriching it with their personal contributions. The internalization of social normative values happens almost indifferently, as they are introjected as important but not because of internal involvement. In this process, the person becomes naive, a template of a passive worldview, devoid of an internal drive. From a life-philosophical perspective, we can say that the person must not become a mere template of the social normative value system of a worldview. On the contrary, they should actively and personally engage with this system of social should-values.

II. However, the second type—the fantasmatical person—must also be approached with caution. The fantasmatical person isolates themselves from any social normative value system and is solely concerned with the relaxation and realization of their own personal fantasm system. In doing so, the individual becomes socially incompatible, as they see no reason to adapt to or engage with established orders. The fantasmatical person thus lives in their own world, isolated from social normative values, which they view only as inhibitors of their personal values and therefore as threats. The resources offered by a social normative value system are completely underestimated here and dismissed as irrelevant. Furthermore, the potential personal growth that can only happen through dialogue is artificially stagnated. Thus, the fantasmatical person exists solely for their own sake, viewing themselves exclusively as an agent without recognizing or acknowledging the significance of social normative values. From a life-philosophical standpoint, we must state that the individual should not rigidly fixate on their own personal fantasm system and private normative values. On the contrary, the person must remain permeable to social should-values.

III. This leads us to the third type in our action-theoretical characterology and life philosophy, the person striving for completion. Since the individual is always in tension between private and social should-values, they must continually resolve this tension. This means they are obliged to create a balance between themselves and their social environment, enriching their personal identity through social normative values while simultaneously contributing to the world as it exists and progresses, thereby enriching or expanding it through their personal actions. This remains a continuous process and is never truly completed. As long as the person lives, they must learn how the world can expand their personal identity and when they need to develop accordingly, as well as how the world can benefit from the individual, allowing them to have a role in the creation of culture. Accordingly, the person is always obligated to strive for this suspended equilibrium. The dialogue with God, who remains always a priori approachable, can help the person update their inner lines of action and thereby diversify their outer actions. The dialogue with God allows the individual not only to form a connection with another person (Stern, 1924), but also to embed inner and outer actions within a larger, holistic context that implies a higher meaning (Zeigan, 2025). The dialogue with God always leads to a divine, meaningful outcome. A successful life can only be aspired to when a balance between the person and the world is achieved, as well as a balance between the person and God. Both the surrounding world and the divine world do not limit but enrich; this is a constructive insight. The dialogue with Boesch has led us to this point, even though Boesch himself did not identify as a philosopher.

References

Ball, H. (2006). *Hermann Hesse: Sein Leben und sein Werk*. Wallstein Verlag.

Boesch, E. E. (1971). *Zwischen zwei Wirklichkeiten: Prolegomena zu einer ökologischen Psychologie*. Huber.

Boesch, E. E. (1980). *Kultur und Handlung: Einführung in die Kulturpsychologie*. Huber.

Boesch, E. E. (1991). *Symbolic action theory and cultural psychology*. Springer.

Boesch, E. E. (1998). *Sehnsucht: Von der Suche nach Glück und Sinn [longing: On the search of joy and meaning]* (1st ed.). Huber.

Boesch, E. E. (2000). *Das lauernde Chaos. Mythen und Fiktionen im Alltag*. Huber.

Boesch, E. E. (2006). *Konnotationsanalyse: zur Verwendung der freien Ideen-Assoziation in Diagnostik und Therapie*. Privater Nachdruck. (Original work published in 1977).

Callaghan, B. L., Graham, B. M., Li, S., & Richardson, R. (2013). From resilience to vulnerability: Mechanistic insights into the effects of stress on transitions in critical period plasticity. *Frontiers in Psychiatry, 4*, 90. https://doi.org/10.3389/fpsyt.2013.00090

Cohn, J. (1914). *Sinn der gegenwärtigen Kultur [the meaning of contemporary culture]*. Meiner.

Danziger, K. (1994). *Constructing the subject: Historical origins of psychological research*. Cambridge University Press.

von Fircks, E. (2024a). Advocating a valid cultural Alltagspsychologie. *Trends in Psychology, 32*(1), 231–250. https://doi.org/10.1007/s43076-022-00231-1

Fircks, E. (2024b). *Learning with William Stern: Personology for future*. IAP.

Fircks, E. (2024c). Making sense of signs: Readjusting William Stern's personological value theory. *Theory & Psychology*. https://doi.org/10.1177/09593543241279264

von Fircks, E. (2024d). Raising questions with Lamiell about future research directions of critical Personalism: A book review essay. *Integrative Psychological and Behavioral Science*, 1–11. https://doi.org/10.1007/s12124-024-09825-z

Fircks, E. (2025). *Baltic-German lifeworlds. Psychography in social context*. IAP.

Fircks, E. von (2025). Understanding the power of mindfulness from a cultural-psychological perspective. *Culture & Psychology, 31*(1), 146–165. https://doi.org/10.1177/1354067X231201387

Freud, S. (1923). *Das Ich und das Es*. Psychoanalytischer Verlag.

Funke, J. (2023). Wundts Irrtum oder: Was ist die bessere Vorbild-Disziplin, Physik oder Biologie? In G. Jütteman (Ed.), *Wie der Mensch sich selbst entdeckte* (pp. 215–226). Psychosozial-Verlag.

Haffner, S. (1978). *Anmerkungen zu Hitler*. Kindler.

Hagemann, N., Tietjens, M., & Strauß, B. (2007). *Psychologie der sportlichen Höchstleistung: Grundlagen und Anwendungen der Expertiseforschung im Sport*. Hogrefe Verlag GmbH & Company KG.
Hanuschek, S. (2016). *Erich Kästner*. Rowohlt Verlag GmbH.
Heiss, R. (1936). *Die Lehre vom Charakter*. Walter de Gruyter.
Kafka. (1995). *Brief an den Vater*. Reclam.
Kafka, F. (2016). Tagebücher, 1910–1923/Franz Kafka. Books on demand. Original work published in 1948.
Klages, L. (1910). *Prinzipien der Charakterologie*. J.A. Barth.
Klages, L. (1926). Zur Ausdruckslehre und Charakterkunde. Gesammelte Abhandlungen. Kampmann.
Klages, L. (1956). *Mensch und Erde*. Kröner.
May, R. (1991). *The cry for myth*. WW Norton.
May, R. (1953). *Man's search for himself*. W. W. Norton
May, R. (1983). *The discovery of being*. WW. Norton & Co.
Picione, R. D. L., & Marsico, G. (2024). The religious nature of the psyche, semiotic mediation, and the evanescence of identity in Liminality. *Integrative Psychological and Behavioral Science, 58*(3), 836–844. https://doi.org/10.1007/s12124-024-09850-y
Scholdt, G. (2011). Innere Emigration und literarische Wertung. In Kanon, Wertung und Vermittlung (pp. 123–144). DE GRUYTER. https://doi.org/10.1515/9783110259964.123.
Stern, W. (1923). *Person und Sache: Die Menschliche Persönlichkeit* (3rd ed.). J.A. Barth.
Stern, W. (1924). *Person und Sache: System des Kritischen Personalismus (Wertphilosophie)*. J.A. Barth.
Stern, W. (1935). *Allgemeine Psychologie auf personalistischer Grundlage*. Martinus Nijhoff.
Thorpe, C. (2019). *Oppenheimer: The tragic intellect*. University of Chicago Press.
Valsiner, J. (2012). *A guided science: History of psychology in the mirror of its making*. Transaction Publisher.
Wanger, S. (2010). Die Altersteilzeit im Zusammenspiel individueller und betrieblicher Einflussfaktoren. *WSI-Mitteilungen, 63*(8), 395–403. https://doi.org/10.5771/0342-300x-2010-8-395
Zeigan, H. (2025). WIE DAS WORT ZUR SCHRIFT WURDE: Geschichten der Bibelentstehung. EVANGELISCHE VERLAGSANSTALT.

Chapter 9
The Power of Symbols: Navigating Existential Themes Through Poetic Dialogue

A Thank You in the Guestbook *(E.E. Boesch, Enno von Fircks translation)*
Nothing remains of the guest but a few delicate lines
On the patient field of a page.
Everything hurries—days and dreams rush by,
The bird flutters with the wind to distant skies,
And only within the cool breadth of this sheet
A thank-you lingers and smiles at you, every now and then.

Cultural psychology is often criticized for failing to provide sufficient applied implications for other sciences or everyday life. Critics argue that it is primarily a theoretical and philosophical discipline that pays little attention to practical matters. However, this superficial impression is somewhat distorted. In recent years, cultural psychological research has undergone a significant shift—moving away from theoretical minutiae and toward applied science. Noteworthy developments include organizational psychology grounded in cultural psychology (Bendassolli, 2017, 2024a, b), pedagogy with cultural psychological foundations (Xu & Tateo, 2022), and psychotherapeutic studies with a cultural psychological focus (von Fircks, 2023a, 2023b). It is striking that these applied studies and writings are heavily impacted by Valsiner's semiotic cultural psychology (Valsiner, 2014, 2019, 2021). However, Boesch's work also holds potential for applied utility (see Chaps. 5 and 6).

The present chapter aims to focus on an aspect of Boesch's cultural psychology that has been underrepresented in general Boesch scholarship—his writings on poetry (Boesch, 2021). Our goal is to provide the general reader with an overview of Boesch's interpretation of poetics, from which we will develop a flexible interpretative framework—a framework that as Boesch explains in the introductory poem stays with the reader and can uplift his general mood ("A thank-you lingers and smiles at you, every now and then"). This framework is intended to assist students and laypersons in interpreting poetry more effectively. Based on my own experience working with young people, I have observed that the interpretation of poetry often presents a challenging area in the education of adolescents and young adults. With the interpretative key

E. Freiherr von Fircks, *Culture of Action*, Theory and History in the Human and Social Sciences, https://doi.org/10.1007/978-3-031-88808-3_9

inspired by Boesch's work, we aim to mitigate these challenges and equip young adults and adolescents with an appealing tool to engage with and appreciate the deeper dimensions of poetics. Poetics must once again become a lived experience (von Fircks, 2022). Enough of the introduction: to bring Boesch's cultural psychology back into greater focus, these explanations are necessary.

The Primacy of Experience in Poetics

The primacy of experience is essential for interpreting a poetic text. It concerns the processing of experiences in a condensed, poetic form. Pfeiffer (1946) emphasizes this point poignantly, stating that poetry and density are closely intertwined. Thus, poetry always involves the condensed processing of experience. For Boesch (2021), this is also the core characteristic of poetry:

> For the poet, if they truly are one, rarely derives their wisdom from specialized expertise but rather from an inner sensitivity to what they experience—a sense, at the same time, for what conditions their experience. Their insight, if one may call it that, springs from the depths of personal experience. Therefore, the poet remains the most subtle medium, one that absorbs, processes, and even shapes the world. Yet this medium, unlike the instruments of science, is ever-changing—sometimes measuring this, sometimes that; at times with subtle clarity, at others impressionistically; sometimes with sharp precision, at others distortingly. (p. 153, author's translation)

This highlights the first characteristic of cultural-psychological poetics: the poet processes the world they experience. However, this processing occurs in written form, meaning it is arranged in a particular way. The structure the poet imposes on their thoughts constitutes a concrete design—a poetic shaping. As we established in the earlier chapters, experience does not occur against an empty backdrop. On the contrary, experiences are always related to a world that is already culturally pre-structured, a world imbued with meaning constructed by others.

Every poetic text reflects this reference to the world—both material and social. Without this world, with all its resistances but also its facilitative aspects, the poet could not create poetry. The poet requires this world not only as an adversary but also as a stimulus. Thus, when we speak of experience—or more precisely, the processing of experience—we must always speak of the experience *of something*. This "something," however, is not a static object; it is always subject to a certain dynamism. The world, in its material and social constitution, is continuously reshaped and restructured in its configuration of meaning according to the needs of individuals—other persons. Consequently, experience must always relate to a changing world.

Likewise, the poet is also subject to transformation. They will reconsider their needs, develop different longings and fears at various stages of life, and require different things as they progress through these phases. Both the poet—as an agent in the process of creating poetry—and the world, which serves as the backdrop for the poet's experience, are subject to the general dynamism of life. Let us, therefore, summarize the first interim conclusion as follows:

1. Poetry processes experience and shapes it into an artistic form. However, this processed experience does not occur in a vacuum but rather references the world with its cultural preconditions, including its material and social resources. These cultural preconditions—as well as the poet's needs—are subject to a certain dynamism, which can continuously alter the interaction between the poet and the object. For an interpretative framework, the following questions are therefore crucial:

    - How is the poet's experience characterized?
    - To what does this experience refer to?
    - How is this reference culturally preconditioned?

Existential Themes Between Content and Form

Poetry unfolds through a process (Pfeiffer, 1946). The poet—similar to the writer in general—struggles with their words, striving to find the right formulations. In this process, the poet writes down words, crosses them out, discards them, replaces them with others, and assesses whether a particular word alters the overall rhythm or tonality of the piece. As scholars, we often overlook this process-oriented nature of poetry. Yet, in this process, every word is crucial, because, as Boesch (2021) asserts, it is about uncovering "inner truths" (p. 154).

What precedes the act of writing a poem? As discussed earlier, it is experience. These experiences often revolve around specific thematic axes that appear archetypal in nature. Poets frequently engage with themes such as love, suffering, death, rebirth, loss, grief, beauty, destruction, war, peace, and similar motifs. What all these thematic nodes share is their existential grounding. These are the themes that individuals confront throughout their lives and to which they must develop a personal stance. However, this stance is not always clear to the individual. It often feels vague, and the person seeks clarity to these nebulous thoughts and feelings.

Before a poem is written, there is often only a vague image or feeling, which crystallizes further during the writing process. One line triggers the next; the subsequent line is connected back to the first; the rhythm is evaluated—does it feel too fast or too slow for the theme? Thus, poetry is not merely about addressing existential themes but about grappling with these themes, seeking to refine thoughts and emotions through the act of writing.

Rarely does someone sit down and write a poem fully formed. That is not the purpose of poetics or writing in general. The purpose of poetics is the artistic processing of a theme or experience, with the aim of transforming or clarifying the initial thoughts that prompted the act of writing in the first place.

> Poems, then, are inner truths—or, more precisely, they seek them. The poet's words aim to give form to the inner world they sense. Left unspoken, this inner world remains a vague shadow, shapeless like an evening mood or the rustling of wind in the forest. Only when it finds words does it seem to become true experience, yet, inevitably, it also becomes untrue. The words, steeped in inherited and external meanings, distort the intended essence. Hence

> the poem. Through the sounds of rhymes and the rhythms of verses, through images and associations, the poem strives to impose its own meaning on the words. It never fully succeeds. The poem can only approximate this inner world of experience—which is, at the same time, a self-experience. And thus, the search never ends; it may lead to further poems, perhaps to music, love affairs, travels, or other actions that externalize the inner in a way that corresponds to its essence. (Boesch, 2021, pp. 154–155, author's translation)

The search for answers to existential questions is thus the central characteristic of poetry. For this reason, poetry is always a *becoming image* (Pfeiffer, 1946)—an image that builds itself up gradually and can, unlike painting, be understood in a genetic sense. In painting, the observer is always confronted with the completed image, not with the individual elements that sequentially accumulate to form a whole in a genetic process. Here is an example from Mathias Claudius and his Advent poem:

Always One More Light
Always one more little light
In the wreath we have wound,
That it may shine so bright
Through the dark hours around.
Two, then three, then four!
What a glow around the wreath,
And so we too shine more,
And the room glows beneath.
And so the world glows
Slowly toward Christmas day.
And He who holds it knows
The blessing on its way!
(Claudius, n.d.)

Poetry is thus always a yearning for answers to certain questions. What is important—and this clearly distinguishes poetry from psychology and philosophy—is that poetry does not represent an intellectual engagement with questions and themes. On the contrary, poetry remains an artistic confrontation with existential questions and must therefore be sharply distinguished from the philosophical treatment of objects. Here too, the search for answers—such as to one's own life in Heidegger's existential philosophy (Peiffer, 1946, 1952)—is central, but the material is not equipped with rhyme, rhythm, and tonality and hence affective.

Many literary scholars describe the poetic unity of content and form in this regard (Jaiser, 2016; Pfeiffer, 1946). A song of mourning is thematically accompanied by a suitable form to give mourning a certain emphasis. A love poem is accompanied by a fast, cheerful rhythm to give the content of love a deeper dimension. However, this is only partially true. While we do not share the opposing view—for example, that form negates content and thus creates something new (Vygotsky, 1976)—we believe that form sharpens the content, for instance, through exaggerations or even imperfect rhymes. The latter stylistic devices powerfully demonstrate that form does not always align with content. On the contrary: sometimes form precedes content, exaggerating it through certain stylistic means.

If we consider an imperfect rhyme, the reader is likely to pause at this exception, perhaps rereading the line and viewing it in a new light. Thus, form enhances content in a way that encourages a deeper engagement with the poetic subject. It does not negate the object; rather, it amplifies personal engagement with it. Truly resonant poems captivate us through their form, illuminating their subject, while philosophical texts sometimes feel cumbersome, requiring sustained concentration to maintain focus—such as in Heidegger's existential philosophy. With this paragraph, we wish to establish a second point:

2. In poetry, the poet seeks answers to existential questions. In this search, the poet desperately looks for the right means to pursue their answer. Often, the poet is prompted by a vague feeling that gives them the impulse to follow this feeling or delicate thought through the act of writing. The poet aims to illuminate their thoughts and feelings in the writing process. The form and rhetorical devices assist the poet in striving for the correct expression that will mediate the answer for them. The form is not meant merely to underscore the content of the poem or, as other theories suggest, to negate it and thus create something new (Vygotsky, 1976); rather, the form is intended to enhance and deepen the personal engagement with the poetic object—and with the experience itself. The focus is not on the unity of content and form—as postulated by Pfeiffer (1946)—but on the deepening of the content through the form. For our interpretive framework, the following questions are relevant:
   - What is the existential theme of the poem?
   - What is the stance of the lyrical "I" toward the theme?
   - How does the initial image evolve?
   - What answer to the existential question is found by the lyrical "I"?
   - How does the form deepen the content, and at which points does this occur?

The Reader and Symbolism

In the previous section, we examined the area of stylistic devices. Here, we want to present a more in-depth approach that also concerns the reader or recipient. This involves the general symbolism that a poem contains. To illustrate this, let us consider the first stanza of Goethe's poem *Willkommen und Abschied*:

> Es schlug mein Herz, geschwind, zu Pferde!
> Es war getan fast eh gedacht.
> Der Abend wiegte schon die Erde,
> Und an den Bergen hing die Nacht;
> Schon stand im Nebelkleid die Eiche
> Ein aufgetürmter Riese, da,
> Wo Finsternis aus dem Gesträuche
> Mit hundert schwarzen Augen sah. (Goethe, 1775, p. 244)[1]

[1] A translation of the poem is given in the appendix.

The phrases *"Die Eiche im Nebelkleid"* (The oak in a foggy dress) and *"die Finsternis mit hundert schwarzen Augen"* (the darkness with a hundred black eyes) are not just personifications, but also metaphors of a very peculiar kind. The image of the oak in a foggy dress vividly demonstrates how nature is perceived as elegant and beautiful. The term *dress*—often worn by women to accentuate certain parts of the body or in response to climatic conditions—reflects the metaphorical representation of nature for the lyrical subject in the poem. Here, nature is described as graceful, enchanting, just as a dress can enchant people or the person wearing it.

However, in the next line, this graceful image is immediately overturned as the size and breadth of the oak are described as terrifying, with the oak being likened to a *giant* (Riese). The giant, in this context, is also metaphorically symbolic, illustrating that the power of nature can also be overwhelming and oppressive. Thus, the poem confronts us early on with duality in the experience, which continues throughout the subsequent stanzas of Goethe's poem. On one hand, the grace of nature is perceived, but it can quickly transform into its opposite—fear, terror, and anxiety. The fluctuating mood is therefore signified by the contradictory metaphors, which also foreshadow the thematic motifs of the love story in the following stanzas.

The *"Finsternis mit hundert schwarzen Augen"* (darkness with a hundred black eyes) further emphasizes the threat posed by nature's power or violence and sets a gloomy foreboding of the upcoming mood. Thus, the mood oscillates between fear and excitement in a peculiar way. The symbolism reveals a further depth of the experience of the lyrical subject—here, the ambivalence between light-hearted excitement and anxious oppression. This tension between contradictory emotions is a common theme in poems of various kinds, aiming to uncover a human inner conflict at the beginning of a search for meaning.

This demonstrates that the particularly graceful poem confronts us with an existential conflict (and *Willkommen und Abschied* is an example of such a conflict), which must be resolved individually, as the person develops a particular stance toward it. The symbolism allows the reader to examine certain verses more closely; indeed, the symbolism may serve as a linguistic stumbling block, prompting the reader to engage more deeply with the poem—such as through repeated reading or by reciting a particular verse aloud. Boesch (2021) states that symbolism and its connotations often resonate with us and trigger our own experiences with the lyrical subject:

> Words can certainly have precise meanings, such as meters or kilograms, which are clearly determinable. However, most words in everyday language, such as *work*, *possess*, or *have*, are less definitive. In poetry, many words serve as markers that can point in various directions—they are symbols. Certainly, they can refer to concrete experiences, but even these often conceal a deeper meaning. For instance, the line *"Wieder ist nach dunkler Nacht heiterhell der Tag erwacht"* from a poem by Hand Roelli may be a simple description, yet the adjectives *dunkel* (dark) and *heiterhell* (brightly clear) point to experiences, and the lines may indicate the relief with which one greets the morning light, or they might even hint at the emergence from a depressive mood. Depending on this, the reader will give the poem a different meaning. This not only points to the multiple meanings of words but also to the inner freedom of the reader, allowing them to fill the external with their own meaning. The linguistic symbolism of the poem touches the creativity of the reader. (p. 164)

It is the symbolism in the poem that opens the dialogue between poet and reader. At the beginning, the question arises: What does this mean? What does the image convey? What is implied? What is the subtext? All these questions clearly demonstrate that symbolism prompts the search for and discovery of another layer of meaning in the text. The Russian literary scholar Bakhtin always claimed that, in contrast to the novel, the poem is monologically conceived (Bachtin, 2010). This statement is certainly valid, as in a poem, there are no different characters and their voices interacting and then leaving the scene. In a poem, there is no intra-polyphony, which is characteristic of a well-crafted novel (Wutsdorff, 1999). However, polyphony is not entirely absent from the concept of a poem. The symbolism in a poem unfolds an inter-polyphony between poet and recipient. Even if the poet has passed away and a direct face-to-face dialogue is no longer possible, other channels of communication remain through which dialogue can take place. I would like to illustrate this with an example. Let us look at a poem by Kafka:

> Perhaps this is not really love either,
> when I say that you are the dearest to me;
> Love is that you are the knife
> with which I rummage within myself. (Kafka, n.d., author's translation)

The metaphor "love as a knife," used to rummage through one's own wounds, is a striking symbol for the power and also the violence of love. A knife is often used to harm someone, to injure their body in a way that causes suffering. However, a knife can also have a positive effect, as when it is used to remove a bullet from the arm or in a surgical procedure. A scalpel, for instance, is a tool used to remove certain things from the body and restore the functionality of the organism. So, once again, we encounter an ambivalent symbolism, just as we did in Goethe's poem *"Willkommen und Abschied."* The knife thus symbolizes both the destructive danger posed by love and the restoration of the person through love's power. The circumstances in an individual's life are crucial in determining whether the symbolism tends more toward one or the other direction.

Returning to the dialogue: We now wish to ask how Kafka might have understood this image in light of his autobiographical background. Kafka—as we have pointed out in a previous chapter—was always in a conflict regarding love, experiencing it as both enriching and expansive, yet also as oppressive, confining, and suffocating. The ambivalence of the symbol thus sparks an intertextual search (von Fircks, 2015), through which we try to illuminate the symbolic image. In this process, we draw on other texts by the author, such as novels, autobiographical writings, and letters between the author and his friends/family. We aim not only to interpret the symbolic image of love on a textual basis, specifically the autobiographical basis, as that would narrow Kafka's reception. On the contrary, we seek to decipher the interconnected intertextual web of the poem, and for this reason, we do not solely rely on autobiographical sources of the author, but also draw on cultural-historical features of the era in which Kafka lived and worked.

Of course, this places us before the figure of Kafka, which seems to crystallize more clearly, thus allowing a deeper, albeit fictional, dialogue with this figure. For example: We know that Kafka worked in an insurance office as a civil servant. Although he carried out his work diligently—according to his colleagues—Kafka still felt inferior (Kafka, 1995, 2016). Only writing gave him the passion that he lacked during the day at the office. At times, Kafka felt trapped. This must also be considered in light of the era's context, with the rising tide of modernization and mechanization, which could transform sleepy towns into noisy, confusing monsters that might suffocate the individual. The oppressive, stifling nature of this is a central feeling in Kafka's grand novels, such as *The Trial* or *The Castle*. Thus, the sense of the individual's annihilation by certain tools (technology, mechanization) of culture and education (authoritarian paternal upbringing) runs powerfully through several of Kafka's texts, illuminating the negative symbolism that is so striking in this short Kafka poem.

With this consideration, we are now facing a figure who has crystallized more clearly through the intertextual understanding of various sources. With this clearly formed figure, we can finally engage in a dialogical interaction. The reader, in the context of their own field of action, goals, needs, and resistances, will direct the dialogue in a particular direction. Questions may arise, such as, from the perspective of my Western-influenced field: How would Kafka characterize the modern internet and the constant connectivity to an invisible world? How would Kafka interpret the freedom of information that accompanies this? How would Kafka view the individual in this society? Critically? What would Kafka say about our alienation from nature? All of these are essential questions that must also be understood in the interpretation of the poem within the reader's field of action, thus enriching the dialogue. In other words: We see that not only two individuals are interacting here—Kafka with his poem and the reader—but also two fields of action are always present in the background, and their differences provide crucial points for dialogical engagement. What the reader may take away from the dialogue with Kafka is the recognition that the self-evidence of modernization and mechanization should be questioned, sharpening the focus on personal passions and needs that should not be compromised by technology. Through Kafka, a reader might feel encouraged not to flee from the world but to create islands within it where they can live and exist for themselves, such as catalyzed by the experience of a walk in the woods.

The poem may not feature different voices coming in and out as Bachtin (2010) suggests, but it still triggers a dialogue between the poet and the reader and can lead to an update of certain needs and goals, as we tried to outline with the Kafka example. Thus, the poem is a dialogical source, triggered by the implementation of symbols, which, by definition, are always ambivalent and therefore resonate with the reader's ambivalence in their existential experiences.

Symbols are therefore designators (Cassirer, 2007) that point to something. As mentioned in the previous section, this pointing function is always anchored in existential themes, representing something universally experienceable. However, Cassirer's expression of the symbol as a designator also clearly demonstrates that the symbol is a pointer—in the classical sense of the word. When I point to a path,

I show a person that something useful might be hidden in that direction. However, the person must actively pursue this search and, if necessary, work with the pointer and attribute meaning to it. Poems are thus existential designators, and because the pointing gesture is always ambivalent, it must also be illuminated dialogically. This illumination process takes place in the intertextual understanding of the life and work of the author, as well as in connection with one's own life. *This is how we understand why people have always turned to poetry: they hope for something from the pointers they seek to decipher in dialogue with the poet.*

The symbol also unfolds a pragmatic use in the life of both the reader and the poet. In Cassirer's sense, the utility of the symbol always occurs against the background of a "what-is" and "what-should-be" state. When I am unsure of how to address a difficult topic, I may point to this topic through symbols. The symbol in use is meant to help the person make the leap from the "is" to the "should-be" state. For example, just yesterday, I used the door frame symbolically. I did not have a bottle opener at hand and opened the bottle on the door hinges instead. The lever function of the hinges was symbolically used for another action—with the goal of achieving a certain "should-be" state.

3. The symbolism in the poem triggers a dialogical engagement between the poet and the recipient. It is precisely the ambivalence of the symbol that enables dialogical engagement and thus also points to the pointing function of the symbol. It is this flexible pointing function that constitutes the primary characteristic of the symbol. Through the symbol, the author points in a certain direction, and this directional hint becomes the impetus for the reader to illuminate this direction in a dialogical interplay with the author. The goal is to decipher the symbolic image intertextually, against the background of one's own life and field of action, as well as the autobiographical field of action of the author and the cultural-historical norms that stand out as background of the poem. In a final step, this continuous, dialogical deciphering releases interesting interaction possibilities between poet and reader, which may lead to a re-examination of one's own field of action and thus to an update of one's own actions as a whole. The temporarily deciphered symbol thus becomes a tool for achieving a desired "should-be" state and, in this way, a semiotic bearer of meaning for the recalibration of one's own actions. For our interpretation key, we need to ask the following questions:

   - Which symbols does the author use? What does the symbol point to? How can the symbol be understood against the cultural-historical background of the time and the autobiographical background of the author?
   - How can the symbol be understood today? What differences emerge in the understanding of the symbol?
   - Which personal actions might the symbol trigger?

Applied Interpretation of the Action-Theoretical Poem Interpretation Key

We have now addressed three important levels of understanding poetry and will apply this to a concrete case. For this, we have chosen the poem *Stufen* (Stages) by Hermann Hesse (Hesse, 1943/2001):

> Wie jede Blüte welkt und jede Jugend
> Dem Alter weicht, blüht jede Lebensstufe,
> Blüht jede Weisheit auch und jede Tugend
> Zu ihrer Zeit und darf nicht ewig dauern.
> Es muß das Herz bei jedem Lebensrufe
> Bereit zum Abschied sein und Neubeginne,
> Um sich in Tapferkeit und ohne Trauern
> In andre, neue Bindungen zu geben.
> Und jedem Anfang wohnt ein Zauber inne,
> Der uns beschützt und der uns hilft, zu leben.
> Wir sollen heiter Raum um Raum durchschreiten,
> An keinem wie an einer Heimat hängen,
> Der Weltgeist will nicht fesseln uns und engen,
> Er will uns Stuf' um Stufe heben, weiten.
> Kaum sind wir heimisch einem Lebenskreise
> Und traulich eingewohnt, so droht Erschlaffen,
> Nur wer bereit zu Aufbruch ist und Reise,
> Mag lähmender Gewöhnung sich entraffen.
> Es wird vielleicht auch noch die Todesstunde
> Uns neuen Räumen jung entgegen senden,
> Des Lebens Ruf an uns wird niemals enden…
> Wohlan denn, Herz, nimm Abschied und gesunde! (Hesse, 1943/2001, p. 407)[2]

The Primacy of Experience in Hesse's *Stufen* Poem

When we consider the first step of our interpretative key, we must first take into account the primacy of experience. In this regard, we are compelled to contextualize the poem. The poem is located in the appendix of Hermann Hesse's *The Glass Bead Game* (Hesse, 1943/2001, p. 404), making it a central part of his major work for which he was awarded the Nobel Prize after World War II. The poem is presented as a work by the protagonist of Hesse's novel, Joseph Knecht. It demonstrates Knecht's playful engagement with poetry, which was not fully endorsed by his boarding school teachers and classmates—in an elite school—but rather reluctantly accepted. Knecht sought to write something burning from his soul through this and other poems. Throughout his life, Knecht faced various challenges, smaller crises that he had to solve either independently or with the help of close friends. But the resolution of these crises, the restoration of a delicate equilibrium, was always intended to

[2] A translation of this poem is also given in the appendix.

remind Knecht not to remain in that balance, for he feared it could lead to stagnation in life.

Knecht was a man of deep sensitivity, who formed strong attachments to his role models, such as his music teacher, the hermit, and his friend Pilnius Designori. He often struggled to detach from these role models and develop a sense of independence—a process that took several decades before Knecht could establish a solid, personal value system, including in dialogue with his own spirituality. Hesse wrote this novel as a utopia, and particularly during the time of National Socialism, he invested considerable energy into the writing process. In a 30-page foreword, Hesse introduces the reader to the necessity of engaging with their own *Glass Bead Game*, that is, the poetic confrontation with their own life (von Fircks, 2023a, 2023b). In the final chapters of *The Glass Bead Game*, Knecht faces his greatest inner conflict. He feels an internal urge to give up his beloved position as Master (Director of the Elite School) and as a Glass Bead Player because he wants to focus on other fields of action. However, the school does not accept his resignation; the statutes and regulations do not allow for the premature departure of the Master, and Knecht must argue his case and fight rhetorically for his departure.

Thus, the poem is situated within a particular experiential framework of Joseph Knecht—namely, the departure from a profession he once loved and passionately practiced, in order to explore new fields of action, to learn new things, and to shed the fatigue that has set in. Culturally and historically, the poem is situated within a context that deals with departure, renewal, pain, and farewell—topics that run through the entirety of Hesse's work.

Hesse's early works were pedagogical, his middle works psychoanalytical, and his later works spiritual and unifying, with each phase of his creative process embracing and actively engaging in this shift. In *The Glass Bead Game*, Hesse brought these various lines of interest together. Politically, Hesse viewed National Socialism as anathema to his beliefs—a totalitarian ideology that saw itself as the most powerful, absolute system, intolerant of any other and actively seeking to abolish them. Therefore, it is not surprising that some literary scholars argue that *The Glass Bead Game* should be understood against the historical backdrop of this context (Hamann, 2005; Malischke, 1999). The poem can also be interpreted as a form of opposition to leading, self-affirming ideologies.

Existential Themes in Hesse's *Stufen* Poem

The existential theme of the poem has been prominently articulated by Hesse. It revolves around departure, renewal, farewell, pain, death, longing, etc. As the image of the "steps" suggests, departure always requires a certain amount of strength from the individual to actively effect the transition. This transition between departure and arrival is strongly ambivalent in nature and threatens the psychological system of the individual. Hesse impressively shows in the poem that this existential departure into a new life phase not only brings suffering and sorrow (the pain of farewell), but also allows the person to grow, as they make new experiences, acquire new virtues, and develop a certain wisdom of life. It thus concerns the conflict between departure and arrival, between leaving and staying, between letting go and accepting the new—an existential theme that deeply resonates with the reader, whether they are a young person or already an adult. Even a child or teenager might identify with this poem, as it addresses the universal process of letting go of old interests, turning to new ones, and seeing what life has in store (e.g., during a school change from elementary school to secondary school, or trying out a new sport, etc.). Hesse addresses a theme here that is not only existential for a specific age group, but is particularly relevant for people in different stages of development. It is therefore no surprise that this poem is discussed, for instance, in the sixth grade (and above) in Switzerland. This pedagogical engagement is highly relevant when children are encouraged to reflect on their own lives in a meta-reflective manner. The idea that every step elevates and expands the person, that it brings something new into their life and enables personal growth, is meant to provide comfort for the sorrow of farewell and the pain of letting go. Moreover, the unknown, the beginning from which everything can unfold, which can be actively steered, unfolds a kind of magic for the lyrical self. It is about accepting this magic and the new, as the person heals and grows from it. Looking at the rhyme scheme of the poem, we see that each stanza follows a different pattern. Just as each step in life brings something new, each stanza introduces a new rhyme scheme, thereby significantly deepening the meaning.

From a metrical perspective, the poem predominantly employs iambic meter with five stresses, and despite the incongruence of the rhyme schemes, there are congruent elements in the rhythm. Hesse thus emphasizes that rhythm and change should not be seen as antitheses, but as parts of a greater whole, which must be meaningfully integrated into life. The feminine cadence as a formal stylistic device also tends in this direction. Hesse, due to his psychoanalytic background, advocated for a harmonious interplay between the masculine and feminine to achieve complete individuation. The lyrical self, in the sense of Joseph Knecht, uses the feminine elements of his self to receive something he had not yet been willing to accept. The form thus deepens the content, urging the bold and decisive acceptance of the new.

The Symbolism in Hesse's *Stufen* Poem

Finally, we can turn to the symbolism of the poem. Here, we want to highlight three key symbols: the life stage, the journey, and the call of life. In the process of ascending to a new stage, the person generally experiences uncertainty. One foot is in the air, about to step down, while the other remains on the previous stage. The departure glimmers through, but it is still uncertain. The person may trip, get hurt, and fall back down the steps. Or they may feel unsure on the new step; the view may not please them, and they may have liked the previous stage better. Yet, the ascent is necessary to avoid staying stuck in the old. Here, the ambivalence of the experience shines through—the stage is both a threat and a possibility at the same time. Hesse's own liberation from authoritarian education, which he portrays in his novel *Unterm Rad* (1906), also subtly echoes here in an intertextual and dialogical manner. Yet, each new stage brings the possibility of self-assertion and self-development, the growth of new virtues and abilities that allow the person to evolve. Therefore, sorrow should be overcome, and the focus should be placed on the virtues to be gained, which will unfold a certain magic. Indeed, magic is a striking symbol—being enchanted shows the profound transformation of the person, which they bring about themselves through courage.

The journey and the call, of course, show that something in the distance beckons, something that corresponds to inner truth and for which it is worth taking on hardships. Travel, as we know from our own experience, is always associated with stress and uncertainty (will everything go well?). Yet, the reward awaits at the end of the journey, not only because the person has expanded their potential and managed the stress, but also because the end of the journey represents the beginning of something new that the person can integrate. Hesse's central message in the poem is to embrace the effort and struggle of departure, not to succumb to the sorrow and grief of the past and what has been left behind, but to move forward with confidence toward the new because it allows us to grow as individuals.

This evokes powerful personal experiences where we have become stagnant in life, have long avoided departure, and thus fallen into stagnation. We might wonder how Hesse himself, or in this case the character Knecht, managed this departure. The answer in *Das Glasperlenspiel* is clear: Even in departure, even in the new, the old is never entirely lost, and parts of old bonds can be made fruitful for the new phase of life without clinging to them. It is not possible, nor desirable, to completely sever ties with previous stages, as they have shaped us significantly; the goal is to integrate them organically into the new life stage. Hesse succeeded in this through his literary work—especially *Das Glasperlenspiel*—and we must ask ourselves if we are capable of achieving the same and whether Hesse or Knecht can serve as examples for us. My personal answer is affirmative; I would love to hear the reader's response dialogically.

The meta-dialogue with Boesch about poetics as such has prompted us to regard poems not only as valuable in themselves but also to highlight the dialogical significance of poetics—a significance that is often neglected such as by Bachtin (2010). I am convinced that with this flexible interpretation guide, poem interpretation can be made personally relevant and especially meaningful. On to the new stage. The time was ripe.

References

Bachtin, M. M. (2010). *The dialogic imagination: Four essays*. University of Texas Press.

Bendassolli, P. F. (2017). Emptiness and work: A meaning-making perspective. *Integrative Psychological and Behavioral Science, 51*(4), 598–617. https://doi.org/10.1007/s12124-017-9382-x

Bendassolli, P. F. (2024a). Why, instead of doing nothing, do we work? A cultural-psychological essay on the foundations of the purpose of working. *Integrative Psychological and Behavioral Science*, 1–21. https://doi.org/10.1007/s12124-024-09832-0

Bendassolli, P. F. (2024b). Work and depression: A meaning-making perspective. *Culture & Psychology, 30*(4), 1060–1078. https://doi.org/10.1177/1354067X241226452

Boesch, E. E. (2021). *Musik, Sprache und die Sehnsucht nach dem Paradies*. Psychosozial-Verlag.

Cassirer, E. (2007). Versuch über den Menschen. Einführung in eine Philosophie der Kultur. Meiner.

Claudius, M. (n.d.). *Lied im Advent*. Retrieved from https://www.gedichte7.de/lied-im-advent.html.

Fircks, E. (2015). Der text im text. Untersuchung der Intertextualität anhand eines praktischen Beispiels. Gymnasium am Löhrtor, Privatdruck.

Fircks, E. (2022). Culture in the seminar room of poetry: Poetic insights for cultural psychology. *Culture & Psychology, 28*(4), 475–490. https://doi.org/10.1177/1354067X221097609

Goethe, J. W. (1775). Willkommen und Abschied. *Iris Vierteljahresschrift für Frauenzimmer, 2*(2), 244.

Hamann, J. (2005). *Inszenierungen von Erinnerung und kollektivem Gedächtnis. Hermann Hesses Das Glasperlenspiel als Gedächtnisroman*. University of Waterloo. https://uwspace.uwaterloo.ca/handle/10012/745

Hesse, H. (1906). *Unterm Rad*. Fischer.

Hesse, H. (2001). *Das Glasperlenspiel*. Suhrkamp. Original work published 1943.

Jaiser, C. (2016). *Poetische Zeugnisse: Gedichte aus dem Frauen-Konzentrationslager Ravensbrück 1939–1945*. Springer.

Kafka, F. (n.d.) *Auch ist das nicht eigentlich Liebe*. Retrieved from https://www.abipur.de/gedichte/analyse/26367-auch-ist-das-vielleicht-nicht-eigentlich-liebe-kafka.html.

Kafka. (1995). *Brief an den Vater*. Reclam.

Kafka, F. (2016). Tagebücher, 1910–1923/Franz Kafka. Books on Demand. Original work published in 1948.

Malischke, A. (1999). Ideal und Wirklichkeit in Hermann Hesses' *Das Glasperlenspiel'*. diplom. de.

Pfeiffer, J. (1946). *Umgang mit Dichtung [dealing with poetry]* (6th ed.). Meiner.

Pfeiffer, J. (1952). *Existenzphilosophie. Eine Einführung in Heidegger und Jaspers* [Exis tential philosophy: An introduction to Heidegger and Jaspers]. Meiner.

Valsiner, J. (2021). *General human psychology*. Springer.

Valsiner, J. (2014). *An invitation to cultural psychology*. Sage.

Valsiner, J. (2019). *Ornamented lives. Advances in cultural psychology*. IAP.

von Fircks, E. F. (2023a). Geštalt terapija izvan potreba: Kako upotreba alata utiče na osobu I okruženje [Gestalt therapy beyond needs: How tool use affects the person and their environment]. *Geštalt Zbornik, 10*(1), 46–59. https://www.ceeol.com/search/article-detail?id=1201024

von Fircks, E. (2023b). Cultural psychological implications of Hermann Hesse's Glasperlenspiel (glass bead game). *Culture & Psychology, 29*(3), 451–473. https://doi.org/10.1177/1354067X221132000

Vygotsky, L. S. (1976). *Psychologie der Kunst [psychology of art]*. Fundus Bücher.

Wutsdorff, I. (1999). Polyphoner Roman und polyphone Prosa: Zur Verwendung des (musikwissenschaftlichen) Begriffs Polyphonie in der Literaturwissenschaft am Beispiel M. Kunderas und S. Richterovás. *Zeitschrift für Slavische Philologie, 58*(2), 423–443.

Xu, S., & Tateo, L. (2022). Drama-in-education for understanding: An investigation from the perspective of cultural psychology of semiotic mediation. *Human Arenas, 5*(3), 389–406. https://doi.org/10.1007/s42087-020-00157-z

Chapter 10
Ongoing Narration in the Works of Hermann Hesse: The Trace of the Nature, You, and Departure Fantasm

Lullaby *(by E.E. Boesch, Enno von Fircks translation)*
Sleep, little child,
the wind is playing at the window.
He is no robber, does no harm,
don't be afraid, my child.
He enters dreams as a fairy,
and over flowers, forest, and night,
he gently carries you into morning light,
sleep, little child.
Sleep, child, still and warm,
along the path stand sorrow/harm—
they will not touch you, my small child.
The night will cover all your tears,
the weary rivers rest,
the sighs are blown away by wind—
When you're grown, you'll hear them sing,
but for now, sleep well, my child!

Boesch's Fascination with Literature/Poetry

Boesch had a strong affinity for literature—in fact, he loved literature and poetry. It is therefore not surprising that Boesch frequently referred to literary and poetic sources in his works to present and elaborate on his cultural psychological theory of action (Boesch, 1991, 2000, 2021). This is also unsurprising, as Boesch, before his career as a school psychologist, developmental worker, and researcher, had expressed a desire to become a writer and even to make a living from his writing. The poems placed at the beginning of each chapter are testimony to his remarkable talent for poetry.

E. Freiherr von Fircks, *Culture of Action*, Theory and History in the Human and Social Sciences, https://doi.org/10.1007/978-3-031-88808-3_10

Boesch (2005, 2021) saw literature as a personal processing of cultural values with which the writer grapples and through which they aim to work out a personal position (Boesch, 2006). We have explored the role of poetry in the previous chapter, but prose also plays a significant role in Boesch's work during certain phases. Here, Boesch (2005, 2021) speaks of ongoing narration, through which the author continually strives to develop a new position concerning simmering conflicts—and also newly arising ones—and works it out. Throughout his life, Boesch was fascinated by the novels of Adolf Muschg (born 1934), a Swiss novelist, and used Muschg's works as a starting point to consolidate his theory of ongoing narration. The ongoing quality of narration is best captured by Boesch's poem Lullaby—where the ongoing stream of consciousness (stimulated through a caregiver) is emphasized.

He enters dreams as a fairy,
and over flowers, forest, and night,
he gently carries you into morning light,
sleep, little child.

What Is Ongoing Narration?

Naturally, psychoanalytic connotations (Boesch, 2021) played a significant role in Boesch's engagement with Muschg's works. But what is ongoing narration? Boesch explains that narration is an unfinished action—yes, the general motivation and potential for action (individually) only ceases with death. Writers often leave more questions than answers in their works; indeed, they struggle to frame a more precise question on moving topics, through which an answer might then more easily be found (Jens, 2020/1988). Ongoing narration, therefore, is a meta-action (secondary action in the sense of Pierre Janet, 1929, 1930) because the initial action in a constantly developing world does not cease or must be updated by the respective person. Thus, this meta-action (narration over the primary action) is necessary to gain clarity about the motifs and techniques of action and to update them, if necessary, for new goals and purposes. With this, we have defined ongoing narration in the sense of Boesch.

Of course, we must make two critiques that Boesch did not consider in his reflections on this literary phenomenon. What are these two critiques? First, Boesch reduces the phenomenon of ongoing narration largely to the autobiographical interpretation of a text—a cardinal error in any literary science. Boesch asks himself what the author might continually tell themselves in ongoing narration (Boesch, 2021, p. 140). In lyric poetry or novels, it is essential that an author can represent something that stands as a counterpoint to their life and cannot be directly derived from it. For example, Kafka's *The Trial* (Kafka, 1925) or *The Castle* (Kafka, 1926) can be read autobiographically—as novels of oppression—due to Kafka's strained relationship with his father (Kafka, 1995; Mitscherlich-Nielsen, 1977). However, such an interpretive approach is reductive. It cannot be denied that Kafka works through oppressive relationships in his novels, yet this does not exhaust the literary creation of the modern author. On the contrary, Kafka's novels must also be read in the context of his time. The overwhelming mechanization and industrialization of labor, which alienates the individual from their own work, the chaotic big city that constantly monitors the individual and prevents them from asserting their own world, and the alienation of God between the individual and the collective are all essential cultural-historical interpretive access points to Kafka's work (Jens, 2020/1988). Therefore, we must approach text interpretation from a holistic perspective, always operating in the tension between individual and collective interpretive paths and utilizing them productively.

It is precisely Boesch's core concern to illuminate this interplay and its significance for the action and development of the plot—and here, the important cultural psychologist is reductive in his argumentation. Thus, we have already partially anticipated our second critique: it is precisely poetry that draws a strict distinction between the lyrical self and the author (poet) and refuses to equate both figures. It is, in fact, even possible for a male poet to present a female lyrical self and let it experience complex emotions. This also refers to the topos of the surreal, the topos of imagination, which plays a prominent role in literature and poetry. Accordingly, the author can engage in a dialogue with the gaps in their own autobiography and make the tension with these gaps the theme of their poetry or literature. Again, a direct derivation from the author's autobiography would not provide fertile ground for text exegesis.

In other words: Boesch (2005, 2021) traced the discovery of the fantasm, its development, and its possible relaxation (a special case, as literature often ends with more questions than it answers) in his ongoing narration, while strongly neglecting the social overdetermination of the narrative material from a mythical perspective. Convinced that literature is more than just personal expression, we must always consider these complementary approaches to a text. This is what we aim to address in the present chapter.

Ongoing Narration in the Novels of Hermann Hesse

We now know from Boesch that he was not only an interested reader of Muschg and his novels but also, for example, of Hermann Hesse. At the end of his chapter on ongoing narration (2005, 2021), Boesch states that a similar analysis—as done for Muschg—should also be extended to other authors, and here he directly mentions the German-Swiss poet and novelist Hermann Hesse (Thomas Mann is also of central interest to Boesch). In the present chapter, we aim to focus on Hesse's ongoing narration—of course, through his novels—and then complement Boesch's perspective by considering the social overdetermination of myths and the autobiographical counterpoint. This will serve to provide a more holistic view of ongoing narration. We will do so by focusing on three core fantasms within the rich work of Hermann Hesse knowing that we will not have exhausted the fantasms in his oeuvre with our analysis.

Fantasms in Peter Camenzind

Hesse should thus be the focal point. Let us begin with his debut novel *Peter Camenzind* (Hesse, 1904). The novel tells the story of Peter Camenzind, a young man from a small mountain village, who aspires to a simple life but is simultaneously searching for personal fulfillment and a higher meaning—for example, through writing or poetry. Peter loses his mother at an early age, and a great source of comfort to him is the beloved nature in which he feels a resonance of his soul's mood. We will now highlight the role of nature for Peter Camenzind in the following:

> In childhood, I feared the Föhn and even hated it. But with the awakening of boyish wildness, I came to love it: the rebel, the eternal youth, the cheeky fighter, and the bringer of spring. It was so magnificent, how it began its wild battle, full of life, exuberance, and hope—storming, laughing, and groaning; how it howled through the ravines, devoured the snow from the mountains, bent the resilient old pines with rough hands, and made them sigh. Later, my love deepened, and I now greeted in the Föhn the sweet, beautiful, overly abundant South, from which streams of joy, warmth, and beauty always flow forth—only to shatter against the mountains and finally bleed out, weary, into the flat, cool North. There is nothing stranger or more exquisite than the sweet Föhn fever that overtakes the people of the mountain lands—especially the women—during Föhn season, robbing them of sleep and gently caressing and arousing all their senses. This is the South, throwing itself again and again, stormily and passionately, into the arms of the austere, poorer North, proclaiming to the snowbound Alpine villages that primroses, daffodils, and almond branches are already blooming by the nearby, purple lakes of Italy. Then, when the Föhn has blown itself out and the last dirty avalanches have melted away, the most beautiful time arrives. Then, on every side, the blooming yellowish meadows stretch upwards into the mountains; the snowy peaks and glaciers stand pure and blissful in their heights; and the lake turns blue and warm, reflecting the sun and passing clouds. All of this alone can fulfill a childhood—and, if necessary, even a life. For all of this speaks loudly and unbrokenly the language of God, as it has never passed through human lips. Whoever has heard it in their childhood will hear its echo throughout their life—sweet, strong, and terrible—and will never escape its spell. Someone who is at home in the mountains may spend years studying philosophy or natural history and clearing away the old God, but when they feel the Föhn again or hear a soft breeze breaking through the woods, their heart will tremble in their chest, and they will think of God and of dying. (Hesse, 2018, pp. 18–19)

What we see in this longer paragraph is the connection to nature that Hesse attributes to his main character. This romantic connection to nature plays a significant role in Hermann Hesse's works and also exhibits some autobiographical traits. We know that Hesse was a passionate gardener (Hesse, 2015) and, among other things, drew central ideas for the design of his novels and characters from observing nature (Faesi, 1974). In other words, what Hesse particularly emphasizes in *Peter Camenzind* (1904) is the dialogical essence of nature, or how humans enter into a dialogical relationship with nature. Of course, this relationship, in contrast to the later novels addressed, is only anticipated here. Hesse portrays the role of nature in *Peter Camenzind* as sacred, meaning-giving, and accompanying—and in this way, a healthy life must engage in dialogue with nature and its forces to seek clarity about itself and its needs. As we also see in the paragraph described, nature is always understood as a reflection or visage of God, and when humans connect with the works of nature, they are also connecting with God. Perceiving nature in its beauty and grandeur and integrating it into one's own delicate soul to enrich life, to enrich one's own decisions, and to ease eternal loneliness is Hesse's core concern in *Peter Camenzind*.

The development of a young person can only proceed as a differentiation and hierarchical process if the person connects with nature and then, logically, with God, seeking engagement. Hesse creates an interesting contrast in the character of Peter C. here. Of course, it is not the case that Peter C. openly criticizes his parents and accuses them of improper upbringing. However, the main character in Hesse's debut novel expresses that nature understands him better and is more capable of educating him than his parents. We would argue that it is precisely this volatility in Peter C. that characterizes him as a protagonist and vividly describes his personality, which he sees confirmed and then elevated through nature. Indeed, this is the very nature of nature: the eternal dynamics, the unpredictability of its moods, its aesthetic, its rhythmic poetry (the ebb and flow of the seasons), which Peter C. also wants to reflect in his own character and see confirmed. Another quote will provide insight into our explanation:

> Mountains, lake, storm, and sun were my friends; they spoke to me, shaped me, and were, for a long time, dearer and more familiar to me than any people or human fates. But my favorites, which I preferred even to the shimmering lake, the sorrowful pines, and the sunlit rocks, were the clouds. Show me, in the wide world, the man who knows the clouds better and loves them more than I do! Or show me the thing in the world that is more beautiful than clouds! They are play and delight for the eyes, they are blessings and gifts from God, they are wrath and the power of death. They are delicate, soft, and peaceful like the souls of newborns; they are beautiful, abundant, and generous like good angels; they are dark, inescapable, and merciless like the messengers of death. They drift in thin silver layers, they sail laughing white with golden edges, they rest motionless in hues of yellow, red, and blue. They creep darkly and slowly like murderers; they rush headlong with roaring speed like frenzied riders; they hang sadly and dreamily in pale heights like melancholic hermits. They take on the forms of blessed islands and the forms of benevolent angels; they resemble threatening hands, fluttering sails, and migrating cranes. They hover between God's heaven and the poor earth as beautiful symbols of all human longing, belonging to both—dreams of the earth in which it presses its stained soul against the pure sky. They are the eternal emblem of all wandering, all searching, yearning, and longing for home. And just as they hang between earth and heaven, wavering, longing, and defiant, so too do the souls of humans hang, wavering, longing, and defiant, between time and eternity. (Hesse, 2018, pp. 18–19)

1. This allows us to identify the first fantasm in Hermann Hesse's work—a fantasm that not only characterizes the protagonist in *Peter Camenzind* but also bears autobiographical traits. Nature, as a fantasm, is depicted by Hesse as a dialogical partner, a companion along the way, and a wise figure who not only provides confirmation and comfort but also helps the individual make decisions and shape their character. Nature, as an educational force, thus becomes one of the central fantasms in Hesse's debut novel, *Peter Camenzind*.

Interpreting nature as a dialogical partner also has to be understood from a mythical (socially overdetermined) perspective. Hesse was a neoromantic (Hill, 1948)—and we will discover the meaning of this term during our analysis of the *departure fantasm* and make it fruitful. The goal of the neoromantics (approximately from 1890 to 1915) was not only departure, wandering, a longing for unknown lands and parts of the soul, but neoromanticism must also be interpreted as a counterpoint to an urban and over-technologized world and city design. This is why Neo-Romanticism is also seen as a counterpoint to modernity and the urban novel. We know from Ernst Bloch's analysis of the *contemporaneity of the non-contemporaneous* (Schmieder, 2017) that this is fundamental for an analysis of urban and technologized life, at least at the turn of the century, from the nineteenth to the twentieth century. The core of this analysis is that many incompatible things occur simultaneously in a large city [of course, the *Neue Sachlichkeit* (New Objectivity) movement, as part of modernism, also focused on this simultaneity technique], portraying the metropolis as fragmented, split, and overwhelming. Topics such as lighting technology (neon signs), the spread of motorized vehicles, waste, dust, and increased noise are all important themes for critics of this modernization—indeed, *Steppenwolf* (Hesse, 1927) can also be read as a fragmentation of the human soul by the city. Nature, as a dialogical partner, thus cleanses the soul—Hesse not only engaged with this myth but also actively shaped and refined it.

The Mentor (You)-Fantasm in Hesse's Peter Camenzind

Another fantasm that prominently appears in Hesse's works is the Mentor-fantasm. We later learn in *The Glass Bead Game*—for which Hesse won the Nobel Prize—that the young Joseph Knecht had the opportunity to attend an elite school because he was recognized and then supported by a mentor who saw his aptitude and potential (Hesse, 1943). This music master played a crucial role in the development of Joseph Knecht's personality in *The Glass Bead Game*. A similar pattern can also be observed in Hesse's *Peter Camenzind*. Here, the young Peter excuses his father at the monastery (his father was unable to serve due to an impediment, so Peter excused him on his behalf), and a monk recognizes Peter's talent and recommends him for higher studies at the school—after personally meeting him and having a conversation. Thus, in Hermann Hesse's works, advancement in knowledge and education is not determined solely by the individual, but is always also dependent on another, on a *Du* (other), who enters into personal contact with the individual. Education, the unfolding of personality, and personal development are not merely the result of personal willingness and effort (one might read this as a cautionary critique against today's startup ideology, which claims that anyone can start a company and become successful if they just try hard enough). Instead, they must be encouraged, challenged, supplemented, and criticized by a *Du (a genuine You)*, so that the person can develop sustainably. For Hesse, *Peter Camenzind* is indeed about reinforcing this insight—the insight that by connecting with a *Du*, a person can outgrow their previous field of action and erase their personal weaknesses.

> Thus equipped and provided with a new garment, I set out on the journey into life. My parents' gifts proved themselves, for ever since, I have stood and walked in the world on my own two feet. Yet something must have been missing, something that neither knowledge nor the ways of the world ever brought me. For even now, as before, I can conquer a mountain, march or row for ten hours, or, if necessary, kill a man barehanded—but I am as far from being a master of life today as I was then. My early, one-sided engagement with the earth, its plants, and its animals allowed few social skills to develop in me, and even now, my dreams are a curious testament to how much I unfortunately incline toward a purely animal existence. I often dream, for instance, that I am lying on the seashore as an animal, usually as a seal, and I experience such an immense sense of well-being that, upon waking, I do not regain my human dignity with joy or pride, but only with regret. (Hesse, 2018, p. 25)

In other words: It is only the rich human connection—through dialogue and shared experiences—that allows the individual to update their self. Without this connection, without the insight that the *Du* (other) provides into another field of action with its techniques and resources, this updating of action potential cannot occur. This is exactly what Peter Camenzind explains in the abovementioned quote: The individual's action potential stagnates and remains static without the richness of human connection. However, everything is development in Hesse's works, and so too must the individual's development be propelled through courageous exchange with a *Du*. In *Peter Camenzind*, this Mentor- and *Du*-fantasm is only intermittently present, or is portrayed in its negation in the early chapters. Peter C. has difficulty making friends and joining a certain community. In short, he becomes lost in smaller infatuations, befriends only superficially, and cannot appropriately resolve conflicts with his classmates. Nevertheless, the longing for a friend, for a genuine *Du*, remains clearly present in Peter C.

> An even stronger and more vivid longing stirred within me: I wanted to have a friend. There was a brown-haired, serious boy, two years older than me, named Kaspar Hauri. He had a confident and quiet manner of walking and being, held his head firmly and earnestly, and spoke little with his peers. For months, I looked up to him with great admiration, followed him on the street, and fervently hoped to be noticed by him. I was jealous of every bourgeois person he greeted and of every house I saw him enter or leave. But I was two grades behind him, and he likely already felt superior to his own peers. Not a single word was ever exchanged between us. (Hesse, 2018, p. 27)

2. In the works of Hesse, the mentor- and *Du*-fantasm plays a dominant role. It is only in human dialogue, in interpersonal, that is, shared experience, that the individual becomes a personality. In other words: The *Du* allows the person to perceive weaknesses in themselves and make additions to their own character formation. In this way, the *Du* is the key to the self-actualization of the personality. Where this key is missing, the person stagnates. One might again interpret this autobiographically in regard to Hesse's own childhood, his perceived misunderstanding by parents and friends, who did not properly assess, confirm, or nurture his delicate nature, mindset, and talent (Ball, 2006).

Of course, we must also embed and contextualize this *fantasm* within the emerging psychotherapy movement. Freud had already worked through his initial difficulties and established a certain autonomy (he was no longer considered just a charlatan in medicine) in the treatment of mental and physical suffering (Puner, 1947). We must also highlight the predecessor, the true founder of psychological analysis, Pierre Janet. His ideas not only prepared the ground for later institutionalized psychoanalysis, but Janet also proposed and implemented an independent definition and treatment approach for his patients (Janet, 1889, 1898)—particularly through the establishment of a human and moral relationship. The human connection was central to that treatment.

Departure, Home- and Longing-for-Foreign Lands Fantasm in Peter Camenzind

With this, we touch upon another important *fantasm* in the works of Hermann Hesse. The *Du* is the key that allows Peter Camenzind's longing for foreign lands to be fulfilled. Peter C. begins his studies in Zurich and vividly describes his fantasm to find a home that suits him. This is crucial for Peter C. In the confined spaces of his childhood home, his sheltered upbringing, and the practical nature of his field of action, Peter C. feels stifled and oppressed. It is not exactly a *fantasm* of oppression, but rather a threatening *fantasm* of the nondevelopment of his very own personality. Here, too, parallels can be drawn with Hesse's own life, as Hesse felt misunderstood throughout his youth—something that culminated in his attempted suicide during his teenage years (Ball, 2006; Schwilk, 2012). Consequently, we see a strong *fantasm* in Hesse's prose—the *fantasm* of home- and longing-for-foreign-lands. In his own homeland, Peter C. feels misunderstood, oppressed, and deprived of his ability to unfold. However, he then moves to Zurich—for the purpose of studying—and it is Zurich that presents a stark contrast to the rural existence of his parents. Yet it is also Zurich that, at this point, better corresponds to the unfolding of Peter C.'s personality. The longing for foreign lands that Peter C. had experienced early in his childhood and youth is also a yearning for a new home that corresponds to him more as a person. Accordingly, we can describe a vivid scene that illustratively traces the tension between home- and longing-for-foreign-lands.

> Having escaped the sober and oppressive air of home, I made great sweeping flights of joy and freedom. If I ever fell short in other aspects of life, I nonetheless richly and purely savored the peculiar, rapturous delight of youth. Like a young warrior resting on the edge of a blooming forest, I lived in blissful restlessness, caught between struggle and play; and like a prophetic seer, I stood at dark abysses, listening to the roaring of mighty rivers and storms, my soul prepared to perceive the resonance of things and the harmony of all life. Deeply and joyfully, I drank from the full chalices of youth, silently endured sweet suffering over beautiful, shyly revered women, and tasted to its depths the noblest joy of youth: a manly, cheerful, and pure friendship. I arrived in a new buckskin suit, with a small chest full of books and other belongings, ready to conquer a piece of the world and to prove to the coarse folks back home as quickly as possible that I was made of different stuff than the rest of the Camenzinds. For three wonderful years, I lived in the same wind-swept garret with a wide view, studying, writing poetry, longing, and feeling all the beauty of the earth warmly surrounding me. Not every day did I have something warm to eat, but every day, every night, and every hour my heart sang, laughed, and wept, full of intense joy, and I held dear life tightly and fervently to my chest. (Hesse, 2018, p. 47)

We can also see here the effort that Hesse portrays, which accompanies the construction of a new home. Home, for Hesse, is not something that passively forms itself, but something that must be actively constructed. This is precisely the essence of the longer quote: finding friends, strengthening these friendships, falling in love, reading what interests you, experiencing the world in its fullness, creating something with your own hands, and building—these are concrete actions that are crucial for a personalistic construction of one's own home. It is precisely the contrast in the previous home, the narrowness, that initiates this longing for foreign lands and then triggers the genuine feeling of home of a life not yet built—a very personal, colored future.

Longing for foreign lands—so we can argue with Hesse and his Peter Camenzind—is by no means a negative, maladaptive, or psychopathological phenomenon. It is precisely this longing that shows the person which needs cannot be fulfilled in certain places (in certain fields of action) and which makes the departure from the old home—a departure from a constricting world—not only necessary but indispensable. We now must clearly distinguish this from a cosmopolitan *fantasm*. It is indeed possible to interpret Peter Camenzind's desire to study in Zurich and pursue his beliefs and talents as a cosmopolitan *fantasm*. However, that would be a grave mistake in our interpretation. Why? Peter C. is not striving for an open society, for ever-changing relationships and romances, for the bustling life of a metropolis, or the wide range of cultural offerings, but Peter C. longs for close friendships with like-minded people, for individuals who can help him grow, for a work that fulfills him (poetry), and in which he can fully express himself. This has little to do with the cosmopolitan hustle and bustle of a large city.

We have now arrived at an interesting point in our examination of Hesse's debut work. At this point, the different *fantasms* converge—united in the figure of Peter C.—and it is precisely this interdependence of the *fantasms* that generates a certain momentum for the further course of the narrative (see also Chap. 1). After Peter C. fell in love unhappily and then had to distance himself from this love (the woman chose someone else), after he lost confidence in his poetry, succumbed to alcohol, and engaged in superficial acquaintances, Peter C. realizes that his new home has once again become too narrow for him, and he must break free—from the patterns that no longer fulfill his needs. This is an interesting point. Here we can see that Peter C. develops as a person, new needs arise for him, which he cannot satisfy appropriately in his environment in Zurich. In other words: his constructed home becomes too narrow for him, no longer aligns with his person, and once again stifles him. Again, Peter C. must break free and build a new home—once again, he is plagued by longing for foreign lands and the yearning for a promising equilibrium. After some time, Peter C. is once again swept back to his hometown, as his father has fallen ill and now requires care. The entire village is affected by severe climatic conditions (Föhn weather) and has to deal with nature-induced devastation. Peter C. decides to put his skills to the service of the community.

> I realized that there was no remedy for this illness and tried at least to work a little. I began tackling the structure of my work, drafted a few studies, and soon recognized that this was not the right time for it. Meanwhile, ominous reports of the Föhn's destruction kept coming

> in from all sides, and even in the village, the distress grew overwhelming. The stream embankments were half destroyed, several houses, barns, and stables had suffered severe damage, and several homeless people arrived from the outlying community. Everywhere there was lamentation and need, and nowhere was there money. It was during these days that, to my fortune, the mayor summoned me to his council chamber and asked if I was willing to join a committee to help alleviate the widespread distress. They trusted me to represent the community's case to the canton and, in particular, to use the newspapers to rally the country to offer aid and contributions. It suited me well to forget my own useless suffering in the face of a more serious and worthy cause, and I threw myself into the task with desperation. In Basel, I quickly gained the support of a few collectors through letters. The canton, as we already knew, had no money and could only send a few relief workers. So, I turned to the newspapers with appeals and reports. Letters, contributions, and inquiries began to pour in, and alongside all the writing, I also had to fight through the municipal council disputes with the stubborn, hard-headed farmers. (Hesse, 2018/1911, pp. 171–172)

It is of crucial importance that Peter C. postpones indefinitely his own literary sketches, which could theoretically be published. What matters more to him is his commitment to the village community. And we see very clearly that Peter C. dedicates his resources to the service of the community, participates in something greater, and finds success in his engagement. Through this commitment, he also forgets past pain and life's hardships (e.g., unrequited loves). Despite his differences in mindset and talent, Peter C. realizes that he has a place in the community, and he can successfully fulfill it with his own character. In this way, his *fantasms* are indeed actualized—it is not necessarily required to change the field of action due to difference, but rather to find ways and means to make one's own mindset and talent fit into the community and continue to develop there.

Equally important is that during this time—in his hometown—Peter C. transcends himself as a person and does not merely engage in intellectual activities. On the contrary: Peter C. also practices manual tasks and realizes that these skills are necessary for the restoration of life and everyday existence, such as when he renovates his father's house. It is precisely in these everyday tasks that he finds friendship and confirmation of his person. In these moments, Peter C. feels that he has arrived—that is how one might interpret it. And yet, a small voice stirs in his chest, whispering to him that the need for departure may come again one day. However, Peter C. no longer experiences this contradiction (the contradiction of longing for home and for foreign lands) as tension, but rather as an inherent part of life.

3. Longing for home and longing for foreign lands are necessary points of tension in every personal life. In this context, the longing-for-foreign lands reveals the narrowness, the suppression of important personal needs that the current field of action can no longer satisfy. This makes departure into a new, personal world (teleology) necessary, as well as the active construction of a new home based on pressing needs. However, this is only a precarious balance (Heiss, 1936) because the person is in constant development, forming new needs, and the new home may become old and narrow. Furthermore, the construction of a home must always be accompanied by the service of one's mindset and talents in relation to the larger challenges of the community. Only then will the person become an integral part of this cultural community, meeting neighbors, friends, and com-

panions, who in turn can serve as door-openers and mentors for future development. One might also read this autobiographically, as Hesse himself repeatedly pondered different careers and vocational desires, pursued them, then abandoned them entirely or after a short period of time, moving on (e.g., the carpentry apprenticeship). Yet we know that Hesse extracted important lessons from each professional experience, lessons that would later play a key role in his novels.

Mythologically speaking, we once again encounter a neoromantic interpretation of Hesse's intertextuality. This is hardly surprising, as we know that Hesse was a connoisseur of Goethe's writings, particularly his romantic period (Hesse, 1999). Indeed, we also know that before Hesse became politically engaged in an anti-militarist and anti-nationalist way, he went through a phase in which he saw a certain value in war (World War I) and even initially enlisted as a volunteer (Schwilk, 2012). This action can certainly be interpreted through a neoromantic lens, a perspective that was also adopted by the expressionists of the time (such as Georg Heym), who viewed war as a means of liberation from stagnation and desolation.

The Development of Fantasms in Hesse's *Siddhartha*

With this, we aim to conclude our analysis of Hermann Hesse's debut work. It is striking how existential themes are already explored and addressed in his early work. It should be noted that Hesse's *Peter Camenzind* contains additional central fantasms that we have excluded from this examination. Decoding all the fantasms in each of Hesse's works would be a task that demands not just a single chapter, but an entire book. Therefore, we wish to proceed to another of Hesse's novels, *Siddhartha* (Hesse, 1922), and explore how this middle work of Hesse addresses and further develops these fantasms—an essential characteristic of ongoing storytelling according to E.E. Boesch (2021).

Let us begin by reviewing the plot of *Siddhartha* (Hesse, 1922): Hermann Hesse's *Siddhartha* narrates the spiritual journey of a young man in ancient India in search of enlightenment. Siddhartha, the son of a Brahman, leaves his wealthy family home because the traditional teachings fail to satisfy him. He lives in asceticism with the Samanas, meets the Buddha, but realizes that enlightenment cannot be achieved by following a teacher, so he decides to walk his own path. He experiences a worldly life, is introduced to love by the courtesan Kamala, and accumulates wealth as a merchant, only to realize that this way of life leaves him empty. At the river, he meets the ferryman Vasudeva, who teaches him to understand the unity of life by listening to and observing the river. In the end, Siddhartha achieves inner peace and shares his realization with his old friend Govinda: wisdom comes through experience and acceptance, not through teaching.

In *Peter Camenzind*, Peter must constantly struggle between himself and his father. His father mocks Peter for wanting to study and leave his hometown. Even later, Peter feels that his father wants to prove that his own viewpoint was correct when Peter eventually returns to his village. In *Siddhartha*, the main character also expresses the desire to break out of his home, his traditional world. This is all the more surprising because Siddhartha is loved by his parents, respected by his acquaintances, and held in high regard within his field of action. Yet, within this superficial harmony, within this floating equilibrium, disharmony breaks in, and Siddhartha expresses his uncertainty—uncertainty about the traditional teachings and customs. Therefore, he wishes to move on, to learn, and perhaps to answer the questions he has about the world and himself through new experiences. Before Siddhartha breaks out of his traditional world and seeks to move forward, he wishes to gain his father's consent. Initially, his father refuses to offer his blessing or moral support, but Siddhartha fights for his undertaking and wants to negotiate with his father. The following scene illustrates Siddhartha's unwavering belief:

> And in the final hour of the night, before the day began, he returned, stepped into the chamber, and saw the youth standing there, appearing tall and strange to him."Siddhartha," he said, "what are you waiting for?""You know.""Will you stand there waiting until it becomes day, noon, evening?""I will stand and wait.""You will grow tired, Siddhartha.""I will grow tired.""You will fall asleep, Siddhartha.""I will not fall asleep.""You will die, Siddhartha.""I will die.""And you would rather die than obey your father?""Siddhartha has always obeyed his father.""Then will you abandon your intention?""Siddhartha will do what his father tells him to do."

The father touched Siddhartha's shoulder. "You will go," he said, "into the forest and become a Samana. If you find bliss in the forest, then return and teach me bliss. If you find disappointment, then come back and let us once again offer sacrifices to the gods together. Now go and kiss your mother, tell her where you are going." (Hesse, 1922, pp. 17–18)

What in *Peter Camenzind* was a conflict between father and son over the interpretation of the departure and the journey, becomes in *Siddhartha* a negotiation between father and son. The son presents his wish to his father with a clear conscience and steadfastness, hoping for his blessing or moral support. In other words, Siddhartha does not simply wish to flee from his social sphere or treat his father and his old homeland disrespectfully. In short: He wants to seek an amicable solution with his father, and this solution should be the blessing and support of his father. Siddhartha underpins his desire to depart by persistently showing his father the importance of this endeavor, not yielding and firmly stating his intention. Through this persistent attitude, the father realizes the central importance of Siddhartha's need to leave and join a new community (the Samanas), even if he does not share their teachings. Thus, we see that the departure fantasm is more conciliatory and no longer comes with a complete break from one's homeland or parents. On the contrary: Siddhartha seeks the support and blessing of his parents, thereby closing one chapter of his life in a reconciliatory manner, a closure that does not exclude the possibility of reestablishing ties with his former homeland in the future. In other words: *Siddhartha* is about reconciliatory farewells, not about desperate ones, as was still the case in *Peter Camenzind*. Closing a life stage or chapter reconciliatorily is still, today, an essential aspect of adaptive development in Gestalt therapy (Polster & Polster, 1974). One might draw autobiographical parallels here—Hesse, who settled in a small village in Baden with his first wife Maria Bernoulli, after a few years began to feel certain tensions in his marriage and embarked on a journey to Asia to create some distance between him and his wife. Later, the marriage was to be dissolved, as both partners—compounded by psychological suffering—drifted apart. Also later, when Siddhartha and Govinda (his closest friend) wanted to leave the Samanas and the elders of the community opposed this wish, Siddhartha was able to convince them through the teachings he had received from the community. Of course, from these two examples—the opening chapters of *Siddhartha* (Hesse, 1922)—we see that Siddhartha, through skill and newly learned techniques, received from those communities he joined voluntarily to learn and grow, was able to convince the elders repeatedly, thus achieving the blessing for himself and his friend Govinda. To close, to depart, to be conciliatory in departure, and to shape this in a socially acceptable way is, therefore, an essential resource (a skill) that the individual can—and must—learn. We can now present the updating of the fantasms as follows:

4. While in *Peter Camenzind* the departure was pursued at all costs, with potential losses (by the protagonist), in *Siddhartha* the departure is socially acceptable and carried out through certain learnable techniques. In other words: The departure is pursued in a reconciliatory manner, without leaving burned earth behind—an important update of the departure fantasms in Hesse's prose.

We already see in this fantasms Hesse's strong engagement with Taoism (Chen, 2007), a spiritual or religious worldview from ancient China. Here, the unity of opposites is postulated, meaning that departure and arrival should not be interpreted as oppositional, but seen as part of a higher whole. Accordingly, departure must not destroy the possibility of a future return, as doing so would cause the overarching whole to lose its meaning.

Mentor-/Du-Fantasmus in *Siddhartha*

We now turn to the next motif in *Siddhartha*, the Mentor/You Fantasm. We previously mentioned Siddhartha's close friend, Govinda. Siddhartha and Govinda left their parental world together and also moved on from the Samanas. In this regard, Govinda has always behaved in a loyal, understanding manner—almost passively submissive to Siddhartha. In other words, Govinda has always submitted to Siddhartha's insights and needs—until a certain point, the friendship was dominated and directed by Siddhartha. Siddhartha was, therefore, the guiding star for Govinda. However, this one-sided friendship was now about to come to an end. When Govinda and Siddhartha left the Samanas, they sought to learn the teachings of Gotama (the Buddha) and be guided by them. Both friends were thirsting for wisdom—they thirsted for answers to burning questions; both the teachings of their parents and homeland, as well as those of the Samanas, had proven inadequate. Once again, the two friends were journeying into the open, into the new, carried forward by their shared friendship—united in similar needs.

Hesse explores the theme of traveling together, embarking on journeys toward new horizons, in many of his works. Indeed, traveling together is a valuable experience for Hesse. This is not surprising, as traveling together reduces some of the uncertainty and ambivalence, as part of one's own homeland and past always remains with them, resonating in the rhythm of time. Yes, when we say that friendship carries both young men, we mean that it provides a sense of security, understanding, acceptance, and even intimacy—essential factors for facing the uncertainties of an unpredictable world. We thus already see a remarkable development of the Mentor-/You-Fantasm in *Siddhartha*. While the Mentor/You relationship in *Peter Camenzind* is still somewhat idealized, romanticized, and longingly distant, in *Siddhartha* the central elements of the fantasm are already relaxed or realized. A certain clumsiness in the main characters is shed, making space for a mature, intimate (albeit platonic) relationship between the young men, who provide each other with support and strength and face their next journeys together, loyally. It is precisely this bond that creates a certain intimacy—a bond that, in a sense, cannot be broken, thereby granting both individuals (Govinda and Siddhartha) stability, courage, and optimism. It is not surprising that Govinda also seeks physical closeness to his friend (again, understood in a platonic sense).

However, the Mentor-/You-Fantasm is not only updated in this regard (in terms of the concrete realization of a genuine relationship rather than idealized longing), but is also fundamentally expanded in other areas. While the relationship between Govinda and Siddhartha remains meaningful and strengthening, we observe a break in their relationship in the third chapter—though this break is not understood in the negative sense of the word, but rather as a necessary rupture, a needed change. Govinda wishes to follow Gotama (the Buddha) and learn from him, while Siddhartha, although impressed by the Buddha's teachings, is not fully convinced by them. Let us now look at a key passage in *Siddhartha* (Hesse, 1922, pp. 35–36).

> Shortly after the Buddha had retired for the night, Govinda turned to Siddhartha and spoke eagerly: "Siddhartha, it is not my place to reproach you. We have both heard the Enlightened One, we have both listened to his teachings. Govinda has heard the teaching and taken refuge in it. But you, my honored friend, will you not also follow the path to liberation? Will you hesitate, will you still wait?" Siddhartha awoke as if from a sleep upon hearing Govinda's words. He looked into Govinda's face for a long time. Then he spoke softly, with a voice free of mockery: "Govinda, my friend, you have now taken a step, you have chosen your path. Always, O Govinda, you have been my friend; always you have walked one step behind me. Often I wondered: will Govinda not one day take a step on his own, without me, from his own soul? Behold, now you have become a man and chosen your way for yourself. May you walk it to its end, O my friend! May you find liberation." Govinda, not fully understanding, repeated with a note of impatience: "Speak, I beg you, my dear friend! Tell me, as it must surely be, that you too, my wise friend, will take refuge in the great Buddha!" Siddhartha placed his hand on Govinda's shoulder. "You have missed my blessing, O Govinda. I will repeat it: May you walk this path to its end! May you find liberation!"

An important Mentor/You relationship must, according to Hermann Hesse's philosophy of life, eventually come to an end. This becomes particularly evident in this passage. Govinda, who had previously stood in the shadow of Siddhartha, passively subordinate and receptive to him, now decides to go his own way—though not without sorrow and loss. This signifies that Govinda has recognized for himself that the teachings of the Buddha and the symbol of the Buddha will satisfy his needs for a time, and he finds himself in a state of suspended equilibrium with Gotama. Naturally, Govinda wishes to explore this equilibrium more deeply—by growing through it and gaining further insights into the world and into himself. However, Govinda hopes that his beloved friend will follow his path. Siddhartha, however, does not find the same equilibrium in Gotama's teachings that Govinda finds, and once again decides to follow his own path and continue his journey.

Hesse thus illustrates that no relationship—no matter how strong or elevated—can take precedence over the individual's search for balance, which is a recurring task, or over the quest for and construction of one's own sense of home, including the home fantasm. The relationship, too, can and must end at certain points because both individuals (Govinda and Siddhartha) must follow their own paths to satisfy their needs and build their personal worlds. To ignore or refuse to heed this deeply personal calling is a cardinal mistake according to Hesse's philosophy of life. Hesse might be suggesting: do not cling to a relationship from which you must inevitably outgrow. Yet, both friends are eventually reunited at the end of Hesse's Indian narrative—Siddhartha, who has found insight into the way, and Govinda, still seeking that very insight. Hesse demonstrates that an end to a relationship often marks the beginning of a wholly personal, autonomous vision of the future—a beginning that may later return to an end, where both parties, both friends (in this case), will meet again, face each other genuinely, and enrich one another.

To further explore this, let us look at the conclusion of *Siddhartha* to gain a clearer understanding of the relationship, and from there, expand our interpretation of the Mentor-/You-Fantasm. First, let us briefly contextualize this scene. Govinda, still thirsty for knowledge and wisdom, is influenced by rumors and seeks out the wise ferryman (who is now Siddhartha), hoping to find answers to his burning, never-ending questions. Before Govinda recognizes Siddhartha, the latter speaks the following:

> Siddhartha said: "What could I possibly have to say to you, Reverend? Perhaps only that you seek too much? That in your seeking, you fail to find?""How so?" asked Govinda."When someone seeks," Siddhartha said, "it easily happens that his eyes see only the thing he is searching for, that he is unable to find anything, unable to absorb anything within himself, because he is always focused solely on the object of his search, because he has a goal, because he is obsessed with that goal. Seeking means: having a goal. But finding means: being free, being open, having no goal. You, Reverend, may indeed be a seeker, for in pursuing your goal, you fail to see many things that are directly before your eyes." (Hesse, 1922, p. 135)

In this passage, we can see how Hesse interprets the content of the I-You relationship (or the You Fantasm) in *Siddhartha*. What does this content look like? What Siddhartha addresses here is the idea that a philosophy of life cannot be taught, but must instead be lived. In other words, Siddhartha is attempting to convince Govinda to stop searching so much and to pay more attention to himself and to his own world. Naturally, he does not want to deny Govinda any sense of purpose—Siddhartha himself is not free from it, and no one ever is (Stern, 1923, 1924). However, it is possible to be surprised by the world and one's social and material overdeterminedness, to enrich one's own sense of purpose, and to spontaneously integrate concrete impulses into one's striving. Siddhartha is trying to show Govinda that a certain directionality in life should not be rigid and fixed, but must always be recalibrated according to the social and material conditions of one's own world. In this sense, the human being is always in development, as every path leads further toward the perfection of one's own person, and thus toward the perfection of the cosmos itself.

Govinda has difficulty understanding and accepting this teaching. It only happens at the moment when Govinda allows himself to kiss Siddhartha's forehead and, in doing so, sees the true face of his old friend. In that moment, he experiences it, and can integrate this experience deeply into himself.

> While Govinda, astonished and yet drawn by great love and intuition, listened to his words, leaned close to him, and pressed his lips to Siddhartha's forehead, something wonderful happened to him. While his thoughts still lingered on Siddhartha's remarkable words, while he was still vainly and reluctantly trying to imagine time being erased, trying to picture Nirvana and Samsara as one, while a certain disdain for the words of his friend was still contending within him with an immense love and reverence, this happened: he no longer saw Siddhartha's face; instead, he saw other faces, many, a long series, a flowing river of faces, hundreds, thousands, which all came and passed, yet all seemed to exist simultaneously, all constantly changing and renewing themselves, yet all were Siddhartha. He saw the face of a fish, a carp, its mouth wide open in endless suffering, a dying fish with glazed eyes—he saw the face of a newborn child, red and full of wrinkles, crying—he saw the face of a murderer, saw him plunging a knife into someone's body—he saw, in the same instant, this criminal kneeling in chains and his head being struck off by an executioner—he saw the bodies of men and women naked in positions and struggles of raging love—he saw corpses stretched out, still, cold, empty—he saw animal faces, of boars, crocodiles, elephants, bulls, and birds—he saw gods, saw Krishna, saw Agni—all these forms and faces were connected in thousands of relationships, each helping the other, loving it, hating it, destroying it, giving birth to it anew. Each was a will to die, a passionate, painful confession of transience, and yet none of them died; they all only transformed, were continually reborn, constantly acquiring a new face, with no time ever seeming to pass between one face and the next. All these forms and faces rested, flowed, created themselves, swam, and merged into one another, and above all, there was constantly something thin, insubstantial, yet existent, drawn over them like thin glass or ice, like a transparent skin, a shell or mask of water. And this mask smiled, and this smile was Siddhartha's smiling face, which Govinda was touching with his lips in this very moment. And, Govinda saw, this smiling mask, this smile of unity over the flowing shapes, this smile of simultaneity over the thousand births and deaths, this smile of Siddhartha's was precisely the same, was the very same quiet, delicate, impenetrable, perhaps kind, perhaps slightly mocking, wise, thousand-fold smile of the Buddha, as Govinda had seen it a hundred times with reverence. So, Govinda knew, the perfected ones smiled in this way. No longer knowing whether there was time, whether this vision lasted a second or a hundred years, no longer knowing whether there was a Siddhartha, whether there was a Buddha, whether there was an "I" and a "you," pierced to the heart as if by a divine arrow, whose wound tasted sweet, enchanted and dissolved in his innermost being, Govinda stood for a little while longer, bowed over Siddhartha's still face, which he had just kissed, which had just been the site of all shapes, of all becoming, of all being. The face was unchanged after the depths of its thousand-foldness had once again closed beneath its surface—it smiled quietly, smiled softly and gently, perhaps very kindly, perhaps very mockingly, exactly as the Buddha had smiled. Govinda bowed deeply. Tears flowed, tears he did not know about, down his old face, and like fire burned the feeling of deepest love, of most humble veneration, in his heart. He bowed deeply to the motionless figure, whose smile reminded him of everything he had ever loved in his life, everything he had ever held dear and sacred. (Hesse, 1922, pp. 144-147)

5. The *Du-Fantasmus* and the mentor relationship are nourished by human contact, which is both physical and mental—psychophysically neutral in nature. Indeed, this psychophysical synchronous contact is an experience in itself. Govinda perceives the life, past, present, and future of Siddhartha in a flow of experiences, and it is this experience—not words—that allows Govinda to grow in his worldview and self-reflection, bringing him closer to enlightenment. Genuine development—concerning one's own philosophy of life or life patterns—can therefore only occur through interpersonal mediation, but this mediation must be understood as an experience and not as an explanation. Explaining this insight destroys the experiential nature of the possible realization. In short: Govinda looks into the personal flow of Siddhartha's being and allows himself to be touched by this experience—for his own enlightenment. Thus, Hesse demonstrates that the *Ich-Du-Fantasmus* must not be intellectualized or talked to death, but must be guided by shared experience and mutual emotional engagement. This is a distinct life philosophy of the author concerning the *Du-Fantasmus* and the potential mentor relationship between individuals.

Once again, mythological connections to Taoism can be drawn. Indeed, in his middle works, psychoanalysis and Taoism were the dominant worldviews and thought systems that impacted Hermann Hesse. This is not surprising—Hesse, after some interpersonal crises (marital), embarked on a journey to Southeast Asia, where he became acquainted with the works of Richard Wilhelm and established a friendly connection with this prominent sinologist (Jung was also friends with Wilhelm). Through this, Hesse—similar to Jung (Barrett, 2022)—was able to develop a peculiar synthesis between psychoanalysis and Taoism, which would shape his later works (*Morgenlandfahrt* and *Glasperlenspiel*). Taoism, we must emphasize again, is known for striving to experience the world profoundly (by Fircks, 2023a, 2023b, 2023c), without interrupting the flow of experience through linguistic mediation. Indeed, for Laozi (2011), language is often an obstacle to the mediation of the experience and life flow that is so significant to Taoism. In brief: The person must immerse themselves in this life flow, not only accepting but actively integrating the polarities of life into their own existence, without artificially fixing one pole in their personality. Only then can virtue arise continuously, which not only benefits the person but can also benefit the entire social system. Hesse—with his psychoanalytic interest—was indeed willing to explore how unconscious polarities can be made conscious, which is what makes his work, as we allow ourselves to repeat, a unique version of Taoism.

The Nature/Cosmos Fantasmus in Siddhartha

Let us now turn to the last fantasm that we must still address in *Siddhartha*. It is the fantasm of the dialogical role of nature or the cosmos. While nature in *Peter Camenzind* still appears very romanticized and full of yearning, serving more as a reflection of the individual's desires and needs, in *Siddhartha*, nature takes on a more independent role as a dialogical partner. This means that nature is no longer just a projection surface for the individual's emotional state, offering confirmation or comfort, but develops an autonomy (yes, the cosmos remains autonomous), which can also confront the person in contrast. To illustrate this, let us quote a scene that reflects the dialogical and autonomous role of the cosmos, the holy spirit, and enriches the person in their growth. Again, a brief contextualization of this scene is necessary: Siddhartha is in a transition. After he has shed the hardships and sufferings of a worldly existence (consumption, sex, greed), he now feels disgusted and aimlessly wanders through the lands, sitting down near a river to reflect on his life. Before he decides to end his life and depart from the world, he hears the holy call of life (the Om) and decides against the plan of suicide. He falls into a deep sleep. After Siddhartha awakens, he sees his friend Govinda, who has happened to pass by the river and is watching over his sleep. After a few words, Govinda leaves Siddhartha, who is now reflecting on his rebirth.

> Now Siddhartha also realized why, as a Brahmin and an ascetic, he had fought in vain against this self. Too much knowledge hindered him, too many sacred verses, too many rules of sacrifice, too much self-mortification. He had been too proud, always the cleverest, always the most zealous, always a step ahead of everyone else, always the learned and spiritual one, always striving to be a priest or wise man. Into this priesthood, into this pride, into this spirituality, his self had crept, there it had entrenched itself and grown, while he had thought to kill it with fasting and penance. Now he saw it, and he saw that the inner voice had been right, that no teacher could ever have redeemed him. That was why he had to go out into the world, had to lose himself in pleasure and power, in women and money, had to become a merchant, a dice player, a drinker, and a greedy man, until the priest and ascetic in him were dead. That was why he had had to endure these ugly years, had to endure the disgust, the teachings, the futility of a barren and lost life, until the very end, until the bitter despair, until even the lustful Siddhartha, the greedy Siddhartha, could die. He had died, and a new Siddhartha had awakened from sleep. He too would grow old, he too would one day have to die—Siddhartha was transient, every form was transient. But today he was young, he was a child, the new Siddhartha, and he was full of joy. These thoughts he pondered, smiling as he listened to his stomach, gratefully hearing the hum of a bee. Cheerfully, he looked into the flowing river—never before had a water pleased him so much as this one, never before had he so strongly and beautifully heard the voice and metaphor of the flowing water. It seemed to him that the river had something special to tell him, something he did not yet know, something that still awaited him. In this river, Siddhartha had wanted to drown himself, in it, the old, tired, desperate Siddhartha had drowned today. But the new Siddhartha felt a deep love for this flowing water and resolved within himself not to leave it again too soon. (Hesse, 1922, pp. 98–99)

In this longer passage, we see very important lessons that elements of the cosmos teach Siddhartha. Here we clearly observe the autonomy, the autotelic nature of the cosmos, which enters into a connection with Siddhartha. Initially, Siddhartha views the river only as a means to the end of suicide—as a way to escape earthly existence. However, the river's role shifts from a passive object, which Siddhartha can simply use for his own purposes, to a genuine dialogical partner that possesses an accountable independence. It is this autotelicity that Siddhartha believes he can learn from (the river of the cosmos). Therefore, it is not surprising that the river has its own voice that speaks to him.

Here, we must first assume that the river takes on a contrastive role as a dialogical partner—in other words: I am not only here for you to use me for your selfish purpose (suicide), but I possess my own voice and worthiness. It is this worthiness (the eternal river, the ever-renewing) that makes Siddhartha recognize the hopelessness and meaninglessness of his plan for suicide, and it updates his position concerning this topic. In other words, Siddhartha looks into the personal world of the cosmos, into the autotelicity of the cosmos and its purposes, and it is this look that allows him to update his own position regarding the value of his own life.

Nature, or better yet, the cosmos, here takes on the fate of the Du-/Mentor-Fantasmus. Hesse vividly demonstrates that nature is not just a projection surface and serviceable to the person—meaning that it is completely dependent on the person—but, through its autotelicity, it matures into a dialogical partner that can confront the person. In this confrontation, the person can then recognize the self-governance of the other, meaning they can look at it and integrate the experience from this observation into their own life. The cosmos thus becomes the Du-Fantasmus in Hesse's *Siddhartha*. Accordingly, it is not surprising that Siddhartha attributes personal qualities to the water and decides to learn from it.

> "I will stay by this river," thought Siddhartha. "It is the same river I once crossed on my way to the world of ordinary people. A kind ferryman guided me then. I will go to him. From his hut, my path once led me into a new life, which has now grown old and dead—may my present path, my present new life, also begin there!" Tenderly, he gazed into the flowing water, into its transparent green, into the crystalline lines of its mysterious patterns. Bright pearls rose from the depths, silent air bubbles floated on the surface, reflecting the blue of the sky. The river looked at him with a thousand eyes—green, white, crystalline, and sky-blue. How he loved this water, how it delighted him, how grateful he was to it! In his heart, he heard the newly awakened voice speaking, and it told him: "Love this water! Stay by it! Learn from it!" Oh yes, he wanted to learn from it, he wanted to listen to it. Whoever understood this water and its secrets, it seemed to him, would also understand much else, many secrets, all secrets. But today, from the mysteries of the river, he saw only one that touched his soul. He saw: this water flowed and flowed, it flowed endlessly, and yet it was always there, always and at all times the same, and yet at every moment new! Oh, if only one could grasp and understand this! He did not fully understand or grasp it but felt only an intuition stirring within him, a distant memory, divine voices. (Hesse, 1922, p. 83)

6. In Boesch's words, we can analyze the written text as follows: Nature possesses a completely independent potential for action; it is, in Hesse's view, a collective potential for action for the individual person who confronts it, and thus an independent person, since only persons have potential for action. The river is the dialogical island through which this potential for action is communicated, through which the individual person can integrate elements of this collective potential into themselves. And since for Hesse, nature carries a divine character, the person can also integrate a divine aspect by incorporating this foreign potential for action into their own potential for action. Siddhartha recognizes this divine character of nature—and actively approaches it. Indeed, one could say that the dialogue with nature as a person (because only with persons can one engage in dialogue) brings life back to Siddhartha and awakens new vitality within him. Again, this is proof of how strongly nature or the cosmos confronts the person as an independent dialogical partner.

Of course, at this point, we once again see the leading role of Taoism. We will refrain from a longer presentation of the fundamental pillars of this interesting religious worldview, as the core of Taoism has already been thoroughly explored. However, one point should still be added. Taoism also emphasizes the inherent connection between humans and nature. Humans have never fully outgrown the biosphere (Stern, 1935); on the contrary, they remain embedded within it. Even modern technology cannot change this natural embedding. When humans speak to nature, they are speaking to themselves. When they speak to themselves, they are also speaking to the cosmos. This connection, this unity with the biosphere, from which there is no escape, is mostly prepared in Siddhartha, but not yet fully realized with determination and ultimate consequence.

Fantasms in Hesse's Narcissus and Goldmund

We now wish to focus on a final novel in Hesse's oeuvre—*Narcissus and Goldmund* (Hesse, 1930). We deliberately set aside *The Glass Bead Game* (Hesse, 1943) in this discussion—certainly the largest and most complex of Hermann Hesse's novels, which has also been most extensively analyzed in the academic reception of Hesse's work. An analysis of this novel would exceed the scope of this chapter—we refer to two articles on *The Glass Bead Game* and its significance for psychology (Hsia, 1970; von Fircks, 2023d). However, in this subsection, *Narcissus and Goldmund* is the primary focus, specifically how Hesse updates the three fantasms we have highlighted. Before proceeding, a brief summary of the internal and external plot of *Narcissus and Goldmund*:

Hermann Hesse's *Narcissus and Goldmund* describes the lives of two friends in the early Middle Ages. Narcissus is a monastic ascetic, while Goldmund is a jovial monk who enjoys life. While Narcissus lives in the monastery, Goldmund ventures out into the world. Goldmund's travels expose him to love, art, suffering, and passion, but he also learns the reality of the endless flow of time. However, rather than being defeated by it, this realization brings him anguish, which ultimately leads him back to Narcissus. The novel shows that although they engage in philosophical debates about intellect and pleasure, they also realize that despite their differences, they complement each other. The novel plays with the tension between spirit and sensuality, order and chaos, as well as duty and freedom.

The Mentor-/Du-Fantasm in *Narcissus and Goldmund*

Before we delve into some passages from Hesse's novel once again, we must first contextualize the situation. Context, after all, seems to be the essential criterion for understanding narratives and their significance (Bruner, 1986, 1996, 1997). Literary studies recognized this crucial phenomenon more quickly and coherently than psychology or sociology. Even today, psychology struggles to include the context in its analysis (Valsiner, 2017). Regardless, here we are. Narcissus quickly rises to the position of an assistant teacher in the monastery, and soon a young, yet very contrasting character, Goldmund, appears in his class. For both students, this encounter is pivotal and profoundly alters the trajectory of their lives. This information is crucial for understanding the Du-/Mentor-Fantasm in Hesse's literary work. It is not only Goldmund who looks up to Narcissus, admiring his sharp intellect, but Narcissus also perceives abilities in Goldmund that complement him as a person. Formally, we once again have a clear dependency relationship. In *Siddhartha*, this dependency is still present (albeit weakened), as Govinda repeatedly encounters Siddhartha, seeking to learn from his wisdom. The enlightened one and the one striving for enlightenment are somewhat trapped in their roles; there is no significant balance or a strong reciprocal friendship, which is so central to the human organism. In *Narcissus and Goldmund*, however, the Du-/Mentor-Fantasm in Hesse's work has been updated. But in what sense? The I/You relationship is less dependent, or in other words, more balanced. This balance is reflected in the fact that Narcissus, as a scholar and teacher, recognizes the qualities of Goldmund and seeks his closeness and friendship to enrich and update his own field of action. Let us take a look at the following passage to examine this further:

> Narcissus's thoughts were far more occupied with Goldmund than Goldmund imagined. He wanted the bright boy as a friend. He sensed in him his opposite, his complement; he would have liked to adopt, lead, enlighten, strengthen, and bring him to bloom. But he held himself back, for many reasons, almost all of them conscious. Most of all, he felt tied and hemmed in by his distaste for teachers or monks who, all too frequently, fell in love with a pupil or a novice. Often enough, he had felt with repulsion the desiring eyes of older men upon him, had met their enticements and cajoleries with wordless rebuttal. He understood them better now that he knew the temptation to love the charming boy, to make him laugh, to run a caressing hand through his blond hair. (Hesse, n.d., p. 25)

Narcissus seeks to help Goldmund grow further through this relationship, not only to perfect Goldmund but also to benefit from this perfection for his own field of action. Narcissus feels no vanity, pride, or arrogance toward Goldmund in light of his higher position in the monastery—it is only a misinterpreted relationship with Goldmund that causes the young scholar some distress. However, even this concern fades soon, as Narcissus comes to understand the warmth and intimacy of their friendship. Before long, these two different partners will not only become friends but also mutual mentors, helping each other fill the gaps in their disposition and talents, assisting the other person's growth.

> But during the first spring of this friendship he ran up against unfamiliar obstacles, unexpected, incomprehensible coolness, frightening demands. It never occurred to him to see himself as the contradiction, the exact opposite of his friend. He thought that only love, only sincere devotion was needed to fuse two into one, to wipe out differences and bridge contrasts. But how harsh and positive this Narcissus was, how merciless and precise! Innocent abandonment, grateful wandering together in the land of friendship seemed unknown and undesirable to him. He did not seem to understand, to tolerate dreamy strolls on paths that led in no particular direction. When Goldmund had seemed ill, he had shown concern, and loyally he helped and advised him in all matters of school and learning; he explained difficult passages in books, opened new horizons in the realm of grammar, logic, and theology. Yet he never seemed genuinely satisfied with his friend, or to approve of him; quite often he seemed to be smiling, seemed not to take him seriously. Goldmund felt that this was not

> mere pedantry, not just the condescension of someone older and more intelligent, but that there was something else behind it, something deeper and important. But he was unable to recognize this deeper something, and this friendship often made him feel sad and lost. Actually Narcissus recognized his friend's qualities only too well; he was not blind to the budding beauty, the vital force of nature in him, his flowering opulence. He was no pedant bent on feeding Greek to a fervent young soul, on repaying an innocent love with logic. On the contrary, he loved the blond adolescent altogether too much, and this was dangerous for him, because loving, to him, was not a natural condition but a miracle. To fall in love was not permitted him; he could not be content with the joyful contemplation of those eyes, with the nearness of this golden light. Not even for a second could he let this love dwell upon the senses. Because where Goldmund felt himself destined for monkish asceticism and a life-long striving for saintliness, Narcissus was truly destined for that life. To him, loving was permitted only in its highest form. Narcissus did not believe in Goldmund's calling to be an ascetic. He knew how to read people more clearly than most, and here love increased his clarity. He recognized Goldmund's nature and understood it deeply, in spite of the contrasts, because it was the other, the lost half of his own. He saw that this nature was armored by a hard shell, by fantasies, faults of upbringing and paternal words; he had long sensed the whole, uncomplicated secret of this young life. He was fully aware of what he must do: reveal this secret to its bearer, free him from the shell, give him back his true nature. It would be hard, and the hardest was that perhaps it would make him lose his friend. (Hesse, n.d., pp. 39–42)

The role of the Mentor-/You-Fantasm is thus defined by what the other person (the living You) has to offer, and how these gifts affect the individual. This relationship is necessarily reciprocal because a person is never fully or completely perfected in their disposition, talents, motivation for action, or potential for action. On the contrary, for Hesse, it is of fundamental importance to experience and recognize the gaps in one's own motivation and potential for action, in order to draw certain conclusions about one's own actions. These conclusions allow for the construction of a more adaptive personal world and, thus, a more adaptive, personalistic future. And because each person tends to have something to offer that the other lacks, this relationship should not be complicated or made impossible by artificial barriers (position, seniority, etc.). The Mentor-/You-Fantasm is, therefore, in *Narcissus und Goldmund*, not only reciprocal but constantly (re)changing and interchangeable—depending on the situation and context. This represents a significant development in comparison to *Siddhartha*, where the roles between the two main characters are more rigid, narrower, and must not be interpreted as interchangeable—teacher and student are predetermined in *Siddhartha*, and the relationship cannot reverse these roles. Only in *Narcissus und Goldmund* is it possible for the roles of mentor and student, or I and You, to shift back and forth. Again, this shows the updating of Hesse's life philosophy: From every personal life, one can and must learn, without societal barriers or inhibitors, and the roles must remain irrelevant.

> But the love that bound him to his friend and their habit of spending much time together had not been fruitless. In spite of the vast differences of their characters, each had learned much from the other. Beside the language of reason, a language of the soul had gradually come into being between them; it was as if, branching off the main street, there are many small, almost secret lanes. Gradually the imaginative power of Goldmund's soul had tracked such paths into Narcissus's thoughts and expressions, making him understand—and sympathize with—many of Goldmund's perceptions and feelings, without need for words. New links from soul to soul developed in the warm glow of love; words came later. That is how, one holiday, in the library, there occurred a conversation between the friends that neither had

> expected—a conversation that touched at the core and purpose of their friendship and cast new, far-reaching lights. They had been talking about astrology, a forbidden science that was not pursued in the cloister. Narcissus had said that astrology was an attempt to arrange and order the many different types of human beings according to their natures and destinies. At this point Goldmund had objected: "You're forever talking of differences —I've finally recognized a pet theory of yours. When you speak of the great difference that is supposed to exist between you and me, for instance, it seems to me that this difference is nothing but your strange determination to establish differences." (Hesse, n.d., pp. 55–56)

Indeed, this excerpt shows that a person can only be touched by a You if they are fundamentally open and ready for it. For this, they must also shed societal conventions (the rigid role relationship in our example) and fully surrender to the relationship. Furthermore, it is essential that the relationship between the two is not overly intellectualized—a clear difference from *Siddhartha*, where the shared search for meaning with and through the other was more prominent, except at the end. Hesse consistently carries out the recalibration of the fantasm—mutual touching only occurs through shared experience, which then crystallizes into insight. Language is more of an obstacle in this process and can distort the relationship; therefore, it must be used carefully. Another important motif in the You-fantasm—particularly visible in *Narcissus und Goldmund*—is the depiction that I and You, You and I, constantly complement each other (and then also alternate), meaning that human relationships progress toward perfection in their polarities. The other often symbolizes this polarity, which, by itself, should not be interpreted as separating but rather connecting. Only through the complementing of the other and through my complementing for them does mutual growth, mutual maturing, and symbiosis become possible. Hesse's own experiences in Southeast Asia, his engagement with Daoism, the I Ching, psychoanalysis, and his friendship with Richard Wilhelm and C.G. Jung surely contributed a great deal to this motif.

7. While the relationship in *Siddhartha* between Govinda and Siddhartha is not depicted as equal or balanced, but rather embedded in a dependency relationship—albeit with a certain degree of exchange—this dependency in *Narcissus und Goldmund* is almost completely dissolved. Despite the age difference and professional disparity, two main characters come to the forefront here, enriching and perfecting each other through their person. Genuine development in *Narcissus und Goldmund* is only possible when both main characters—free from rigid role obligations—face each other as individuals, observe their mutual dependency and that of their personal worlds, and integrate the results of this experience into the future direction of their lives.

Longing/Departure Fantasm in Narcissus and Goldmund

Of course, now we want to explore the longing and departure fantasm. First, a contextualization. Goldmund is tasked by the abbot of the monastery to accomplish something. For this, he must leave the protected space of the monastery and venture into the rough world outside. In the process, Goldmund meets a woman who has a magical effect on him. They kiss and grow closer; Goldmund falls in love with the homeless woman and promises to return to her in the evening/night. For Goldmund, this is a sign to finally leave the monastery and begin his own journey. He explains this to his friend, Narcissus, as follows.

> Timidly Goldmund touched his friend's shoulder. "You have just said it. Although this time you didn't say it well, Narcissus, not accurately. It is altogether different. I was out in the fields, and I fell asleep in the heat, and when I woke up, my head was resting on the knees of a beautiful woman and I immediately felt that my mother had come to take me home. I did not think that this woman was my mother. Her eyes were brown and her hair was black; my mother had blond hair like mine. This woman didn't look in the least like her. And yet it was my mother, my mother's call, a message from her. It was as though an unknown beautiful woman had suddenly come out of the dreams of my own heart and was holding my head in her lap, smiling at me like a flower and being sweet to me. At her first kiss I felt something melt inside me that hurt in an exquisite way. All my longings, all my dreams and sweet anguish, all the secrets that slept within me, came awake, everything was transformed and enchanted, everything made sense. She taught me what a woman is and what secrets she has. In half an hour she aged me by many years. I know many things now. I also suddenly knew that I could no longer remain in this house, not for another day. I'm going as soon as night falls." Narcissus listened and nodded. (Hesse, n.d., pp. 112–113)

Of course, Goldmund's focus in this scene is not on sex (or only secondarily on sex). Rather, it is about his wish to integrate the archetypal qualities of the mother into himself. Here, we see how deeply Hesse was influenced by C.G. Jung and his theory of the collective unconscious and archetypal psychology—naturally combined with an intriguing spirituality from Southeast Asia that amplifies this archetypal theory (Barrett, 2022). This is unsurprising since we know that Hesse underwent psychotherapy with Jung and one of Jung's students (Schwilk, 2012). The psychoanalytical knowledge made accessible to Hesse through this therapy found both direct and indirect expression in the middle and later works of the poet and author Hermann Hesse (Baumann, 1997).

Returning to the archetype of the mother: the archetypal image of the mother conveys certain symbolic qualities that strike Goldmund as peculiar and that he wishes to experience and integrate (individuation as a process). Goldmund, who is closer to the feminine archetype (due to his sensuality, emotionality, and prioritization of experience) than to the masculine archetype (spirit, rationality, reason), now wants to follow the call of his nature and flee from the newfound constriction of the monastery, seeking a state of dynamic balance.

Certainly, Goldmund had, for a long time, adhered more to the masculine archetype in his life and personality and often found himself in difficulties and conflicts with teachers and fellow students due to the qualities of the feminine archetype. Here, Goldmund is set to break free and walk his own path—to Narcissus's delight. For Goldmund, the aim is not only to experience the feminine archetype through encounters with people and different places but also to differentiate and realize his innermost tendencies and abilities, which lean more toward the general feminine archetype.

Thus, homesickness and wanderlust fall into the broader tension of the masculine and feminine archetypes, a tension that every individual must resolve and integrate personally (Jung, 1997a, 1997b, 1997c, 1997d). The goal is to find a balance by the reciprocal and contextual integration of each archetype. In other words, for Hesse—and, by extension, for the departure and wanderlust fantasm—it is essential to integrate the polarity of life within oneself. For this, one must actively seek out the polar ends of experience to cultivate this capacity within oneself.

Only when the ability to manage polarity is fully developed can an individual respond to challenges and resistance with a broad repertoire of actions and interpret them meaningfully without being shattered by them. Here, masculine and feminine are no longer simple opposites but parts of a higher whole (part of personal life itself), which must be adaptively integrated and cultivated within one's self-image.

This is why Narcissus reacts not with anger at Goldmund's departure but rather with understanding and approval. Finally, Narcissus promises not to forget Goldmund and to pray for him—their friendship will endure.

Ultimately, this perspective provides us with interesting insights into how certain chapters of life should be approached. Hesse demonstrates that chapters in life must come to an end once a person has fully satisfied their needs in one place and developed essential skills that will serve them later in life. However, once this is deemed complete, the individual must move on and seek further fulfillment. Fulfillment is never simply given but always to be pursued.

8. The fantasm of departure in Narcissus und Goldmund: In *Siddhartha*, the departure was socially acceptable, with the possibility of a reunion always shimmering in the background (albeit under different circumstances). Similarly, in *Narcissus und Goldmund*—as previously in *Peter Camenzind* and *Siddhartha*—departure is portrayed as absolutely necessary and unavoidable. However, in *Narcissus und Goldmund*, the departure is preceded by the realization that something essential is missing within the person. What is missing is not discovered during the journey but is instead actively prepared and made tangible through shared experiences, meditation (contemplation), and dialogue. These practices provide the individual with clues and reference points for the journey's next destinations. Through genuine self-study, the individual can acknowledge the gaps in their personal philosophy of life. Based on these recognized gaps, they can then prepare their next actions. The departure into the unknown world becomes less disorienting—because a rough goal is already outlined through contemplation, which prepares the individual for potential fractures, resistance, and difficulties along the way. In short, the fantasm of departure is differentiated and deepened by the individual even before the journey begins. Through certain techniques—such as meditation or introspection, the purposes, goals, and means of the journey become more clearly defined, providing a foundation for more appropriate actions. This process does not imply that the individual embarks on the next stages of their journey with a fully formed goal; however, they possess a broad understanding of what they need and the direction they should orient themselves toward.

The Nature Fantasm in Narcissus and Goldmund

Let us now examine the last fantasm in our reflections on Hermann Hesse's Narcissus and Goldmund: the nature fantasm. To shed light on this topos, we will analyze a peripheral scene in the novel. In this scene, Goldmund is instructed by a teacher at the monastery on pharmaceutical matters and is tasked with venturing beyond the monastery walls to gather a specific plant. Goldmund agrees and sets out into the countryside to fulfill this mission. Here, we once again see nature initially portrayed as something with instrumental value, serving the self-interests of human beings. However, this functional view of nature soon gives way to its inherent autonomy, which becomes essential for the individual to discover a dialogical partner in nature and the cosmos.

> Goldmund thanked him for the most welcome assignment to pick flowers for a couple of hours rather than sit in the classroom. To make his joy complete, he asked the stablemaster to let him take the horse Bless, and soon after lunch he led the animal from the stable. It greeted him enthusiastically; he jumped on and galloped, deeply content, into the warm, glowing day. He rode about for an hour or more, enjoying the air and the smell of the fields, and most of all the riding itself, then he remembered his errand and searched for one of the spots the father had described to him. He found it, tethered the horse in the shade of a maple, talked to it, fed it some bread, and started looking for the plants. There were a few strips of fallow land, overgrown with all kinds of weeds. Small, wizened poppies with a last few fading petals and many ripe seed pods stood among withering vetch and sky-blue chicory and discolored knotweed. The heaps of stones between the two fields were inhabited by lizards, and there, too, stood the first, yellow flowered stalks of John's-wort; Goldmund began to pick them. After he had gathered a sizable bunch, he sat down on a stone to rest. It was hot and he looked longingly toward the shadowy edge of the distant forest, but he didn't want to go that far from the plants and from his horse, which he could still see from where he sat. So he stayed where he was, on the warm heap of stones, keeping very still to see the lizards who had fled come out again; he sniffed at the John's-wort, held one of its small leaves to the light to study the hundred tiny pin pricks in it. Strange, he thought, each of these thousand little leaves has its own miniature firmament pricked into it, like a delicate embroidery. How strange and incomprehensible everything was, the lizards, the plants, even the stones, everything. Father Anselm, who was so fond of him, was no longer able to pick his John's-wort himself; his legs bothered him. On certain days he could not move at all, and his knowledge of medicine could not cure him. Perhaps he would soon die, and the herbs in his pantry would continue to give out their fragrance, but the old father would no longer be there. But perhaps he would go on living for a long time still, for another ten or twenty years perhaps, and still have the same thin white hair and the same funny wrinkle-sheaves around the eyes; but what would have become of him, Goldmund, in twenty years? Oh, how incomprehensible everything was, and actually sad, although it was also beautiful. One knew nothing. One lived and ran about the earth and rode through forests, and certain things looked so challenging and promising and nostalgic: a star in the evening, a blue harebell, a red-green pond, the eye of a person or of a cow. And sometimes it seemed that something never seen yet long desired was about to happen, that a veil would drop from it all; but then it passed, nothing happened, the riddle remained unsolved, the secret spell unbroken, and in the end one grew old and looked cunning like Father Anselm or wise like Abbot Daniel, and still one knew nothing perhaps, was still waiting and listening. (Hesse, n.d., pp. 102–104)

Goldmund initially sought out nature not for its own sake, but to perform a service for the monastery. In doing so, nature or the cosmic order appeared to enter into a subordinate relationship with Goldmund—a dynamic that, in *Siddhartha*, had seemingly already been transcended. We explicitly use the term "seemingly," for this subordinate relationship between nature and humanity remains fundamentally resolved. Despite the fact that Goldmund ventures into nature solely to fulfill Father Anselm's request, nature's inherent autonomy and self-regulating order quickly come to the forefront. Nature draws Goldmund into its vibrant, colorful allure, even diverting him from his initial task—a diversion that Goldmund neither resists nor even perceives as such. Instead, he is drawn into a dialogical relationship with nature.

The stones warm him, lizards scurry across the ground, and the vivid colors of the herbs deeply move him. Ultimately, Goldmund feels so at ease that he falls into a restorative sleep. Here, nature appears as a cosmic order that, through its autonomy and intrinsic laws, provides humans with rest, renewal, and profound experiences. Not coincidentally, it is amidst nature that Goldmund contemplates Father Anselm, the fate of the monastery, and his own existence. Such reflection is only possible because nature stands before Goldmund as an autonomous and self-regulating dialogical partner. Furthermore, the fact that nature, in its autonomy and will, leaves so many open questions is perceived by Goldmund as a mandate to understand nature and the cosmic order more deeply—ultimately with the goal of gaining greater clarity about himself.

9. The autonomy and self-regulating essence of nature remain consistent from *Siddhartha* to *Narcissus and Goldmund*. In this regard, no significant updates to the fantasm are evident. Even in Hesse's final major work, *The Glass Bead Game* (1943), the quality of this fantasm remains largely untouched. However, *Narcissus and Goldmund* illustrates this fantasm in greater detail. One might concur with Lotman (1990) that once a style is achieved, it is further elaborated and consolidated as the artist progresses. A similar pattern applies to the nature fantasm in Hesse's oeuvre. The notion that nature helps refine the purposes and goals of the individual is vividly depicted in *Narcissus and Goldmund*, particularly in the scene discussed above. It demonstrates how the individual benefits and grows from continuous dialogue with the cosmos as separated and united force. Indeed, this represents a perfect symbiosis between humanity and nature—a testament to the idea that humanity should not outgrow nature but instead integrate it into themselves for its own sake.

With this, we conclude our analysis of Hesse's narrative progression as illustrated through three of his novels. The summary of the various fantasms, their content, and their development is presented in Table 10.1.

We are convinced that the table, along with our subsections in each respective section, clearly traces the development of these fantasms and illustrates how Hesse differentiated them over the years, ultimately developing a sharper life philosophy based on these meaning patterns. Certainly, Hesse continues to captivate his readers because he is an existential author who encourages them to find autonomous answers to the fantasms of the main characters (which then become Hesseian myths)

Table 10.1 Development of three Fantasms in Hesse's major works (Novels)

Fantasm	Peter Camenzind	Siddhartha	Narcissus & Goldmund	Developmental law
Nature Fantasm	Nature as a formative fact—still unspecialized and undeveloped. Nature as a projection surface of one's own mood and life situation	Nature emerges as an independent agent—it provides the individual with a collective potential for action that, when integrated, differentiates the person and stimulates growth. Nature appears as a dialogical partner	Nature appears as a dialogical partner with a certain autonomy and self-regulating order in relation to the individual—yet, nature is not presented as separative, but as integrative, depicting a human/ nature symbiosis	Progressive autonomy of nature and the symbiotic role between human and nature
Departure Fantasm	The person wants to escape the suppression and repression of their own needs and create a meaningful future (home)	Departure is initiated because new needs emerge within the person's psychological system—and an appropriate environment must be created to satisfy these needs. Yet, the departure is executed in a socially acceptable way, leaving no burnt bridges behind. The possibility of a reunion remains	Departure into a new world (wanderlust) must be prepared through self-study (immersion, meditation, reflection) so that potential difficulties and obstacles encountered in the creation of a personally meaningful world can be anticipated	Departure becomes more socially acceptable and anticipated, rather than impulsively executed
I/You/ Mentor Fantasm	The person needs the other, the genuine "You", in order to grow into a full personality. Otherwise, the person will stagnate in their self-development	The I/You relationship can only allow personality development through the mutual experience of both individuals. The genuine I/You relationship is characterized not by intellectuality but by a phenomenological component	The I/You relationship fosters growth in the individual when the opposites and differences of the genuine other are not seen as weaknesses or separative, but rather as something complementary and necessary. This complementarity can be integrated into the person through experience and practice	The I/You relationship is enriched when it is understood not only as a dialogical partner but as an experiential interplay of complementarity—enhancing personal integrity through the genuine "You"

and to embark on the path of paradoxical perfection—a lifelong process. In this sense, Hesse has also prefigured the notion of lifelong development in a specific way, which underscores his ongoing relevance. With Boesch, his ongoing narration, and our minor revision (which also includes the social overdetermination of the context), we have succeeded in analyzing the development of the fantasms in the interplay of autobiography and certain modes of thinking—certainly with a peculiar preview of Hesse's life philosophy, which appears so profound and appealing to readers of various kinds.

References

Ball, H. (2006). *Hermann Hesse: sein Leben und sein Werk*. Wallstein Verlag.

Barrett, L. (2022). A Jungian approach to coaching. *Routledge*. https://doi.org/10.4324/9781003167846

Baumann, G. (1997). Hermann Hesses "Demian" im Lichte der Psychologie C.G. Jungs. In V. Michels (Ed.), *Materialien zu Hermann Hesse "Demian"* (Vol. II, pp. 331–351). Suhrkamp.

Boesch, E. E. (1991). *Symbolic action theory and cultural psychology*. Springer.

Boesch, E. E. (2000). *Das lauernde Chaos. Mythen und Fiktionen im Alltag*. Huber.

Boesch, E. E. (2005). *Von Kunst bis terror: Über den Zwiespalt in der Kultur [from art to terror: On the cleavage of culture]*. Vandenhoeck & Ruprecht.

Boesch, E. E. (2006). *Reime und Bilder*. Privatdruck.

Boesch, E. E. (2021). *Musik, Sprache und die Sehnsucht nach dem Paradies*. Psychosozial-Verlag.

Bruner, J. (1986). *Actual minds, Possible worlds. The Jerusalem-Harvard Lectures*. Harvard University Press.

Bruner, J. (1996). The culture of education. *Harvard University Press*. https://doi.org/10.2307/j.ctv136c601

Bruner, J. (1997). Sinn, Kultur und Ich-Identität. .

Chen, Z. (2007). Hermann Hesse-ein vom taoistischen Gedankengut durchtränkter Geist. *Literaturstraße. Chinesisch-deutsche Zeitschrift für Sprach-und Literaturwissenschaft, 8*, 217–231. https://doi.org/10.11588/litst.2007.0.39970

Faesi, R. (1974). Hermann hesses Glasperlenspiel [Hermann Hesse's glass bead game]. In V. Michels (Ed.), *Materialien zu Hermann Hesse Das Glasperlenspiel* (1st ed., pp. 7–25). Suhrkamp Verlag.

von Fircks, E. F. (2023a). The dialogical self of Taoistic dynamics: How scientists and practitioners can trigger the discovery of a harmonious self. *Integrative Psychological and Behavioral Science*, 1–25. https://doi.org/10.1007/s12124-023-09807-7

von Fircks, E. (2023b). Understanding the process of Taoistic-informed mindfulness from a Meadian perspective. *Integrative Psychological and Behavioral Science*, 1–20. https://doi.org/10.1007/s12124-023-09805-9

von Fircks, E. (2023c). Understanding the power of mindfulness from a cultural-psychological perspective. *Culture & Psychology.*, article 1354067X231201387. https://doi.org/10.1177/1354067X231201387

von Fircks, E. (2023d). Cultural psychological implications of Hermann Hesse's Glasperlenspiel (glass bead game). *Culture & Psychology, 29*(3), 451–473. https://doi.org/10.1177/1354067X221132000

Heiss, R. (1936). *Die Lehre vom Charakter*. Walter de Gruyter.

Hesse, H. (1904). *Peter Camenzind*. Fischer Verlag.

Hesse, H. (1922). *Siddhartha: Eine indische Dichtung*. S. Fischer Verlag.

Hesse, H. (1927). *Der Steppenwolf*. S. Fischer Verlag.

Hesse, H. (1930). *Narcissus und Goldmund*. S. Fischer Verlag.
Hesse, H. (1943). *Das Glasperlenspiel*. Fretz & Wasmuth Verlag.
Hesse, H. (1999). *Dank an Goethe: Betrachtungen, Rezensionen, Briefe*. Insel Verlag.
Hesse, H. (2015). *Freude am Garten*. Suhrkamp Verlag.
Hesse, H. (2018). *Peter Camenzind* (30th ed.). Suhrkamp.
Hesse, H. (n.d.) Narcissus and Goldmund (transl. Ursule Molinaro). https://bcsenglish.weebly.com/uploads/3/8/6/6/38668167/hermann_hesse_-_narcissus_and_goldmund.pdf.
Hill, C. (1948). Hermann Hesse als Kritiker der bürgerlichen Zivilisation. *Monatshefte, 40*(5), 241–253. http://www.jstor.org/stable/30164681
Hsia, A. (1970). Hesses esoterisches Glasperlenspiel. *Deutsche Vierteljahrsschrift für Literaturwissenschaft und Geistesgeschichte, 44*(2), 354–362. https://doi.org/10.1007/bf03375189
Janet, P. (1889). *L'automatisme psychologique*. Félix Alcan.
Janet, P. (1898). *Névroses et idées fixes*. Félix Alcan.
Janet, P. (1929). L'évolution psychologique de la personnalité (Tome 1). Cahine.
Janet, P. (1930). L'évolution psychologique de la personnalité (Tome 2). Cahine.
Jens, W. (2020, April 14). *Ich bin Ende oder Anfang—Franz Kafka (Vortrag)* [Video]. YouTube. https://www.youtube.com/watch?v=ANuqa8t4J0c. Original work published 1988.
Jung, C. G. (1997a). Die chinesische Weltsicht [the Chinese worldview]. In J. J. Clark (Ed.), *C.G. Jung und der östliche Weg* (pp. 101–108) Patmos paperback.
Jung, C. G. (1997b). Yin und Yang: Die Einheit der Gegensätze [Yin and Yang: The unity of opposites]. In J. J. Clark (Ed.), *C.G. Jung und der östliche Weg* (pp. 109–114) Patmos paperback.
Jung, C. G. (1997c). Chinesische Alchemie und Psychische Individuation. In J. J. Clark (Ed.), *C.G. Jung und der östliche Weg* (pp. 114–162) Patmos paperback.
Jung, C. G. (1997d). Ein Gespräch mit dem I Ging [a conversation with the I Ging]. In J. J. Clark (Ed.), *C.G. Jung und der östliche Weg* (pp. 162–187) Patmos paperback.
Kafka, F. (1925). *Der Prozess*. Verlag Die Schmiede.
Kafka, F. (1926). *Das Schloss*. Kurt Wolff Verlag.
Kafka. (1995). *Brief an den Vater*. Reclam.
Puner, H. W. (1947). Freud: His life and his mind, a biography. Howell, Soskin.
Laozi, T. (2011). *Tao Te Ching*. Mauad Editora Ltda.
Lotman, Y. (1990). *Universe of the mind: A semiotic theory of culture*. Indiana University Press.
Mitscherlich-Nielsen, M. (1977). Psychoanalytischer Bemerkungen zu Franz Kafka. *Psyche, 31*(1), 60–83.
Polster, E., & Polster, M. (1974). *Gestalt therapy integrated: Contours of theory and practice*. (2 ed.). New York: Brunner/Mazel.
Schmieder, F. (2017). Gleichzeitigkeit des Ungleichzeitigen: Zur Kritik und Aktualität einer Denkfigur. *Zeitschrift für kritische Sozialtheorie und Philosophie, 4*(1–2), 325–363. https://doi.org/10.1515/zksp-2017-0017
Schwilk, H. (2012). *Hermann Hesse. Das Leben des Glasperlenspielers*. Piper Verlag.
Stern, W. (1923). *Person und Sache: Die Menschliche Persönlichkeit [The Human Personality]* (3rd ed.). J.A. Barth.
Stern, W. (1924). *Person und Sache: System des Kritischen Personalismus (Wertphilosophie) [the philosophy of values]*. J.A. Barth.
Stern, W. (1935). *Allgemeine Psychologie auf personalistischer Grundlage*. Martinus Nijhoff.
Valsiner, J. (2017). *From methodology to methods in human psychology*. Springer.

Dialogues in Cultural Psychology: Between Boesch, Bruner, and Valsiner (Postface)

We have reached the conclusion of our scientific treatise. I must admit, this has been an exciting journey for me. Writing a book is always a challenge. Is there enough material to support the various chapters? How can I go beyond existing authors and present something novel? Where are the research gaps, and are my attempts to address them sufficient, or do I need to make improvements somewhere? Not to mention the formal challenges every academic writer faces. Yet, gratitude prevails. This journey into both familiar and uncharted territory has helped me gain a deeper understanding of Boesch, better comprehend certain aspects of his theory, and even offer revisions in some areas.

Regrettably, readers often only see the finished product—the completed manuscript. However, this manuscript is preceded by numerous drafts, rejections, and improvements, as well as dialogues with other scholars, which must be acknowledged.

Boesch in Dialogue with Valsiner

Jaan Valsiner, for example, read all the chapters and provided profound feedback, which significantly informed the revisions. This dialogue was enriching, not only for me as a psychologist and as an individual seeking new perspectives, but it also indirectly resonates throughout this book. For instance, Valsiner encouraged me to expand the theological considerations in Chap. 7 beyond Christianity. A scientific work that explicitly documents and reflects this type of dialogue would be a valuable complement to the many finalized manuscripts researchers engage with in their work.

Valsiner himself engaged in dialogue with Boesch, visiting him in his home in Saarbrücken and receiving private reprints of his works, which he then shared with me. The symbolism of this dialogical transmission is striking. I must reiterate: we must make this dialogue explicit. Valsiner has developed an entirely autonomous

E. Freiherr von Fircks, *Culture of Action*, Theory and History in the Human and Social Sciences, https://doi.org/10.1007/978-3-031-88808-3

cultural psychology (Valsiner, 2007, 2014, 2019, 2021), which, alongside Boesch's symbolic action theory and Bruner's narrative cultural psychology (Bruner, 1986, 1996, 1997), has had—and continues to have—a significant influence on many psychologists and scholars. It is no coincidence that some scholars refer to a distinct Valsinerian cultural psychology.

Valsiner, for example, introduces the concept of hyper-generalized fields of meaning, which either inhibit or enhance everyday actions, and the personal expansion of social patterns of meaning which are schematized (termed pleromatization as process). The pleromatization process is an interesting semiotic regulation of meaning-making, which accounts for fullness in meaning-making (Valsiner, 2006, 2007). For example, a person experiences an overwhelming sunset and feels close to God (This must be God or only God can create something so beautiful) not knowing how to describe this full and rich experience guiding human experience—it can be felt and experienced. Hence, the pleromatization process is superimposed on the schematization process (sunset, God which serve as schematized denominators of meaning) but which in this crucial moment are transformed into something higher and hyper-generalized—the hyper-generalized semiotic regulator of meaning that defies verbal description but becomes then a central hierarchically organized regulator of action expanding to different contexts as well (Valsiner, 2019). Of course, from this pleromatization process, further schematization can be derived. The sunset experience becomes symbolic for God's mercy, benediction, closeness among many other possible denominators which in itself is a schematized outlook from the pleromatic process. If the person then focuses on the closeness to God within experience, an actualized hyper-generalized meaning-making pattern might emerge transforming the initial pleromatization and schematization process. Which Valsiner describes as the interdependence of schematization and pleromatization (Valsiner, 2006). With Boesch, we might argue that the hyper-generalized meaning-making pattern—coming into being through pleromatization—is a fantasm but Boesch omits the process how to get to the fantasm (genetic perspective). To put it short, the cultural psychologies of Valsiner and Boesch are related (fantasm as hyper-generalized meaning-making complex), and it is up to us to decipher these connections while also drawing attention to their differences (genetic vs. nongenetic perspective). These differences, in particular, could foster an intriguing meta-dialogue about central concepts in cultural psychology.[1]

If we were to include Bruner in this dialogue, we could further identify intertextual commonalities—analogies—as well as differences—contrasts—that could help fill research gaps within both frameworks or even strive for a synthesis of their theories and practices. Based on the themes explored in this book, concrete questions—regarding the interplay of Boesch and Valsiner—could include:

1. How would an analysis of dream experiences and dream interpretation look within Valsiner's cultural semiotics? How does this differ from our guiding action-theoretical questions for dream experiences and interpretations? To offer

[1] The comparison is not exhausted with the notions of fantasms and pleromatization.

the reader a brief insight into our fertile comparison, here are some preliminary answers to the question at hand. Dreams, as a whole, operate through various hyper-generalized meaning-making complexes that have been pleromatized. Within the therapeutic encounter, the therapist may activate and amplify specific elements from this hyper-generalized meaning-making field, which are then schematized through verbal language into more precise, point-like signs. For instance, the therapist might invite the patient to reexperience a particular scene from the dream, focusing on a specific meaning complex derived from it (schematization). Within the therapeutic process, the therapist might further encourage the patient to reengage with the scene or adopt a different stance toward a dialogical partner in the dream. For example, if the patient lacked the courage to express a particular sentence within the dream, the therapeutic encounter could provide a space where the patient is supported to articulate it. This process then feeds back into the pleromatization and the actualization of the hyper-generalized meaning-making field within the hierarchy of semiotic regulation.

2. What would a guide for the interpretation of poetry look like through the lens of Valsinerian cultural semiotics? What is the role of the symbol in Valsiner's cultural semiotics? This question is inherently ambivalent, as it requires distinguishing between the symbolic meaning of poetry for the poet and for the reader—two perspectives that may diverge significantly. For the poet, poetry emerges as a response to profound experiences they have undergone, experiences they wish to share with others. The act of writing poetry is thus preceded by a pleromatic experience—an engagement with existential themes, as previously discussed—with which the poet wrestles, unsure of what these experiences mean for them. Through poetry, they seek clarification, striving to access the hyper-generalized meaning-making field by carefully arranging words in a way that might illuminate their understanding of both the world and themselves. Poetry, uniquely, operates in the liminal space between verbal and nonverbal meaning-making, bridging the gap between generalized and hyper-generalized meaning fields (corresponding to Level 3 and Level 4 in Valsiner's framework). The words in a poem are schematized products derived from fragments of hyper-generalized meaning fields, leaving the poet in an ambivalent state. They recognize that their words can only approximate the hyper-generalized experience but can never fully encapsulate it. This perpetual tension—the inability to fully mirror the ineffable experience through language—explains why poets continue writing and are often inspired by their own creations. The process of self-understanding, as well as the attempt to make sense of the world, necessarily remains incomplete. For the reader, poetry serves a different but equally profound purpose. Readers typically turn to poetry not as a pastime but as a means of seeking answers to deeply existential questions in their own lives. A poem, in its entirety, may evoke the emergence of a hyper-generalized feeling field, potentially initiating a pleromatization process. This occurs because the poem can reveal previously unnoticed facets of experience (e.g., reinterpreting death not as a loss but as liberation), thereby transforming the reader's hyper-generalized meaning-making patterns. Through this transformation, new schematized

products may emerge, brought to life by the pleromatization process. In this regard, Boesch and Valsiner's perspectives align closely. Boesch posits that individuals turn to poetry to navigate fantasms and myths, while Valsiner's framework allows us to more precisely analyze the poet's ambivalent experience—rooted in the tension between Level 3 and Level 4. This approach highlights the dynamic interplay of meaning-making fields and the enduring incompleteness that characterizes both the creation and reception of poetry.

3. How could a philosophy of life and characterology be constructed based on Valsiner's cultural semiotics? What guidelines for a successful life can be derived from this framework? How do these compare with our Boeschian philosophy of life? This question is relatively straightforward, as Valsiner has developed his theory of semiotic mediation by drawing upon linguistics and philosophy, with additional concepts borrowed from biology (e.g., promoter and inhibitor signs). The question can also be approached from a cultural-historical perspective. Consider the experiences of National Socialism (NS) and Communism. These ideologies function as systems of promoter and inhibitor signs—predominantly inhibitor sign systems—that mediate human meaning-making processes from the bottom up (from Level 3 to Level 4 in particular). However, such ideologies remain schematized products. However, these schematized products—if they are used repeatedly and with a strong emphasis (as was done in NS and Communism), then those signs can trigger affective meaning-making (hyper-generalized) in line with the ideology. Hence, the strategy to overflood the periphery of an action field in Communism and NS so that people might use these schematized products as primary products for their higher affective meaning-making. This perspective was complemented by NS and Communism forbidding some novelists and poets to share and disseminate their work as those oeuvres would have allowed people to have alternative schematized products to superimpose their affective hyper-generalized meaning-making upon—allowing the actualization of meaning-making to come into being. From this perspective, it becomes evident that a life philosophy and character development rooted in Valsiner's theory of semiotic mediation fundamentally relies on the pleromatization process—or more precisely, on the means that initiate and sustain this process. In essence, Valsiner's life philosophy could be summarized as follows: At all stages of development, engage actively with literature and poetry, immerse yourself in dialogue with individuals who bring diverse experiences, and embrace those experiences—not by fully submitting to them but by using them as a stimulus to enrich your own experiential landscape. Be vigilant against ideologies, as they aim to overly promote your hyper-generalized meaning-making process. It is especially the distrust of ideology and the search for genuine stimulants of meaning-making and its actualization that unites Boesch with Valsiner.

These are just a few examples of questions that could play a pivotal role in establishing a new meta-theoretical foundation for cultural psychology. They could also expand Valsiner's cultural semiotics into new, uncharted, and applied areas of psychology and related disciplines.

Boesch in Dialogue with Bruner

Let us briefly return to another significant figure in cultural psychology: Jerome S. Bruner (1915–2016) and his narrative cultural psychology. According to Bruner (1996, 1997), humans do not passively accept the world and its dynamics but instead perceive the world through meaning-mediating narratives. These narratives can also be made fruitful for one's self-identity. For instance, nature may be interpreted as an expression of divine grace and beauty if a person has grown up within a theological narrative and actively identifies with it, while another person might view nature as indifferent, perhaps even as a dumping ground for waste. Thus, one person may treat nature as a nearly sacred space, while another might treat it with indifference, using it as a site for waste disposal.

However, what I find lacking in narrative cultural psychology is the dynamic aspect of narratives. In other words, narratives are constantly evolving; people do not remain stagnant in their narratives. A pacifist today may passionately advocate war in 10 or 20 years and even participate in it. Likewise, someone who is indifferent to environmental concerns today could become an environmental activist in just a few months or years. Our narrative life paths are often disrupted, and these disruptions lead to further use and development of the narrative.

Linear narratives that simply consolidate and provide people with enduring security, recognition, love, and other qualities are rare. Because the world is inherently dynamic, narratives are continually disrupted and rewritten by individuals and groups. A natural disaster, for example, may transform an ecological skeptic into an advocate for environmental protection, provided the person realizes their dependence on the biosphere and its diversity.

The degree to which we actively integrate a narrative into our self-concept thus depends on the needs we have developed, which are themselves continuously evolving. In Boesch's terminology, fantasms (Fantasmen) and the way we interact with myths are also tied to our central needs—an area of overlap between Boesch and Bruner that calls for comparative research or a meta-theory of cultural psychology. Yet, Bruner's narrative cultural psychology lacks precisely this personalistic component, which Boesch expresses in the concept of fantasmen.

In my terms, a narrative is a conglomeration of various normative ideals (Soll-Werte). Its meaning is socially overdetermined, compelling individuals to adapt to certain normative values. However, individuals do not always passively accept narratives; they often attempt to enrich or co-construct them. When a narrative is personally appropriated and enriched through personal meaning, it ceases to be merely a narrative. Instead, it is transformed into something more personal—perhaps even into an individual philosophy of life.

Furthermore, Bruner (1996, 1997), in his monographs exploring the significance of narrative, draws on scientifically grounded research on heroism. In narratives, we are often confronted with certain figures or groups who believe in a specific cause. These personal convictions then encounter a social context, which is often antagonistic to those convictions—or the context collapses, requiring the heroic figure to restore order. In this process, the hero embarks on a journey during which they acquire new qualities and transform the social order they save.

From *Harry Potter* to *Frozen*, the story of the hero is a recurring theme, illustrating the appeal of such universal meaning mediators. However, if we examine the story of Harry Potter (written by J.K. Rowling), Harry does not save the world single-handedly. On the contrary, his profound connections with friends (Ron Weasley and Hermione Granger), beloved mentors (Dumbledore, and later Snape), and his remaining family (Lupin and Sirius) enable him to develop crucial skills to confront darkness and evil in the world.

In my view, Harry's capacity for action is expanded and enriched by the intersecting collective potentials for action—fostered on various dialogical "islands" of mediation. Thus, the hero myth in *Harry Potter* is effectively refuted. Instead of perpetuating the hero myth, we should adopt a more collective perspective, one that highlights how Harry Potter developed into a powerful and effective wizard through various interactions—something he could not have achieved alone.

I am convinced that our theory of co-constructed action potential better captures the essence of the Harry Potter saga than the common interpretations of the hero myth. This also demonstrates that narratives, in their social overdetermination, often underestimate the role of collectives and cooperation.

From these preliminary considerations, we see the value and fertile ground that a hypothetical dialogue between Bruner, Valsiner, and Boesch could yield. More research in this area is essential.

The Future of Cultural Psychology

A journey always offers a glimpse into the next journey, indicating the next steps to take. Our aim in this reflection was to outline these possible next steps through dialogue. Improving individual cultural psychologies is only possible through building bridges, that is, through dialogue. We have set the pillars; constructing them into a bridge will require cooperation. At the end of this triadic bridge, we might even glimpse a meta-cultural psychology. Let us strive earnestly to develop this together.

Appendix A: English Translation of Welcome and Farewell by Edgar Alfred Bowring (1874)

Welcome and Farewell (Goethe)

Quick throbb'd my heart: to horse! haste, haste
And lo! 'twas done with speed of light;
The evening soon the world embraced,
And o'er the mountains hung the night.
Soon stood, in robe of mist, the oak,
A tow'ring giant in his size,
Where darkness through the thicket broke,
And glared with hundred gloomy eyes.

From out a hill of clouds the moon
With mournful gaze began to peer:
The winds their soft wings flutter'd soon,
And murmur'd in my awe-struck ear;
The night a thousand monsters made,
Yet fresh and joyous was my mind;
What fire within my veins then play'd!
What glow was in my bosom shrin'd!

I saw thee, and with tender pride
Felt thy sweet gaze pour joy on me;
While all my heart was at thy side,
And every breath I breath'd for thee.
The roseate hues that spring supplies
Were playing round thy features fair,
And love for me—ye Deities!
I hoped it, I deserved it ne'er!

E. Freiherr von Fircks, *Culture of Action*, Theory and History in the Human and
Social Sciences, https://doi.org/10.1007/978-3-031-88808-3

But, when the morning sun return'd,
Departure filled with grief my heart:
Within thy kiss, what rapture burn'd!
But in thy look, what bitter smart!
I went—thy gaze to earth first roved;
Thou follow'dst me with tearful eye:
And yet, what rapture to be loved!
And, Gods, to love—what ecstacy!

Appendix B: STEPS by Hermann Hesse and Translated by Gonzalez-Trejos)

Steps

As every blossom wilts and youth must yield
To age, so too life's stages bloom and fade,
Each virtue, every wisdom, once revealed,
Must find its hour, then softly be unmade.
The heart must, at each call of life, be still
Prepared to part, begin anew, and grow,
With courage, free of sorrow, free of will,
To join with other bonds we do not know.
For every start holds magic, deep and bright,
That shields us and sustains us in the night.
We're meant to pass through chambers, light and wide,
Not cling to one as though it were our home.
The world-spirit would not have us confined—
He lifts us step by step, compels us on.
Just when we're nestled in a life too deep,
Secure, familiar, still—then comes the drift;
Only the soul prepared to rise and leap
Can shake the weight of habit, make the shift.
And even death's dark hour may yet unfold
New spaces, youthfully, before our gaze.
Life's call will never cease, nor will it hold—
So, heart, farewell—and heal in brighter days!

E. Freiherr von Fircks, *Culture of Action*, Theory and History in the Human and Social Sciences, https://doi.org/10.1007/978-3-031-88808-3

Appendix C: Sterben und Auferstehn (Matthias Claudius)

Du Menschenkind, sieh um dich her ...
Und weißt du eine Lehre,
Die größer und die tröstlicher
Für uns hienieden wäre? –

Dort, wo die Siegespalmen wehn,
Ist Sein nur, ist kein Werden,
Kein Sterben und kein Auferstehn,
Wie hier bei uns auf Erden.

Dort freun sie ewig ewig sich,
Ist ewig Licht und Friede,
Das Leben quillt dort mildiglich
Aus sich, und wird nicht müde.

Doch dieser Unterwelt ist nicht
Solch glorreich Los gegeben;
Hier ist ohn Finsternis kein Licht,
Und ohne Tod kein Leben.

Der Löwe liegt und fäult und schwellt –
Dann geht vom Fresser Speise;
Der Same in die Erde fällt
Und stirbt – und keimt dann leise.

Und die Natur ein Spiegel ist;
Es wird darin vernommen:
Was deinem Geist du schuldig bist
Soll er zum Leben kommen.

Willst du wahrhaftig glücklich sein,
Auf festem Grunde bauen;
Mußt du den Dornenweg nicht scheun,
Der Rosenbahn nicht trauen.

E. Freiherr von Fircks, *Culture of Action*, Theory and History in the Human and Social Sciences, https://doi.org/10.1007/978-3-031-88808-3

Einst war ein großer Mann bedacht
Uns darin einzuweihen,
Und führte durch die lange Nacht
Das Volk zum Fest der Maien.

Drum spare dir viel Ungemach,
Du Menschenkind, und höre,
Und denke der Verleugnung nach,
Und jener großen Lehre.

In uns ist zweierlei Natur,
Doch ein Gesetz für beide;
Es geht durch Tod und Leiden nur
Der Weg zur wahren Freude.

References

Bruner, J. (1986). *Actual minds, possible worlds. The Jerusalem-Harvard lectures*. Harvard University Press.
Bruner, J. (1996). *The culture of education*. Harvard University Press. https://doi.org/10.2307/j.ctv136c601
Bruner, J. (1997). *Sinn, Kultur und Ich-Identität*. Carl Auer.
Valsiner, J. (2006). The overwhelming world: Functions of pleromatization in creating diversity in cultural and natural constructions. In: *International Summer School of Semiotic and Structural Studies* (Conference Paper, pp. 1–21).
Valsiner, J. (2007). *Culture in minds and societies* (pp. 276–299). Sage.
Valsiner, J. (2014). *An invitation to cultural psychology*. Sage.
Valsiner, J. (2019). *Ornamented lives. Advances in cultural psychology*. IAP.
Valsiner, J. (2021). *General human psychology*. Springer.

E. Freiherr von Fircks, *Culture of Action*, Theory and History in the Human and Social Sciences, https://doi.org/10.1007/978-3-031-88808-3

Index

E. Freiherr von Fircks, *Culture of Action*, Theory and History in the Human and Social Sciences, https://doi.org/10.1007/978-3-031-88808-3

The manufacturer's authorised representative in the EU is Springer Nature Customer Service Centre GmbH, Europaplatz 3, 69115 Heidelberg, Germany. If you have any concerns regarding our products, please contact ProductSafety@springernature.com

Printed and bound by CPI Group (UK) Ltd, Croydon, CR0 4YY
30/09/2026
02225569-0003